# Aerodynamics of Wind Turbines

## A Physical Basis for Analysis and Design

*Sven Schmitz*
Department of Aerospace Engineering
The Pennsylvania State University
PA, USA

WILEY

*Registered Offices*
John Wiley & Sons, Inc., 111 River Street, Hoboken, NJ 07030, USA
John Wiley & Sons Ltd, The Atrium, Southern Gate, Chichester, West Sussex, PO19 8SQ, UK

*Editorial Office*
The Atrium, Southern Gate, Chichester, West Sussex, PO19 8SQ, UK

For details of our global editorial offices, customer services, and more information about Wiley products visit us at www.wiley.com.

Wiley also publishes its books in a variety of electronic formats and by print-on-demand. Some content that appears in standard print versions of this book may not be available in other formats.

*Library of Congress Cataloging-in-Publication Data*

Names: Schmitz, Sven, 1976- author.
Title: Aerodynamics of wind turbines : a physical basis for analysis and
    design / Sven Schmitz, Department of Aerospace Engineering, Pennsylvania
    State University.
Description: First edition. | Hoboken : Wiley, 2020. | Includes
    bibliographical references and index.
Identifiers: LCCN 2019024050 (print) | LCCN 2019024051 (ebook) | ISBN
    9781119405610 (paperback) | ISBN 9781119405597 (adobe pdf) | ISBN
    9781119405641 (epub)
Subjects: LCSH: Wind turbines–Aerodynamics.
Classification: LCC TJ828 .S37 2019 (print) | LCC TJ828 (ebook) | DDC
    621.4/5–dc23
LC record available at https://lccn.loc.gov/2019024050
LC ebook record available at https://lccn.loc.gov/2019024051

Cover Design: Wiley
Cover Images: © YaiSirichai/Shutterstock, © nito/Shutterstock,
© travellight/Shutterstock, Cover figure courtesy of Sven Schmitz

Set in 10/12pt WarnockPro by SPi Global, Chennai, India

Printed and bound by CPI Group (UK) Ltd, Croydon, CR0 4YY

10  9  8  7  6  5  4  3  2  1

**Aerodynamics of Wind Turbines**

*To my son Andreas.*
*Never lose your spirit of invention. You can change the world.*

# Contents

# About the Author

Sven Schmitz is an Associate Professor in the Department of Aerospace Engineering and The Institutes of Energy and the Environment (IEE) at The Pennsylvania State University. He is an expert in rotary wing aerodynamics, with particular emphasis on wind turbines and rotorcraft. He has authored 30 peer-reviewed journal publications and more than 60 conference papers and presentations. His research program embraces the areas of wind turbine aerodynamics and rotorcraft aeromechanics. Current activities include improvements to wind turbine blade-element momentum theory, wind farm wake modeling, icing on wind turbines, rotor hub flows, rotor active control, ship air-wake modeling, and future concepts for transonic commercial aircraft. At Penn State, he developed and maintains the XTurb code, a wind turbine design and analysis software that he integrates in his teaching and research.

Sven Schmitz grew up in Effeld (Germany), a small and beautiful village on the German-Dutch border in the lower Rhineland, approximately 60 km west of Düsseldorf and north of Aachen. In the Fall of 1996, he enrolled in the Engineering program at the Rheinisch-Westfälische Technische Hochschule (RWTH) Aachen where he graduated in 2002 with a Diploma degree. From 2002 to 2006, he was a Ph.D. student at the University of California (UC) Davis under the guidance of Professor Jean-Jacques Chattot. From 2006 to 2010, he was a postdoctoral researcher and project scientist at the U.S. Army Aero Flight Dynamics Directorate (AFDD) at the NASA Ames Research Center in Moffett Field, CA and at UC Davis. During this time, he also worked as a Computational Fluid Dynamics (CFD) consultant for wind energy applications with General Electric (GE) Global Research. In 2010, he joined the faculty of Aerospace Engineering at The Pennsylvania State University. He continues to publish and teach in the areas of wind turbine aerodynamics and rotorcraft aeromechanics.

# Preface

The vision for the book was to develop a self-contained and affordable unique text with focus on the aerodynamics, scaled design and analysis, and aerodynamic optimization of horizontal-axis wind turbines. It is not a systems-engineering text on wind energy, which distinguishes the book from other available texts. On the contrary, the author was encouraged by many colleagues over the past several years to develop a well-integrated and focused account on the blade aerodynamics of horizontal-axis wind turbines. The technical content is based on lecture notes developed by the author at the senior-level undergraduate and graduate level. A further unique aspect of the book is the close integration of the text with the wind turbine design and analysis software, XTurb, developed and maintained by the author at The Pennsylvania State University. The XTurb code is essentially a teaching and research tool used since 2011 by the author in his graduate course and by Penn State students participating in the U.S. DOE Collegiate Wind Competition. The XTurb examples in the book add a "hands-on" component, thus enhancing the learning experience to readers and resulting in a deeper and more complete understanding of the subject matter. This gives readers from interdisciplinary backgrounds in the area of wind energy the opportunity to fully absorb and understand the design principles and governing concepts in blade aerodynamics through the text and independent analyses using XTurb.

Chapter 1 concerns a brief description of horizontal-axis wind turbine development, with particular considerations for the history of aerodynamics and its impact on the design evolution of wind turbines. In addition, the reader is introduced to the atmospheric boundary layer and the wind resource. Chapter 2 is a classical account on momentum theory for horizontal-axis wind turbines, with some differences and additions compared to most texts. Chapter 3 covers classical Blade Element Momentum (BEM) theory. A special section includes a detailed description of root and tip loss factors used in today's BEM-type methods and a discussion of their respective limitations. As far as BEM solution techniques are concerned, the presentation of classical work is accompanied by a description of various numerical techniques of solving the BEM equations (new in this book), followed by a complete description of models for the turbulent wake state (also not in any other text). A simplified BEM theory is described as a classical means to introduce simplified dependencies of the effect of design parameters on power coefficient, with the later subsection containing multiple examples of XTurb analyses and associated input decks. Chapter 4 begins with an introduction to thin-airfoil theory and the foundations of viscous airfoil flow. This is followed by a brief historical review on wind turbine airfoil design and airfoil design criteria along the blade radius. The chapter concludes with a catalog of wind turbine airfoils. Chapter 5 focuses on introducing unsteady aerodynamics occurring on horizontal-axis wind turbines, with subsections describing yaw effects, tower interaction, and dynamic stall. A big portion of this chapter is devoted to a comprehensive review on rotational augmentation and available stall-delay models (also new in this text). Chapter 6 concerns vortex-wake methods and starts off by a comprehensive introduction to lifting-line theory and describes the basics of computing induced velocities from planar and vortical wake sheets. A unique subsection on prescribed wake methods (not in any other text) includes additional XTurb examples. A subsection on free-wake methods includes classical descriptions and introduces the reader to the numerical problem of vortex cores to avoid wake singularities, including recent advances on singularity-free-wake methods. Chapter 7 gives an introduction to advanced computational methods including Computational Fluid Dynamics (CFD), hybrid CFD methods, and an introduction to recent advances in actuator-type methods for wake modeling (i.e. actuator disk, actuator line, actuator surface). Chapter 8 introduces the reader to principles of (scaled) wind turbine design and optimization. The technical content draws from the author's experience in designing scaled experiments in both wind-/water tunnels for wind turbine and rotorcraft applications. This is a very unique chapter of the book and one that is vitally important to the wind energy community at large. The XTurb software is used extensively in this chapter and brings the material of previous chapters into an overarching context to the reader, thus allowing a deeper understanding of the book material by considering the difficult and non-unique aspects of scaled blade design and optimization. Supplementary files including XTurb input decks are available on a respective Wiley website.

The primary intended audience consists of senior undergraduate students and graduate students in MSc. and Ph.D. programs at universities, focusing on coursework and research in wind energy in engineering, atmospheric science, and meteorology. Further audiences are instructors and university professors, as well as practicing engineers and scientists in industry and national laboratories. Readers are expected to have basic knowledge of incompressible flows through coursework or work experience. The book is written at an intermediate level using college-level algebra and analysis. The book can be used as a primary text for courses on wind turbine aerodynamics and/or as an affordable instructional aid for a multitude of courses on wind energy systems and power engineering in the international wind energy community.

In essence, this is a book "for Students, written by a student," as the author sees himself as a continuous learner. I hope that you will find the book helpful to your careers. Wind energy has the potential to playing a major role in battling climate change by powering the world with a clean and renewable source of energy. Never give up, not as long as you have strength left.

January 2019

*Sven Schmitz*
State College, PA

# Acknowledgments

A number of individuals have assisted the author and contributed to the book in different ways. I want to start by thanking my beloved wife Cristina for her patience and encouragement while writing the manuscript. I am especially grateful to my parents Helmut and Irmgard Schmitz who have always supported me in the path I have taken far away from home. I am also thankful to my Ph.D. advisor Professor Jean-Jacques Chattot at the University of California Davis who educated me in wind turbine aerodynamics and numerical methods.

Many thanks are directed at The Pennsylvania State University for allowing me to spend my sabbatical leave at home. Furthermore, I am grateful to Professor Carlo L. Bottasso of the Technical University of Munich (TUM) and Professor Jens. N. Sørensen of the Technical University of Denmark (DTU) for hosting me a week each and their many valuable suggestions. The discussions with many colleagues at TUM and DTU were inspiring and helpful in organizing some of the book content.

I also acknowledge very much the help of my dear colleague Professor Mark D. Maughmer for numerous discussions on aerodynamics during my time at Penn State and for checking many subtle details thoroughly in the book. Furthermore, I am thankful to the former Aerospace Department Head Professor George A. Lesieutre for his encouragement in developing a graduate course focused on wind turbine aerodynamics, as well as colleagues Dr. Susan W. Stewart and Professor Dennis K. McLaughlin without whose dedication there would not be a continuing graduate certificate in wind energy in the department.

Special thanks to Dr. Scott Larwood of the University of the Pacific and Dr. Mike P. Kinzel of the University of Central Florida for reading the entire manuscript. Scott and Mike represent a portion of a diverse readership with different background and experience; their many questions and comments helped clarifying derivations, figures, and the narrative. I would also like to thank Mr. Dan Somers of Airfoils Inc. in Port Matilda, PA for his review and suggestions on the history of airfoil design for wind turbines over the past 30 years. In addition, comments and suggestions by Dr. Niels Troldborg of Risø National Laboratories on advanced computational methods and actuator line modeling are very much appreciated.

Last but not least, I want to thank my students in wind energy for their commitment to hard work, inspiration, and passion for wind energy as a viable source of renewable energy. The future lies with our students who will advance wind energy to the next generation. I would like to close by thanking the U.S. Department of Energy (DOE), the National Science Foundation (NSF), the University Corporation for Atmospheric Research (UCAR), and industry for their generously supporting my students at Penn State.

# Abbreviations

| | |
|---|---|
| 1P | One Per revolution |
| 2-D | Two-Dimensional |
| 3-D | Three-Dimensional |
| ABL | Atmospheric Boundary Layer |
| ACE | Actuator Curve Embedding |
| AEP | Annual Energy Production |
| ALM | Actuator Line Model |
| AoA | Angle of Attack |
| AR | Aspect Ratio |
| BEM | Blade-Element Momentum |
| BEMT | Blade-Element Momentum Theory |
| CF | Capacity Factor |
| CFD | Computational Fluid Dynamics |
| COE | Cost of Energy |
| DTU | Technical University of Denmark |
| DUT | Delft University of Technology |
| ECN | Energy Research Center of the Netherlands |
| FS | Full Scale |
| HAWT | Horizontal-Axis Wind Turbine |
| HVM | Helicoidal Vortex Model |
| IEC | International Electrotechnical Commission |
| KJ | Kutta–Joukowski |
| LCOE | Levelized Cost of Energy |
| LIR | Low Induction Rotor |
| MEXICO | Model rotor EXperiments In COntrolled conditions |
| MS | Model Scale |
| NACA | National Advisory Committee for Aeronautics |
| NASA | National Aeronautics and Space Administration |
| NREL | National Renewable Energy Laboratory |
| PIV | Particle Image Velocimetry |
| PSU | Penn State University |
| rpm | Revolutions Per Minute |
| RWT | Research Wind Turbine |

| SST | Shear Stress Transport |
|-----|-----|
| TKE | Turbulent Kinetic Energy |
| TS | Tollmien–Schlichting; Technical Specifications (Chapter 8) |
| TUM | Technical University of Munich |
| UIUC | University of Illinois Urbana-Champaign |
| VWM | Vortex Wake Method |

# List of Symbols

Note: This list includes the most relevant symbols used in the book, and omits some symbols that are unique to particular subchapters.

## English

| | |
|---|---|
| $a$ | axial (flow) induction factor |
| $a'$ | angular (flow) induction factor |
| $adv_i$ | (vortex) sheet advance ratio (Chapter 6) |
| $A$ | actuator/rotor disk area |
| $A_0$ | entrance area of streamtube (Chapter 2); Fourier coefficient in airfoil theory (Chapter 4) |
| $A_1$ | exit area of streamtube |
| $A_n$ | Fourier coefficients in airfoil/wing theory (Chapters 4, 6) |
| $A_t$ | airfoil cross-sectional area (Chapter 4) |
| $b$ | wing span |
| $B$ | blade number |
| $c$ | blade chord; Weibull scale parameter (Chapter 1) |
| $\bar{c}$ | mean aerodynamic chord |
| $c_0$ | wing mid-span chord |
| $c_d$ | section (profile) drag coefficient |
| $c_{d,min}$ | minimum section (profile) drag coefficient |
| $c_l$ | section lift coefficient |
| $c_{l,max}$ | maximum section (airfoil) lift coefficient |
| $c_{l,\alpha}$ | airfoil lift-curve slope |
| $c_m$ | section moment coefficient about quarter-chord location |
| $c_{m,0}$ | section moment coefficient about leading edge |
| $c_p$ | pressure coefficient; specific heat constant (at constant pressure) |
| $C(k)$ | complex Theodorsen function |
| $c_D$ | wing drag coefficient |
| $c_{D_i}$ | wing induced drag coefficient |
| $c_L$ | wing lift coefficient |
| $C_P$ | power coefficient |
| $C_{P,max}$ | maximum power coefficient |
| $C_Q$ | torque coefficient |
| $C_T$ | thrust coefficient |

| | |
|---|---|
| $d$ | spacing between semi-infinite planar vortex sheets; airfoil camber |
| $dr$ | incremental width of blade element (or rotor annulus) |
| $dA$ | incremental disk-annulus area $(2\pi r \cdot dr)$ |
| $dA_c$ | wetted area of infinitesimal blade element $(c \cdot dr)$ |
| $dD$ | incremental/sectional profile drag force |
| $dF_N$ | incremental/sectional (or blade-element) normal force – i.e. force normal to local airfoil/section chord line towards upper airfoil/section surface |
| $dF_T$ | incremental/sectional (or blade-element) tangential force – i.e. force parallel to local airfoil/section chord line towards leading edge of airfoil/section |
| $dL$ | incremental/sectional lift force |
| $d\dot{m}$ | incremental mass-flow rate |
| $dQ$ | incremental torque |
| $dQ_B$ | blade-element torque, incremental torque (one blade) |
| $dP$ | incremental power |
| $dT_B$ | blade-element thrust, incremental thrust (one blade) |
| $D$ | rotor diameter; (profile) drag force |
| $D_i$ | induced drag |
| $e$ | Oswald efficiency factor |
| $f()$ | functional relationship |
| $\boldsymbol{f}_{N,m}$ | elemental rotor/blade force vector (Chapter 7) |
| $F$ | total loss factor – i.e. product of local root- and tip-loss factor $(F_R \cdot F_T)$ |
| $F()$ | objective function (Chapter 8) |
| $F(k)$ | real part of Theodorsen function |
| $\vec{F}$ | airfoil force vector $(D\vec{i} + L\vec{k})$ |
| $Fr$ | Froude number $(V_0/\sqrt{gR})$ |
| $F_C$ | centrifugal force $(m\Omega^2 R)$ |
| $F_{Co}$ | Coriolis force $(-2m\Omega v_{rel})$ |
| $F_1$ | tip correction factor applied to section normal-/tangential force coefficients (Section 3.4) |
| $F_B$ | Prandtl's approximate loss factor to account for a finite number of blades (Section 3.4) |
| $\boldsymbol{F}_P$ | volumetric body-force vector (Chapter 7) |
| $F_R$ | root-loss factor |
| $F_T$ | tip-loss factor |
| $g$ | gravitational acceleration; coefficient or function used in tip corrections (Section 3.4) |
| $G(k)$ | imaginary part of Theodorsen function |
| $H$ | shape factor of airfoil boundary layer $(\delta^*/\Theta)$ |
| $\vec{i}$ | unit vector in the $x$-direction |
| $I$ | turbulence intensity; mass moment of inertia |
| $\vec{j}$ | unit vector in the $y$-direction |
| $k$ | integer; shape parameter for Weibull distribution (Chapter 1); reduced frequency (Chapter 5); specific turbulent kinetic energy (Chapter 7) |
| $\vec{k}$ | unit vector in the normal/vertical $z$-direction |

| | |
|---|---|
| $l$ | turbulent mixing length |
| $L$ | lift force; Monin–Obukhov length scale (Chapter 1) |
| $Lo$ | Lock number ($c_{l,\alpha}\rho c R^4 / I$) |
| $m$ | mass |
| $\dot{m}$ | mass flow rate |
| $M_0$ | pitching moment about leading edge |
| $Ma$ | (tip) Mach number ($\lambda V_0 / a_0$) |
| $n$ | integer; coordinate normal to surface (element) |
| $\vec{n}$ | outer unit normal vector (along contour or control-volume surface) |
| $p$ | fluid static pressure |
| $p_0$ | ambient static pressure |
| $p_\infty$ | freestream static pressure |
| $p(U)$ | probability density function |
| $P$ | actuator/rotor/turbine power |
| $\overline{P}$ | average wind/turbine power |
| $P_{Rated}$ | rated wind/turbine power |
| $q_s$ | surface heat flux |
| $Q$ | rotor torque |
| $r$ | radial blade coordinate (radial location) |
| $\vec{r}$ | marker position vector (Chapter 6) |
| $r'$ | normalized radial blade coordinate ($r/R$) |
| $r_c$ | viscous core radius (Chapter 6) |
| $r_j$ | radius of vortex filament $j$ emanating from blade trailing edge in HVM (Chapter 6) |
| $\vec{r}'_j$ | initial position vector (or vector to starting point) of vortical trailer in HVM (Chapter 6) |
| $r_{Root}$ | (blade) root cut-out radial location |
| $R$ | blade radius; specific gas constant (air) |
| $Re$ | Reynolds number |
| $Re_c$ | chord-based (blade section) Reynolds number ($\rho V_{rel} c / \mu$) |
| $Re_D$ | Reynolds number based on rotor diameter ($2\rho V_0 R / \mu$) |
| $s$ | coordinate tangential to surface (element) |
| $S$ | wing area |
| $t$ | time; airfoil thickness (Chapter 4); blade thickness distribution (Chapter 8) |
| $t^*$ | eddy turnover time |
| $T, T(z)$ | actuator/rotor thrust force; absolute temperature |
| $u$ | uniform axial ($x$-direction) velocity across actuator/rotor disk |
| $u_1$ | uniform axial ($x$-direction) velocity across streamtube exit plane |
| $u_i$ | axial induced velocity (Chapter 6) |
| $u, u(t)$ | instantaneous wind speed along the $x$-direction |
| $u', u'(t)$ | fluctuating wind speed along the $x$-direction; perturbation velocity in the $x$-direction |
| $u^*$ | horizontal velocity scale (friction velocity) in boundary layer |
| $U, U(z)$ | mean wind speed in the $x$-direction (mean value of $u(t)$) |
| $\overline{U}, \overline{U}(z)$ | average wind speed in the $x$-direction (typically taken over a period of 10 minutes or 1 hour) |

| | |
|---|---|
| $U_e$ | edge velocity of airfoil boundary layer |
| $U_\infty$ | freestream speed (referring to airfoil flows) |
| $\vec{V}_{rel}$ | fluid-parcel velocity vector relative to rotating system |
| $V_0, \overline{U}(z_{hub})$ | wind speed (average) at hub height |
| $V_{In}$ | cut-in wind speed |
| $V_{Out}$ | cut-out wind speed |
| $V_{Rated}$ | rated wind speed |
| $V_{rel}$ | (local) inflow/relative velocity |
| $\vec{V}(\vec{r}, t)$ | marker advection velocity (Chapter 6) |
| $w'$ | fluctuating wind speed along the $z$-direction; perturbation velocity in the $z$-direction |
| $w^*$ | vertical velocity scale in boundary layer |
| $w_i$ | angular induced velocity (Chapter 6) |
| $w_T$ | downwash in Trefftz plane |
| $w_W$ | downwash at lifting line (or wing) |
| $x$ | downwind/streamwise coordinate (fixed and rotating axis systems) |
| $x_{a.c.}$ | chordwise location of airfoil aerodynamic center (measured from leading edge) |
| $x_{c.p.}$ | chordwise location of airfoil center of pressure (measured from leading edge) |
| $y$ | lateral (across-wind) coordinate with respect to vertical axis in fixed axis system; wing-span coordinate; radial (along blade length) coordinate in rotating axis system |
| $z$ | vertical coordinate (upwards positive) in fixed axis system |
| $z_0$ | surface roughness |
| $z_i$ | height of capping inversion; boundary-layer height/depth |

**Greek**

| | |
|---|---|
| $\alpha$ | angle of attack – i.e. angle between (local) flow incident on the airfoil/blade and the blade chord line; (wind-shear) power-law exponent |
| $\overline{\alpha}$ | mean angle of attack |
| $\dot{\alpha}$ | time rate-of-change of angle of attack |
| $\alpha_0$ | airfoil zero-lift angle of attack |
| $\alpha_g$ | wing setting angle (or geometric angle of attack) |
| $\alpha_e$ | effective angle of attack $(\alpha_g + \alpha_i)$ |
| $\alpha_i$ | induced angle of attack |
| $\alpha_{Stall}$ | deep-stall angle (of attack) |
| $\alpha_W$ | wing setting angle at mid-span |
| $\beta$ | blade pitch angle – i.e. angle between local chord line and rotor plane $(\beta_{Twist} + \beta_0)$ |
| $\beta_0$ | collective blade pitch angle $(\beta_{Tip} - \beta_{Twist}(R))$ – i.e. the collective blade pitch angle is chosen such that when applied to a built-in twist, results in the desired blade tip pitch angle. |
| $\beta_{Tip}$ | blade tip pitch angle, $\beta(R)$ – i.e. angle used to describe "pitch setting" of blade |

| | |
|---|---|
| $\beta_{Twist}$ | built-in (manufactured) twist |
| $\gamma$ | vorticity distribution along airfoil camber line (Chapters 4, 5); rotor/turbine yaw angle (Chapter 5) |
| $\gamma_w$ | wake vorticity (Chapter 5) |
| $\Gamma$ | (bound) circulation on airfoil (and along wing or blade) |
| $\Gamma_B$ | circulation of rotor with $B$ blades |
| $\Gamma_{max}$ | maximum circulation |
| $\Gamma_\infty$ | circulation of rotor with infinite number of blades |
| $\Gamma()$ | gamma function (Chapter 1) |
| $\Gamma_d$ | adiabatic lapse rate |
| $\delta$ | boundary-layer thickness (Chapter 4); relative drag increase (Chapter 6) |
| $\delta^*$ | displacement thickness |
| $\Delta c_l$ | viscous lift-coefficient correction (Chapter 4) |
| $\Delta c_p$ | local change/difference in pressure coefficient from upper to lower airfoil surface |
| $\Delta_{grid}$ | grid spacing |
| $\Delta p$ | pressure discontinuity/jump at actuator disk |
| $\Delta t$ | time step |
| $\Delta\alpha$ | amplitude of angle-of-attack oscillation; range of angle of attack |
| $\delta\Gamma$ | trailing vorticity |
| $\Delta\delta_{TE}$ | difference between upper and lower boundary-layer thickness at airfoil trailing edge ($\delta_u - \delta_l$) |
| $\Delta\Gamma$ | viscous circulation correction (Chapter 4) |
| $\varepsilon$ | specific turbulent energy dissipation rate; Gaussian projection radius (Chapter 7) |
| $\eta_{N,m}$ | Gaussian projection function (Chapter 7) |
| $\Theta$ | transformed chordwise coordinate in airfoil frame (Chapter 4); momentum thickness (Chapter 4) |
| $\Theta_0$ | potential temperature on the ground (z = 0) |
| $\phi$ | local blade flow angle ($\alpha + \beta$) – i.e. angle between local flow incident and rotor plane |
| $\kappa$ | von Kármán constant; thrust constraint parameter (Chapter 8) |
| $\lambda$ | tip speed ratio |
| $\lambda_r$ | local speed ratio ($r/R \cdot \lambda$) |
| $\Lambda$ | Lagrange multiplier (Chapter 8) |
| $\mu$ | air (or fluid) dynamic viscosity |
| $\mu_t$ | turbulent eddy viscosity |
| $\nu$ | air (or fluid) kinematic viscosity |
| $\Xi$ | blade geometry parameter ($\lambda\sigma'r'c_l$) |
| $\rho$ | air (or fluid) density |
| $\rho_\infty$ | freestream air (or fluid) density |
| $\sigma$ | rotor solidity (integral of local solidity along blade radius) |
| $\sigma'$ | local solidity ($Bc/2\pi r$) |
| $\sigma_u$ | standard deviation of wind speed in the $x$-direction |
| $\sigma_v$ | standard deviation of wind speed in the (across-wind) $y$-direction |
| $\sigma_z$ | standard deviation of wind speed in the (vertical) $z$-direction |
| $\tau$ | shear stress; time constant; characteristic time |

| | |
|---|---|
| $\tau_0$ | wall shear stress (on boundary-layer surface) |
| $\tau_w, \vec{\tau}_w$ | skin friction (or wall shear stress) on airfoil |
| $\tau'_{xz}$ | turbulent shear stress in the $x$-direction (normal vector in the $z$-direction) |
| $\varphi(k)$ | phase angle of Theodorsen function |
| $\varphi_j$ | phase angle of initial position vector $\vec{r}'_j$ (Chapter 6) |
| $\psi$ | azimuthal (or circumferential) angle – i.e. blade angle in rotor plane with respect to tower axis |
| $\omega$ | wake angular speed (or wake swirl, added wake rotation, Chapter 2); vorticity (Chapter 4); angular frequency (Chapter 5); turbulent energy frequency (Chapter 7) |
| $\omega_N$ | natural frequency |
| $\vec{\omega}$ | vorticity vector ($\vec{\nabla} \times \vec{U}$) |
| $\Omega$ | disk/rotor angular speed (or rotor speed); simply-connected domain (Chapter 4) |
| $\vec{\Omega}$ | angular velocity vector |
| $\partial\Omega$ | boundary of simply connected domain |
| $\vec{\nabla}$ | nabla operator ($\partial/\partial x\vec{i} + \partial/\partial y\vec{j} + \partial/\partial z\vec{k}$) |

## Subscripts

| | |
|---|---|
| 0 | reference condition (upstream or wall/surface); reference angle; reference to airfoil leading edge |
| 1 | reference condition (downstream); reference in tip loss correction |
| $\infty$ | farfield/freestream condition; reference to infinite number of blades |
| – | normalized quantity in airfoil boundary layer |
| $A$ | aerodynamic |
| $a.\,c.$ | aerodynamic center |
| $B$ | blade; reference to finite number of blades |
| $Bo$ | bound |
| $c$ | chord; core (Chapter 6) |
| $crit$ | critical |
| $c.\,p.$ | center of pressure |
| $C$ | reference to arbitrary point/location; centrifugal |
| $CO$ | Coriolis |
| $d$ | drag; adiabatic |
| $e$ | effective |
| $D$ | reference to rotor diameter; drag |
| $fs$ | fully separated |
| $hub$ | hub height |
| $He$ | helix |
| $i$ | integer; inversion; induced |
| $inv$ | inviscid |
| $iter$ | iteration |
| $j$ | integer |
| $k$ | integer |

| | |
|---|---|
| $l$ | lift; lower |
| $m$ | integer; moment |
| $n$ | integer index used for fourier coefficients |
| $N$ | integer; normal; natural |
| $opt$ | optimum |
| $p$ | pressure; pitching |
| $P$ | power; reference to arbitrary point/location |
| $r$ | radial coordinate along blade; reference quantity |
| $rel$ | relative (at blade section) |
| $rot$ | rotor entrainment |
| $s$ | surface |
| $S$ | small scale |
| $t$ | turbulent; thickness |
| $T$ | thrust; tangential; Trefftz |
| $TE$ | trailing edge |
| $u$ | reference to wind speed in the (streamwise) $x$-direction; upper |
| $U$ | utility scale |
| $v$ | reference to wind speed in the (lateral) $y$-direction |
| $w$ | reference to wind speed in the (vertical) $z$-direction; wake |
| $W$ | wing; wake |
| $x$ | vector component in the $x$-direction; streamwise variable |
| $Xtr$ | chordwise location of laminar-turbulent transition (measured from leading edge) |
| $y$ | vector component in the $y$-direction |
| $z$ | vector component in the $z$-direction |

## Superscripts

| | |
|---|---|
| $*$ | reference to velocity scale in boundary layer |
| $'$ | normalized/dimensionless; fluctuating component; per unit length/width; slope (first derivative) of function; perturbation |
| $-$ | time average |
| $\sim$ | normalized |
| $rem$ | remainder |
| $st$ | static |

## About the Companion Website

This book is accompanied by a companion website:

**www.wiley.com/go/schmitz/wind-turbines**

The website includes:

- X Turb Files

Scan this QR code to visit the companion website.

XTurb executable & User's manual can be downloaded from:

**https://www.rotoraero.psu.edu/xturb-psu/**

# 1

# Introduction: Wind Turbines and the Wind Resource

*Energy is the single most critical challenge facing humanity.*
                                  – Richard Smalley (Nobel Laureate, Chemistry)

## 1.1  A Brief History of Wind Turbine Development

The stewardship of energy, food, and water is the most critical challenge facing humanity. The realization that (i) our planet's resources are finite, (ii) an adverse impact of humanity on climate and the environment is undeniable, and (iii) planet Earth might be hosting $10^{10}$ people within the next couple centuries can affect individuals in different ways. Some of us may try to forget, or simply ignore the truth in order to not be consumed by worry and the apparent lack of responsibility for the future to come. Others are in denial, and are so intentionally to maximize personal power and profit. But then again others may decide "to do something" through science, education, and/or service to the community and at whichever scale they are capable of and comfortable with. Every contribution, be it small or large, is essential to growing and teaching future generations in mindfulness and responsibility to our planet and the life it hosts.

### 1.1.1  Why "Wind Energy"?

For this purpose, it is always important to have "the facts right" and to be able to explain to anybody and at any time the "Why" a particular subject is important and impactful to the stewardship of energy, food, and water. The author asks every student at the beginning of his graduate course to prepare a townhall talk on "Why Wind Energy?" The reader is also encouraged to think of 20 second, 2 minute, and 20 minute explanations suitable for the public as to the importance of wind energy. The following bullets form a simple list of facts that distinguish wind energy from practically all other sources of energy and are helpful in developing these pep talks.

**Why "Wind Energy"?**

- Free energy source
- No emissions
- No water use

*Aerodynamics of Wind Turbines: A Physical Basis for Analysis and Design,* First Edition. Sven Schmitz.
© 2020 John Wiley & Sons Ltd. Published 2020 by John Wiley & Sons Ltd.
Companion website: www.wiley.com/go/schmitz/wind-turbines

- Scalable
- Wind energy = less dependence on fossil fuels

First of all, wind energy is a *free* source of energy, and some wind resource is available anywhere on the planet. Hence "wind" is not a rare resource whose harvest requires mining, transportation, and other costly logistics and environmental impacts. Thus, by virtue of its nature, wind energy is unlikely to generate conflicts and trade inequalities as other forms of energy do. Second, wind turbines produce energy with *zero* emissions. Consolidated life cycle greenhouse gas emissions associated with turbine manufacturing, installation, operation, and decommissioning are one of the lowest among all sources of energy, second only to nuclear power (Logan et al. 2017; NREL 2013). Third, and becoming an issue of growing importance, is that wind energy does *not* require any use of water. This is a major advantage of wind produced energy with respect to the stewardship of energy, food, and water as it does *not* use this increasingly valuable resource. Fourth, wind energy is *scalable*. Wind enables producing energy over several orders of scales, from small-scale powering phones or single light bulbs to entire neighborhoods by a single large utility-scale wind turbine. Last, but not least, any energy produced by wind reduces our dependence on fossil fuels, a fact with far-reaching consequences on global climate and politics.

Though all of these are compelling facts in support of wind energy, one has to also address its primary weakness. The "Wind" is *intermittent*, that is, it is neither constant in strength nor does it occur all the time. This may present challenges in energy supply but nothing that cannot be not solved through appropriate storage systems at smaller scale and/or smart power grid systems at larger scale where a projected "lack of wind" can be compensated for by other sources of energy. It is therefore important to realize that wind energy cannot easily supply 100% of power and at all times, though it nearly does so in some smaller countries (e.g. Denmark), but that wind energy is definitely already a notable competitor to coal, gas, and oil in some large industrial countries (e.g. Germany) (GWEC 2018).

### 1.1.2 Wind Turbines Then and Now

Windmills have existed in many forms for two millennia and have been used primarily for grinding grain and pumping water. Today, wind turbines have converged to the most popular types of either two- or three-bladed upwind horizontal-axis machines that generate electrical power over a range of scales. The interested reader is referred to other literature for a more complete description of the history of wind energy. For example, the work of Golding (1977) covers the history from the ancient Persians to the 1950s, with an emphasis on turbine design. A briefer description can be found in Eldridge (1980). An example of a historical review of the more recent history of wind energy system can be found in Spera (2009), while Eggleston and Stoddard (1987) give emphasis to some of the key wind turbine components in use today. In the following, a brief history of windmill and horizontal-axis wind turbine (HAWT) technology is presented.

#### 1.1.2.1 The Windmill – Hero of Alexandria (First Century CE)

Hero of Alexandria, also known as Heron of Alexandria, is believed to have lived in the early first century CE. He was a mathematician and engineer and is considered to be one of the most influential experimentalists of antiquity. His work on *pneumatics* is considered to contain the first reference to a windmill, though there has been some debate about the windmill device described by Hero, see Drachman (1961) and Shepherd (1990). Nevertheless, H.P. Vowles, one of the primary scholars on the subject, concludes that Hero's description of a windmill powering a pipe organ is indeed supported by the technology knowledge of his time, see Vowles (1932). It is interesting to see that the next reference to a windmill dates back to somewhere between the fifth and ninth century CE and the region of Seistan (today, eastern Iran) where vertical-axis windmills were used for grinding grain. In this concept, rectangular sails were attached to a vertical shaft to drive a grinding stone, see Vowles (1932) for reports by Al Masudi. These drag-type devices are still in use today.

### 1.1.2.2  1200s–1300s – Post Mills and Tower Mills

In Northern Europe, windmills made their first appearance in the 1200s in England but had probably already been introduced in the tenth and eleventh centuries, see Vowles (1932). It is unknown whether or not the general concept was developed in Europe or imported from the Middle East. Vowles (1932) suggests that it is plausible the Vikings actually picked up the conceptual idea on their voyages and brought it back home. It is also not well understood how the transition from a vertical-axis to a horizontal-axis machine came about, one possible explanation being the realization that horizontal-axis machines are less prone to break in high winds. It is reasonable to think, however, that this significant transition was accompanied by a parallel design evolution of rigging on ships, that is, replacing square (drag-device) sails by concepts such as an upwind tacking (lift-device). The post mill, developed in the 1200s, was typically a four-bladed configuration mill mounted on a central post, thus being a horizontal-axis machine. The motion of the horizontal shaft was translated to rotation about a vertical axis to turn a grindstone using wooden cog-and-ring gears. In the 1300s, the Dutch refined the design to Tower Mills by affixing the standard post mill to the top of a multi-story tower, with separate floors devoted to grinding grain, removing chaff, storing grain, and living quarters for the windsmith and his family. The main disadvantage of both the post and tower mill designs was that the entire structure had to be oriented manually into the wind by pushing a large lever at the back of the mill, which was no easy task for the windsmith. By the 1800s, the European windmill had reached a high level of sophistication, with later smock mills being able to yaw only the upper part with the rotor into the wind, see Figure 1.1 for an example.

The European windmills were widely used until the late 1800s for grinding grain and pumping water but eventually fell victim to the Industrial Revolution and were replaced by coal powered steam turbines. At this point, the windmill blades had acquired some features of an airfoil shape, including a limited amount of blade twist. The power output of some machines could be adjusted by a mechanical control system.

### 1.1.2.3  1700s – John Smeaton

Scientific testing and performance evaluation started in the mid-1700s in England through the work of John Smeaton (1724–1792), see Smeaton (1759), the founder of the first windmill testing apparatus seen in Figure 1.2.

Smeaton conducted experiments using his apparatus and found three basic rules concerning wind turbine performance.

**Figure 1.1** European smock mill near Waldfeucht (Germany). Source: Photo credit: Michael Küsters.

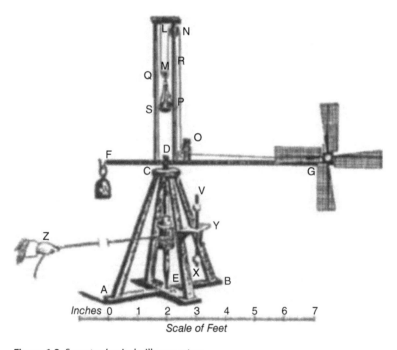

**Figure 1.2** Smeaton's windmill apparatus.

- *Rule 1.* There is an optimum ratio of the blade tip to wind speed (known today as the tip speed ratio)
- *Rule 2.* The maximum torque is proportional to the wind speed squared
- *Rule 3.* The maximum power is proportional to the wind speed cubed

**Figure 1.3** American water-pumping fan mill (U.S. Department of Agriculture).

The third rule means, in particular, that a twice-as-fast wind speed allows for an eight-times increase in power. All three basic rules will be derived mathematically in Chapter 2.

### 1.1.2.4   1800s – Windmills in the American West
While European windmills were increasingly replaced by steam turbines, an evolution of the "fan mill" occurred in the American West. Figure 1.3 shows a typical fan mill used for pumping water on ranches and railroads.

The fan mill had an easy and cheap design that allowed unattended operation. A downstream vane worked as a passive control system to quickly yaw the turbine into the wind and had hinges to furl the rotor in high winds. Its high-solidity low-speed design generated a high torque ideal for pumping water.

### 1.1.2.5   Late 1800s – Wind in Transition (Mechanical – Electricity, Drag – Aerodynamic Principles)
The transition from mechanical to electrical power emerged with the invention of the electrical generator in the late 1800s, and it seemed reasonable to try and turn a generator using a suitable wind turbine design. Indeed, turbine design became increasingly influenced by concurrent advances in aerodynamic principles. This transition is illustrated in Figure 1.4, with the earliest example being the high-solidity low-speed Brush postmill (built in 1888 by Charles Brush in Cleveland, Ohio), with a rotor diameter of 17 m and a rated power of approximately 12 kW, thus essentially being a large fan mill design.

In contrast, Poul La Cour in Denmark built the first wind turbines based on aerodynamic design principles between 1891 and 1918, with rated power between 10 and 35 kW and lighter low-solidity higher-speed rotors enabled by using actual airfoils along

(a)                                                                                (b)

**Figure 1.4** Wind in transition: Brush postmill (a) and Poul La Cour in front of his experimental wind turbines (b). Source: Loaned from the Poul La Cour Museum.

the blade span, see Figure 1.4. As such, La Cour's turbines clearly illustrate turbine design principles in transition.

### 1.1.2.6 1900s–1950s – Wind Turbines across Scales (kW – MW)

In the early twentieth century, other small wind turbine generators were developed, most notably those by Marcellus Jacobs. These were essentially three-bladed propeller-based designs driving a DC generator. Having a rotor diameter of only approximately 4 m, these small-scale turbines were ideal designs for residential-scale electricity generation in the 1.8 – 3.0 kW range. The Jacobs turbines, developed by the Jacobs Wind Electric Company, were used primarily for lighting farms in the American Midwest. Their demise came with increasing rural electrification in the 1930s, though the Jacobs turbines are still the baseline for modern small-scale wind turbines in use today. In Europe, Johannes Juul (a student of Professor Poul La Cour) was a pioneer in AC power generation and developed the first wind turbine to generate AC power in the 1950s. The Gedser wind turbine was a three-bladed wind turbine rotor with a rated power of 200 kW. It is noteworthy that Johannes Juul also invented the tip brake, see the photograph of a Gedser turbine in Figure 1.5. A small two-bladed design rated at 35 kW was developed by F.L. Smidth in Denmark and the second turbine to produce AC power. This upwind turbine was stall-regulated, used modern airfoils, and is considered to be a trendsetter for modern wind turbines.

Completed in 1941, the Smith-Putnam wind turbine was the first MW-scale wind turbine. It was designed by P.C. Putnam and manufactured by the S. Morgan Smith Company in Vermont, USA. The turbine had two blades and a diameter of 53.3 m, achieving a rated power of 1.25 MW, see Figure 1.5 for a photograph. Each blade weighed approximately 8 tons. The Smith-Putnam turbine suffered a blade failure after only a few hundred hours of operation, resulting in the project being abandoned entirely in 1945 as materials were deemed not advanced enough for the scale, and resources were not available right after WWII. The interested reader is referred to *Power from the Wind* (Putnam 1948) for a truly fascinating summary of Putnam's engineering ingenuity. A breakthrough in materials research was done in Germany by Ulrich Hütter in the 1950s, who developed a new approach of using elastic fiberglass blades on a

(a)                                                    (b)

**Figure 1.5** Wind across scales: Gedser turbine, 200 kW (a) and Smith-Putnam turbine, 1.25 MW (b).

downwind teetering two-bladed hub, see Dörner (2002). These experimental turbines later became the baseline prototypes for more modern configurations after the oil crises in the 1970s, though until then, wind generated electricity became dormant in the competitive energy market of expanded use of coal and nuclear power.

### 1.1.2.7  1970s–2000s – Modern Utility-Scale Wind Turbines (>1 MW)

The slow resurrection of wind energy came along with an environmental movement that began to advocate cleaner sources of energy. Many people became aware of the environmental consequences associated with fossil-fuel-based industrial developments through books such as *Silent Spring* (Carson 1962) and *Limits to Growth* (Meadows et al. 1972). This combined with the oil crises in the 1970s resulted in several administrations of industrial countries investing in basic research and technology in the area of wind power generation. A prime example is that of the California Wind Rush, particularly along the Altamont Pass area in California, that inspired people around the world. While investment in the United States has been very intermittent, for example federal tax credits initiated by the Carter administration were withdrawn by the Reagan administration, Europe has seen a more continuous technology development and deployment of wind power. It is only over the last decade that Asia has realized the competitive technology of wind power, first investing in wind power deployment and more recently in research and development. At the same time, the US continues an up-and-down investment and incentive trend in various states.

Modern onshore wind turbine designs have converged to three-bladed upwind HAWTs, see Figure 1.6. An upwind configuration, that is, rotor located upwind of the tower, requires a yawing control system (as an upwind rotor is statically unstable),

(a)                                        (b)

**Figure 1.6** Examples of modern wind turbines: (a) a Siemens 2.3 MW turbine at the National Wind Technology Center in Boulder, CO, USA (with Dr. Matt Churchfield and the author in June 2012) and (b) a wind turbine array near Kirchhoven, Germany. Source: Photo credit: Pankaj K. Jha and Michael Küsters.

while a downwind configuration does not require a yawing mechanism (and is statically stable), though it operates in the tower wake for some portion of the rotor azimuth. A photograph of a modern utility-scale wind turbine is shown in Figure 1.6, along with a wind turbine array used to generate community-scale power.

At this stage, onshore turbine development is somewhat limited by transportation and therefore blade length and tower height, rotor tip speed due to noise constraints, and the fact that many of the best wind sites have been taken, though they are scheduled for replacement from the 2020s. The offshore market, on the other hand, has been very attractive, particularly in Northern Europe where a rather shallow sea and good wind resource allowed the development of large-scale wind farms, for example Horns Rev off the coast of Denmark and the Lillgrund wind farm in southern Sweden.

### 1.1.3   Influence of Aerodynamics on Wind Turbine Development

Fundamentally, the extraction of power from the wind is rooted in basic aerodynamic principles. Before introducing the mathematical description and derivation of the theory in Chapters 2 and 3, a very brief history of aerodynamics is presented, with focus on discoveries pertinent to wind turbine aerodynamics. A good reference on the complete history of aerodynamics is that from Anderson (1997).

*250 BCE* According to Archimedes, a difference in pressure is required to set a fluid in motion; likewise, a fluid flow is decelerated by exerting an adverse pressure gradient (i.e. increasing pressure in streamwise direction) on the fluid. We will use this mechanism in Chapter 2 when introducing the actuator disk concept.

*1490.* Leonardo da Vinci derives the important "continuity relation" (or mass balance) for an incompressible fluid and observes that the resistance force acting on a solid body immersed in a fluid stream is proportional to the frontal area of the body, an important finding that we will make use of in streamtube analysis in Chapter 2.

*1600s.* The experimental and theoretical contributions by Galileo, Mariotte, Newton, and Huygens proved that the resistance force is proportional to the fluid density and the flow speed squared, the basis for blade-element forces as derived in Chapter 3.

*1700s.* Relevant works to the field of wind energy today are those by Euler and Smeaton. Euler derived the energy principle (also known in its simpler form today as "Bernoulli's Equation") from his incompressible and inviscid momentum equation in differential form. The rules of Smeaton (Section 1.1.2) concern basic mechanical relations of rotor torque and power and their proportionality to the flow speed squared and cubed, respectively. These relations are very relevant in basic momentum theory as we will see in Chapter 2.

*1800s.* The fundamental equations (mass, momentum, and energy) for an incompressible and inviscid fluid are now extended to the viscous flow equations by the combined contributions of Navier and Stokes, known today as the Navier–Stokes equations. These form the most general set of differential equations in incompressible flow until today and form the basis for boundary-layer theory and are solved numerically on today's computers in the field of computational fluid dynamics (CFD), more on this appears in Chapters 4 and 7.

*Early 1900s.* The field of applied aerodynamics is initiated by the circulation theory of lift due to Kutta and Joukowski and Prandtl's boundary-layer and lifting-line theories. This was followed by Munk's thin-airfoil theory, Goldstein's propeller theory, the axial momentum and energy considerations according to Betz, and, most notably, the work of Glauert in the 1930s that forms the basis for blade-element momentum (BEM) theory still in use today and expanded in Chapter 3.

*Mid-1900s – Today.* The field of rotor aerodynamics has seen continued advances and cross-disciplinary discoveries in both areas of wind turbine aerodynamics and rotorcraft aeromechanics. Airfoil design methodologies have been developed, resulting in high-performance airfoils designed for specific operating conditions along the blade span. Advances in CFD have further propelled full solutions to the Navier–Stokes equations around complex blade/rotor configurations. Research data generated by experiments have assisted in detailed verification and validation studies of both BEM and CFD codes, with both having seen continued developments. Highly-resolved experimental and CFD data are helpful in improving faster design tools based on BEM theory, which are still used as the primary design tools due to their inherently fast computational speed. The triangle of BEM, CFD, and experiments has been exceedingly valuable in advancing aerodynamics in general and particularly the area of wind turbine aerodynamics. Nevertheless, many flow phenomena are not well understood, for example, tip effects discussed in Chapter 3, unsteady effects and rotational augmentation as presented in Chapter 5, and continued basic research is necessary on advanced computational methods (Chapter 7), as well as design of scaled wind turbine rotors and optimization methods (Chapter 9).

### 1.1.4 Design Evolution of Modern Horizontal-Axis Wind Turbines

The evolution of wind turbine scale from the KW to MW range has gradually progressed since the 1980s, see Figure 1.7, with the rated power being approximately proportional to the turbine diameter squared (see Chapters 2 and 3), depending on particular design and application. Figure 1.7 displays the evolution and indicates the overall dimensions of the largest turbine rotor over time.

It is impressive to see that the largest wind turbines in use today have rotor diameters larger than the wingspan of an Airbus A380. A breakout view of the nacelle showing the key mechanical components of a modern upwind utility-scale wind turbine is shown in Figure 1.8, with the primary components described in brief next.

- *Nacelle.* The housing on top of the tower that contains all key components, in particular gearbox and generator
- *Hub.* Structure outside the nacelle to which rotor blades are attached. The hub rotates with the low-speed shaft at typically $5 - 20$ rpm
- *Low-speed shaft.* Connects the hub and the gearbox
- *Gearbox.* Transfers torque from low-speed shaft to high-speed shaft
- *High-speed shaft.* Delivers power to the generator at $\sim$1500 rpm (gearbox transmission ratio is $\sim 1:100$)
- *Generator.* Designed to handle a time-varying power source, different from other conventional uses of motors/generators
- *Yaw mechanism.* Driven by electric motors to rotate nacelle into wind direction. Controlled by data from wind vane on the rear top of the nacelle
- *Anemometer and wind vane.* Measures wind speed and direction. Data used to operate yaw, pitch, and variable-speed mechanisms and to stop turbine when wind is too low or high

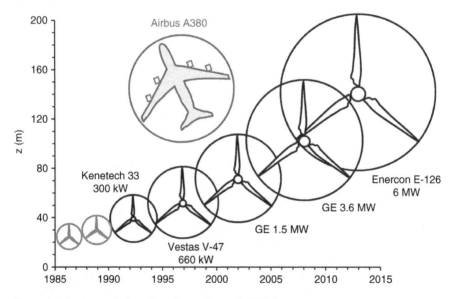

**Figure 1.7** Design evolution of modern utility-scale HAWTs.

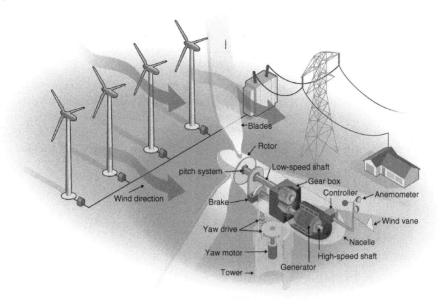

**Figure 1.8** Key components of a modern utility-scale wind turbine. Source: Published with permission from the National Renewable Energy Laboratory.

- *Control panel.* Continuously monitors wind conditions and turbine. Activates control mechanisms and sends alarm messages to control station
- *Cooling unit* (not shown). Contains electrical fan and radiator to cool gearbox oil

The following section gives an introduction to the wind resource, the power curve of a typical utility-scale wind turbine, and basics on the statistical description of wind data and estimating wind power production.

## 1.2   Wind Resource Characterization

The sun is the original source of all wind power and energy on Earth. The global origins of "Wind" arise from both uneven solar heating and the effects of the Earth's rotation as a function of latitude. Both mechanisms cause pressure differences over the Earth's surface, thus driving the largest wind scales, that is, the "Trade Winds," predominantly in the horizontal direction coming from the East closer to the equator and changing direction as "Westerlies" north/south of the horse latitudes (Hiester and Pennell 1981). Friction on the Earth's surface retards the flow close to the ground, but its effect vanishes at the edge of the atmospheric boundary layer (ABL) between 0.5 and 2.0 km above the surface. Above the ABL, essentially a frictionless geostrophic flow balances inertial forces due to large-scale motions, pressure-gradient forces, and the Coriolis force due to the rotation of the Earth. The global wind patterns of trade winds and westerlies are perturbed due to the Earth having considerable surface variations ranging from land/sea distributions to flat and mountainous terrain. Because the sea acts primarily as a sink of solar radiation, flow patterns arise at the land/sea interface and due to variable surface roughness and radiation absorption over land. These, when occurring over the global scale, contribute to secondary climatic circulation patterns such as hurricanes, monsoons, and cyclones. Tertiary circulations refer to local flow patterns such as thunderstorms, land-/sea breezes, mountain-/valley winds, and local tornadoes. While the tertiary circulations are of much smaller scale compared to the climatic-scale winds, they are still of a large scale with respect to wind turbine aerodynamics as they are driven by scales larger than the boundary-layer height/depth. Small-scale motions within the ABL, on the other hand, do affect wind turbines and wind farms and include thermals, wakes, dust devils, and low-level jets. A brief summary of the primary wind scales is given next:

- *Climatic scale* (100 km – 10 000 km). Trade Winds, Westerlies, Monsoons, Cyclonic Storms, Hurricanes
- *Large scale (100 m – 100 km).* Thunderstorms, Land-/Sea Breezes, Mountain-/Valley Winds, Tornadoes
- *Small scale (1 m – 100 m).* Thermals, Dust Devils, Wakes, Low-level Jets, Gusts

Figure 1.9 shows a schematic of the aforementioned spatial scales and their respective temporal scales, ranging from minutes to practically a complete year (Spera 2009). This, as we will discuss in more detail in Chapter 5 with focus on unsteady aerodynamics, has important implications on the interaction of atmospheric scales with blade-section aerodynamics.

Also shown in Figure 1.9 are approximate ranges of spatial-/temporal scales that affect wind resource assessment, wind site selection, and wind turbine design. It becomes clear that only the smaller scales on the order of the rotor diameter and smaller (<100 m) are the primary scales that affect wind turbine design. It is important to realize that the large scales affecting wind site selection range over temporal scales from minutes (short-term)

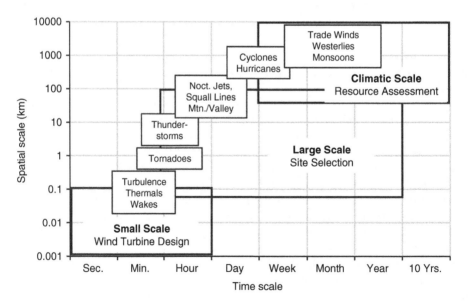

**Figure 1.9** Atmospheric motion – time and space scales.

to multiple years (inter-annual). Similar to the primary spatial scales listed previously, one can categorize the temporal scales coarsely as:

- *Inter-annual (years).* In general, meteorologists advise taking at least five years of data at a given site to have a reliable average wind speed, a difficult (and sometimes not feasible) task when assessing a new wind site with sparse data records.
- *Annual (days to years).* Seasonal wind variations are common over most parts of the world; for example, the Great Plains having spring maxima, while winds in the Eastern US peak during the winter.
- *Diurnal (minutes/hours to days).* There can be large wind variations caused by differential solar heating during day and night, leading to different states of atmospheric stability. In general, the largest diurnal variations occur around summer and the lowest in winter, with large site variation between flat and mountainous terrain.
- *Short-term (minutes).* Usually defined as intervals of 10 minutes or less, include gusts (minutes) and turbulence (seconds to minutes). Effect on wind turbine aerodynamics through maximum load and fatigue prediction, structural modes, control, and power system response.

A gust is a discrete event in a turbulent wind field and is also associated with, for example, a high-speed turbulent region (or streak) within the ABL with a time scale up to several minutes. Figure 1.10 shows an example of simulated short-term wind variations. Statistical analyzes of time series such as shown in Figure 1.10 can extract gust amplitude, rise/lapse time, and maximum wind-speed variation over the event.

### 1.2.1 Wind Resource – Available Power in the Wind

The wind resource (or available power in the wind) can be best described in terms of how much wind power, $P$ [W], crosses a reference area, $A$ [m$^2$], as a function of a representative wind speed, $V_0$ [m s$^{-1}$]. We can use the continuity equation from basic fluid

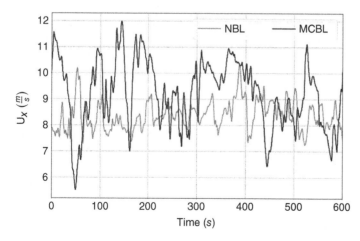

**Figure 1.10** Example of simulated short-term wind variations, *OpenFOAM ABL-LES*. (NBL = Neutral Boundary Layer; MCBL = Moderately-Convective Boundary Layer). Source: Reproduced with permission from Adam Lavely.

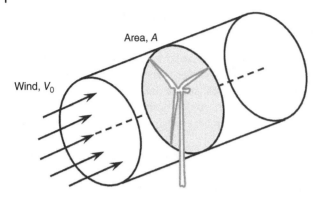

**Figure 1.11** Flow at wind speed, $V_0$, through area, $A$.

mechanics and compute the mass flow rate, $\dot{m} = dm/dt$ [kg s$^{-1}$], at a reference area as:

$$\text{Mass Flow Rate: } \dot{m} = \frac{dm}{dt} = \rho V_0 A \tag{1.1}$$

where $\rho$ [kg m$^{-3}$] is the air (or fluid) density, see schematic in Figure 1.11. The wind power, $P$, is the kinetic energy per unit time crossing area, $A$:

$$\text{Wind Power: } P = \frac{1}{2}\frac{dm}{dt}V_0^2 = \frac{1}{2}\rho V_0^3 A \tag{1.2}$$

Thus, the wind power per unit area, $P/A$ [W m$^{-2}$], becomes:

$$\text{Wind Power Density: } \frac{P}{A} = \frac{1}{2}\rho V_0^3 \tag{1.3}$$

The wind power density as defined in Eq. (1.3) is used to characterize the wind resource at a given site based on a mean wind speed, $V_0$.

Some important relationships can be inferred from Eqs. (1.2) and (1.3). First, both wind power and wind power density are proportional to the wind speed cubed, that is, $\sim V_0^3$, as was found experimentally by Smeaton (Rule 3). Furthermore, wind power is proportional to the flow cross-sectional area, that is, $\sim A$, and hence proportional to the diameter squared, that is, $\sim D^2$, for a circular area, $A$. These are two of the most important findings concerning the scaling of wind power. Table 1.1 lists some computed values for $P/A$ over a typical operating range of $V_0$.

The data presented in Table 1.1 reveal the significant influence of the wind speed, $V_0$, on the power available in the wind. If annual wind-speed averages are known for a given wind site, the wind power density can be used to generate "wind maps" for large areas as a function of height. For example, reference turbine hub heights of 50 and 80 m are typical to estimate wind turbine power production. Due to the cubic dependence on $V_0$, more accurate power estimates can be obtained if hourly wind-speed averages are available. For qualitative magnitude evaluations, a wind resource with $P/A < 100$ W m$^{-2}$ is considered "low," $P/A \approx 400$ W m$^{-2}$ is "good," and $P/A > 700$ W m$^{-2}$ is "great" for a candidate wind site.

Figure 1.12 shows a qualitative power curve, $P$ versus $V_0$, for a modern utility-scale, pitch-controlled, HAWT. For comparison, the wind power $P \sim V_0^3$ according to Eq. (1.2) is also shown in Figure 1.12. In general, a wind-turbine power curve is divided into three regions:

**Table 1.1** Wind power density, *P/A*, for steady uniform wind ($\rho = 1.225$ kg m$^{-3}$).

| Wind speed, $V_0$ (m s$^{-1}$) | Wind power density, $P/A$ (W m$^{-2}$) |
|---|---|
| 5.0 | 76.6 |
| 10.0 | 612.5 |
| 15.0 | 2067 |
| 20.0 | 4900 |
| 25.0 | 9570 |

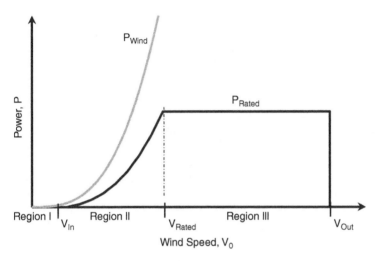

**Figure 1.12** Power curve of a pitch-controlled HAWT.

*Region I ($V_0 < V_{In}$)*. At wind speeds less than the cut-in wind speed, $V_{In}$, the flow around the turbine blades does not generate sufficient torque to accelerate the rotor against friction in the bearings and its mass moment of inertia. The wind turbine does not produce any power in Region I.

*Region II ($V_{In} < V_0 < V_{Rated}$)*. For wind speeds between the cut-in wind speed, $V_{In}$, and the rated wind speed, $V_{Rated}$, the wind turbine produces power $\sim V_0^3$ and to first order half of the total wind power, *P*, at least for large utility-scale machines. Typically, the rotor blades operate at constant pitch angle in Region II. In contrast, the rotor speed, $\Omega$, is variable in Region II and increases linearly with wind speed, $V_0$, thus keeping a close-to-optimal ratio between rotor tip speed and wind speed (known as the rotor tip speed ratio, $\lambda$). The rated wind speed, $V_{Rated}$, defines the end of Region II at which the rotor produces the rated generator power, and hence rated wind turbine power.

*Region III ($V_{Rated} < V_0 < V_{Out}$)*. For wind speeds higher than the rated wind speed, $V_{Rated}$, the power produced by the wind turbine needs to be controlled as it exceeds generator capacity. Pitch-controlled wind turbines actively feather (or pitch) the blades into the wind, thus reducing blade-section angle of attack and sectional force/torque. Consequently, the wind-turbine power no longer follows the $\sim V_0^3$ trend, but is now controlled to stay at the rated power, $P_{Rated}$. The end of Region III is marked by the cut-out speed, $V_{Out}$, at which the rotor needs to be decelerated safely without exceeding structural load limits.

In the following sections, we will learn more about the relationship between Regions II and III of a wind turbine power curve, the actual wind distribution about a mean wind speed, and the resulting power production estimates.

## 1.2.2 Basic Characteristics of the Atmospheric Boundary Layer

The ABL (or planetary boundary layer) is the lowest part of the troposphere, its height varying between 0.5 and 2.0 km depending on the daytime, weather, and location. In general, the wind speed is zero close to the ground and increases with height, also called the vertical wind shear. Knowledge of the wind conditions as a function of time and height above the ground are essential for siting wind projects and predicting wind turbine power production. Figure 1.13 shows some basic characteristics of the ABL. One can see, in particular, that the ABL is composed of the mixing layer, the surface layer, and the roughness layer (including the viscous sublayer).

The mixing layer (or outer layer) is the largest part of the ABL, roughly starting at a height of $z > 100z_0$ where $z_0$ is the surface roughness. In the mixing layer, turbulent momentum and energy are transported upward/downward depending on the particular weather event. In a highly simplified manner, the mixing layer can be thought of as providing momentum and energy for the turbulent energy cascades down to the surface. The largest turbulent eddies in the mixing layer are $>100$ m. In the surface layer ($10z_0 < z < 100z_0$), on the other hand, wind shear dominates, creating new scales of energy/momentum transported further along the inertial subrange of the turbulent energy cascade. Note that the depth of the surface layer is only about 10% of the total boundary-layer height/depth, $z_i$. Finally, the roughness sublayer ($z < 10z_0$) bridges the

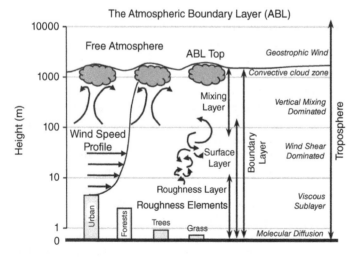

**Figure 1.13** Basic characteristics of the ABL.

**Figure 1.14** Vertical wind profiles – effect of stability state.

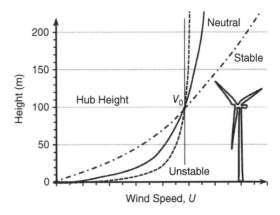

lower part of the surface layer and the viscous sublayer, characterized by molecular diffusion processes and essentially dissipating the cascaded energy/momentum at the smallest (Kolmogorov) scales of $1 - 2$ mm.

Note that modern utility-scale wind turbines have hub heights around 100 m (see Figure 1.14); hence some interaction can be expected between rotor aerodynamics and turbulent scales in the atmospheric surface layer. We proceed by writing a time-varying turbulent velocity, $u(t)$, as the sum of a mean velocity, $U$, and a fluctuating component, $u'$, as:

$$\text{Turbulent Velocity}: u(t) = U + u'(t) \tag{1.4}$$

The mean wind speed is typically averaged over a relatively short time period of either 3 minutes, 10 minutes, and up to 1 hour, depending on wind site and turbine selection. As a general rule-of-thumb, the time period chosen for averaging is longer than the characteristic time of turbulent velocity fluctuations, $u'(t)$. Hence the mean wind speed, $U$, at a given height above ground can be computed according to:

$$\text{Mean Velocity}: U = \frac{1}{\Delta t} \int_0^{\Delta t} u \, dt \tag{1.5}$$

Note that the time average of the fluctuating component, $u'(t)$, is equal to zero. In the following subsection, we assume an ABL of neutral stability state, that is, without surface heating/cooling, and derive some basic relations for the vertical wind shear.

### 1.2.2.1 Steady Wind Speed Variation with Height

Let us denote the steady wind speed as a function of height, $z$, by $U(z)$. When referring to the wind speed, $V_0$, in wind turbine design and analysis, we actually refer to the wind speed at turbine hub height, $z_{Hub}$, as:

$$\text{Wind Speed at Turbine Hub Height}: V_0 = U(z_{Hub}) \tag{1.6}$$

*Logarithmic Wind Profile (Log Law), U(z)* The logarithmic wind profile can be derived using either eddy-viscosity theory, similarity theory, or mixing-length theory (Wortman 1982). The dominant turbulent shear stress in the horizontal direction (normal vector in the $z$ direction) is

$$\text{Turbulent Shear Stress}: \tau'_{xz} = -\overline{\rho u' w'} = \mu_t \frac{\partial U}{\partial z} \tag{1.7}$$

where $u'$ and $w'$ are turbulent velocity fluctuations in the horizontal ($x$) and vertical ($z$) directions, and $\mu_t$ is the turbulent eddy viscosity, a model for small-scale turbulence based on the argument that turbulence acts as an increased viscosity. One such model is the simple mixing-length model where $\mu_t$ is defined as:

$$\text{Mixing-Length Model}: \mu_t = \rho l^2 \left| \frac{\partial U}{\partial z} \right| \tag{1.8}$$

In Eq. (1.8), $\rho$ is the air density and $l = \kappa z$ is the mixing length, assuming a smooth surface, with $\kappa = 0.4$ being the von Kármán constant. For a negligible horizontal pressure gradient, $\partial p / \partial x \approx 0$, it can be inferred from the momentum equation that the turbulent shear stress approximately equals the wall shear stress, $\tau_0$. Combining these relations, we obtain the following differential equation for $U(z)$

$$\frac{dU}{dz} = \frac{1}{\kappa z} \sqrt{\frac{\tau_0}{\rho}} \tag{1.9}$$

which can be integrated easily by separation of variables. We thus obtain

$$\text{log law}: U(z) = \frac{u^*}{\kappa} ln\left( \frac{z}{z_0} \right) \tag{1.10}$$

where $u^* = \sqrt{\tau_0/\rho}$ is the definition of the "friction velocity" and $z_0 > 0$ is a "surface roughness length," depending on the ground topology. Table 1.2 lists values for the surface roughness length, $z_0$, for various types of terrain taken from Garratt (1994) and Manwell et al. (2009). Note that $z_0$ is not the actual height of the respective roughness.

The log law of Eq. (1.10) can be used primarily for two applications: (i) Both the friction velocity, $u^*$, and surface roughness height, $z_0$, can be calculated if measurements of the mean wind speed, $U$, are available at two heights, $z$, for an assumed logarithmic profile; and (ii) the log law can be used to extrapolate the wind speed known at a reference height

**Table 1.2** Surface roughness length, $z_0$, for various natural surfaces.

| Surface | $z_0$ (m) |
| --- | --- |
| Very smooth ice | 0.00001 |
| Calm open sea | 0.0002 |
| Blown sea | 0.0005 |
| Snow surface | 0.003 |
| Grass (thin – thick) | 0.001–0.05 |
| Rough pasture | 0.01 |
| Crops (wheat – corn) | 0.025–0.064 |
| Woodland (trees, savannah) | 0.4 |
| Forests (pine – tropical) | 0.32–2.2 |
| Suburbs | 1.5 |
| Tall buildings (Downtown) | 3.0 |

that is, $U(z_r)$, to another height above the ground using the following relation:

$$\text{log -law extrapolation}: \frac{U(z)}{U(z_r)} = \frac{\ln(z/z_0)}{\ln(z_r/z_0)} \tag{1.11}$$

For example, Eq. (1.11) can be used to extrapolate meteorological wind data at $z = 50$ m to $z = 80$ m to estimate the wind power density for a higher turbine hub height.

***Exponential Wind Profile (Power Law), U(z)***  The power law is probably the simplest model for a vertical wind-speed profile as it only requires one reference point and a power law exponent, $\alpha$. The power law is mostly written in its extrapolation form as:

$$\text{Power-Law Extrapolation}: \frac{U(z)}{U(z_r)} = \left(\frac{z}{z_r}\right)^{\alpha} \tag{1.12}$$

A baseline from turbulent boundary-layer theory over a flat plate (Schlichting 1968) is $\alpha = 1/7$. In practice, however, values for $\alpha$ are highly variable depending on the wind site and atmospheric stability state. A number of correlations have been proposed for the power-law exponent, $\alpha$, in the literature. For example, Counihan (1975) proposes a correlation as a function of the surface roughness height, $z_0$, while Justus (1978) suggests that $\alpha$ is a function of the wind speed, $U$, measured at a given reference height, $z_r$. More recently, researchers at NASA proposed $\alpha$ equations based on both reference wind speeds and surface roughness (Spera 2009). In this context, it is imperative to understand that the wind power density in Eq. (1.3) is $\sim V_0^3$, hence extrapolation errors associated with either an erroneously assumed logarithmic wind profile or uncertainty in the power-law exponent can have a significant effect on a projected wind power density at a different height. The reader may quantify the associated uncertainties in wind power density by using the previous equations.

### 1.2.2.2 Turbulence and Stability State

In general terms, turbulence refers to velocity fluctuations in the wind speed with time scales of typically 10 minutes or less, the largest eddy scales containing most of the energy and the smallest scales progressively less energy. Turbulence is not a conserved quantity but a dissipative process. Hence turbulence decays and disappears, unless there are active processes to generate and/or sustain it. In the context of flat terrain, there are two primary mechanisms that generate and sustain turbulence: (i) Wind shear – viscous stress between predominantly horizontal velocity layers generates vorticity (rotation of fluid particles), leading eventually to shear instabilities and turbulence of various scales; and (ii) buoyancy – surface heating and associated vertical temperature gradients cause large-scale updrafts, forming turbulent eddies on the order of the largest-scale motions down to the viscous dissipation scale.

Turbulence is best described by statistical methods depending on the application. One of the main turbulence quantities is the turbulence intensity, $I$, defined as:

$$\text{Turbulence Intensity (I)}: I = \frac{\sigma_u}{U} \tag{1.13}$$

In Eq. (1.13), $U$ is again the mean wind speed and $\sigma_u$ is the standard deviation about the mean wind speed, given in its sampled form through

$$\text{Standard Deviation (u)}: \sigma_u = \sqrt{\frac{1}{N-1}\sum_{i=1}^{N}(u_i - U)^2} \tag{1.14}$$

where $N$ is the number of discrete samples and $u_i$ is the $i$-th sample of the horizontal velocity measurement. The turbulence intensity, $I$, is one measure of how turbulent a given velocity field is as a function of height and atmospheric stability state. Sometimes, the variance, $\sigma_u^2$, is used instead of the standard deviation, as an overall statistic of the "gustiness" of a flow. Similar definitions can be made for $\sigma_v^2$ (cross-flow) and $\sigma_w^2$ (vertical flow). Note that turbulence is said to be isotropic when $\sigma_u^2 = \sigma_v^2 = \sigma_w^2$, which is practically never the case for any wind turbine application. Another statistical measure of turbulence is the "turbulent kinetic energy," $TKE$, per unit mass defined as:

$$\text{Turbulent Kinetic Energy (TKE) : } TKE = \frac{1}{2}(\sigma_u^2 + \sigma_v^2 + \sigma_w^2) \tag{1.15}$$

TKE is produced at the scale of the boundary-layer depth/height and cascades from the large eddies through the inertial subrange down to the small eddies where molecular viscosity dissipates the energy into heat.

**Atmospheric Stability State** The stability state of the atmosphere is characterized primarily by the effective surface heat flux, $q_s$ [Km s$^{-1}$], with heat flux from the ground to the air considered being positive ($q_s > 0$) and from the air to the ground being negative ($q_s < 0$). In general, we talk about an atmosphere of "neutral" stability for $q_s = 0$, while an "unstable" atmosphere is characterized by $q_S > 0$ and a "stable" atmosphere by $q_s < 0$, respectively. Note, however, that the nature of turbulence in a boundary layer is not only characterized by the surface heat flux, $q_s$, and its associated buoyant mixing processes or lack thereof, but essentially by the interactions between both mechanisms of turbulence, that is, wind shear and buoyancy associated with surface heat flux. A measure of this interaction and hence turbulence nature in a given boundary-layer flow is the Monin–Obukhov (MO) length scale, $L$, defined as:

$$\text{MO Length Scale : } L = -\frac{u^{*3}}{\kappa \frac{g}{\Theta_0} q_s} \tag{1.16}$$

In Eq. (1.16), $u^* = \sqrt{\tau_w/\rho}$ is the friction velocity with $\tau_w$ being the wall/surface shear stress, $\kappa$ is the von Kármán constant, $g$ is the gravitational acceleration, $\Theta_0$ is the potential temperature on the ground ($z = 0$), and $q_s$ is again the surface heat flux. In simple terms, the MO length scale, $L$, is a measure of turbulent shear production over buoyancy and can be also interpreted as the height in the surface layer below which shear production of turbulence exceeds buoyant consumption. Note that $L \to \pm\infty$ for the "neutral" case, $L < 0$ for an unstable atmosphere, and $L > 0$ under stable conditions. For atmospheric stability classification, the MO similarity parameter, $-z_i/L$, is typically used where $z_i$ is the height of the "capping" inversion based on the potential temperature, $\Theta$, or simply boundary-layer depth/height. The potential temperature, $\Theta$, is defined according to

$$\text{Potential Temperature [K] : } \Theta(z) = T(z) + \Gamma_d z \tag{1.17}$$

where $\Gamma_d$ is the adiabatic lapse rate, see Section 1.2.2.3. One can think of the potential temperature being a thermodynamic analogue to the hydrostatic pressure. To a good approximation, the potential temperature, $\Theta$, is constant across the mixing layer. The "inversion" of the potential temperature at height $z_i$ is described by a strong gradient $d\Theta/dz > 0$ and does indeed act as a lid or cap to motions in the ABL. If turbulence tries to push an air parcel from the mixing layer upward into the free atmosphere, its potential temperature would be colder than the surrounding environment and the air parcel would hence fall back down into the mixing layer. In other words, the air in the free atmosphere (starting at $z_i$ up to the end of the troposphere) is unmodified by turbulence. As for the different stability states, surface heating ($q_s > 0$) in an unstable atmosphere occurs for $dT/dz > \Gamma_d$, while surface cooling ($q_s < 0$) in a stable atmosphere is due to $dT/dz < \Gamma_d$. In the following, let us note some basic characteristics of each boundary-layer stability state:

*Neutral ABL ($-1 \ll -z_i/L \ll 1$).* A neutral atmosphere is characterized by moderate to strong winds, with little or no radiative heating or cooling from the surface, that is, no buoyancy. Neutral stability typically occurs during transition from a stable to unstable atmosphere in the morning or an unstable to stable atmosphere in the evening, though it can persist longer under overcast conditions where a release of latent heat can cause increased convective fluxes and hence wind. Turbulence is generated and sustained by wind shear only.

*Unstable ABL ($-z_i/L > 1$).* An unstable atmosphere is seen typically for light winds and a surface that is warmer than the air, for example sunny days and fair weather. However, unstable conditions can also occur when cold air blows over a warmer surface either during the day or at night. Both shear and buoyancy effects contribute to the turbulent nature of the unstable ABL. The range of $1 < -z_i/L < 10$ is called a "moderately-convective" state that is typical of an afternoon boundary layer over land and a calm sea (for the lower $-z_i/L$), while $-z_i/L \to \infty$ can be considered the convective limit where shear-driven turbulence has practically no role anymore in the nature of the turbulent flow.

*Stable ABL ($z_i/L > 1$).* The stable atmosphere occurs typically at night with fair weather and clear skies. Winds are light and the surface is colder than the air. Note, though, that this can also occur when warm air blows over a colder surface during the day. The stable boundary layer suppresses buoyancy effects, turbulent mixing, and turbulence altogether. Hence turbulence is either weak or non-existent in a true stable layer above the ground. As a result, winds are fairly low close to the ground, though effects such as "low-level jets," seen, for example, in the Rocky Flats (Colorado), can inject themselves into a stable atmosphere and persist for long distances.

Figure 1.14 shows qualitative examples of vertical wind profiles under neutral, unstable, and stable conditions. Note that all wind profiles are normalized to the same reference wind speed, $V_0$, at turbine hub height. In general, the neutral wind profile follows a logarithmic distribution, the unstable wind profile has an exponential power-law relationship with height, and the stable wind profile has a more log-linear form, or small power exponent $\alpha$, which is a result of suppressing turbulence. Hence a stable wind profile is more laminar and can be thought of resembling more classical solutions such as Couette flow between two horizontal planes subject to a streamwise pressure gradient. Considering the variations in speed across the difference in height swept by the rotor blades in Figure 1.14, it becomes clear that not only does the total wind power density flowing through the rotor vary with the atmospheric stability state, but the total wind speed variation across the rotor differs significantly between the various stability states. This certainly has implications on time-varying wind-turbine blade loads, not only due to turbulence but also the mean wind shear over the rotor azimuth itself.

*Diurnal Variation of Boundary-Layer Stability*   The stability state of the atmosphere changes due to the daily (diurnal) cycle of radiative heating and cooling. In general, steady winds in the ABL are slower than the equilibrium geostrophic wind, which acts as a boundary condition to the boundary-layer dynamics below the inversion height, $z_i$, for any given weather. For a typical sunny day and stable weather conditions, that is, constant geostrophic wind, the mixing layer thickens as the day progresses due to increasing buoyancy as a result of radiative heating. This unstable (shear + buoyancy) boundary layer reaches its maximum thickness by mid-afternoon. Approaching sunset, the boundary layer becomes more neutral (shear-driven), until stable conditions suppress mixing processes altogether (no/low turbulence). Neutral conditions are reached again around sunrise and duration varies depending on the surface absorption/radiation properties in conjunction with air temperature. The ABL then transitions again to an unstable stability state as radiative surface heating increases. It is important to understand that a neutral boundary layer over land may only occur for relatively short periods of time during a typical diurnal cycle, which has to be taken into account when estimating wind turbine power production.

*Integral Turbulence Scales Relevant for Wind Turbines*   The average time over which fluctuations in the wind speed are correlated with each other can be calculated from an autocorrelation function of wind speed data and is the primary integral time scale of the turbulent flowfield. The primary integral scale is a measure of the cycle time of the largest eddies. The associated turnover time, $t^*$, of the largest eddies, can be approximated by $t^* = z_i/u^*$ in the neutral case ($u^*$ being the friction velocity) and $t^* = z_i/w^*$ in the convective (unstable) case where $w^* = \sqrt[3]{gq_s z_i/\Theta_0}$ is a vertical velocity scale. Note that the large-eddy turnover time can be significantly larger than 10 minutes in the neutral case, while being around 10 minutes in a moderately-convective case (Jha et al. 2015) due to enhanced vertical mixing.

As a general rule-of-thumb, a utility-scale wind turbine rotor with a hub height of around 80 m ($\approx 0.1 z_i$) experiences atmospheric turbulent eddies with time scales on the order of one minute and horizontally elongated eddies that are on the order of 500 m in length and 100 m across vertically (i.e. on the order of the rotor diameter). These "gusts" are typically fairly coherent and associated with low/high-speed regions (or streaks) in the atmospheric surface layer. Considering an operational rotor speed of 12 rpm (rpm = revolutions per minute) for a modern utility-scale wind turbine, this means in particular that wind turbine blades "sweep" through a turbulent eddy spanning the rotor disk area over about $8 - 12$ revolutions, which has a time-varying effect on rotor power but not necessarily on unsteady aerodynamics effects at spanwise blade sections, see Chapter 5 for a more detailed discussion on quasi-steady and unsteady aerodynamics along the radius of a wind turbine blade. As turbulence cascades down the inertial subrange, turbulent length scales become closer to the order of the blade chord length and are of higher frequency and therefore more likely to affect unsteady aerodynamic effects; however, the energy contained in those eddies is significantly smaller, following Kolmogorov's $-5/3$ law. In addition, turbulence on the order of the blade chord length is also affected by wake turbulence downstream of a wind turbine and affects the inflow conditions to inner turbines located in an array.

It is important to understand that atmospheric boundary-layer flow as introduced briefly in this section is practically decoupled from the otherwise close-to-inviscid geostrophic wind scale (mesoscale) and global flow patterns (macroscale) above the ABL and that the grid size of mesoscale weather prediction models is on the order of the height of the surface layer itself. In this context, atmospheric boundary-layer flow is a branch of what is called micrometeorology, and meso-microscale coupling is an active area of research, with challenges in understanding the range of disparate scales from the mesoscale down to the boundary-layer microscale, also called the "terra incognita" (Wyngaard 2010).

### 1.2.2.3 Atmospheric Properties (Troposphere)

In the troposphere ($z < 11.2$ km), pressure, density, and temperature follow well-known relations. The most basic relations derive from fluid statics and thermodynamics. For a quiet atmosphere in hydrostatic equilibrium, the hydrostatic relation for the static pressure, $p$, reads in differential form:

$$\text{Hydrostatics} : dp = -\rho g dz \tag{1.18}$$

The variation of temperature, $T$, with height can be obtained from the adiabatic lapse rate. The adiabatic lapse rate can be obtained easily from the hydrostatic relation and the 1st law of thermodynamics (energy equation), which reads in its general differential form:

$$\text{1st Law of Thermodynamics} : dq = c_p dT - \frac{1}{\rho} dp \tag{1.19}$$

Equation (1.19) can be simplified for adiabatic flow, $dq = 0$, and by using Eq. (1.18) to obtain the adiabatic lapse rate as:

$$\text{Adiabatic Lapse Rate} : \Gamma_d = -\frac{dT}{dz} = \frac{g}{c_p} \tag{1.20}$$

In Eq. (1.18), the gravitational acceleration is $g = 9.81$ m s$^{-2}$ and the specific heat constant (standard atmosphere) equals $c_p = 1.005$ kJ/(kg K). In general, the adiabatic lapse rate is defined as the rate of change of temperature with height. Using the Earth values for $g$ and $c_p$ in Eq. (1.20), the adiabatic lapse rate becomes $\Gamma_d \approx 1$ K/100 m; however, based on definitions of the international standard atmosphere using mean temperatures at the surface and at 10.8 km height, the adiabatic lapse rate is typically referred to as $\Gamma_d \approx 0.66$ K/100 m. In other words, the temperature in the troposphere (i.e. <11.2 km height) linearly decreases by 0.66 K every 100 m at practically a constant potential temperature, $\Theta(z)$, according to Eq. (1.17). This can be found from:

$$\text{Temperature (neutral) [K]} : T(z) = T_0 - \Gamma_d z \tag{1.21}$$

Here, $T_0$ is the surface temperature. Knowledge of the vertical temperature profile, $T(z)$, is required in order to determine the respective vertical profiles of pressure, $p(z)$, and density, $\rho(z)$.

***Atmospheric Pressure and Density*** The variation of atmospheric pressure, $p$, with height is governed by the hydrostatic equation in Eq. (1.18) for a calm atmosphere. Using the "ideal gas law," that is,

$$\frac{p}{\rho} = RT \tag{1.22}$$

where $\rho$ is the air density, $T$ [K] is the absolute temperature, and $R = 287$ J/(kg · K) is the specific gas constant for air, a separable ordinary differential equation for $p(z)$ can be found as:

$$\text{Pressure ODE} : \frac{dp}{p} = -\frac{g\,dz}{R(T_0 - \Gamma_d z)} \tag{1.23}$$

Subsequent integration of Eq. (1.23) from the surface to height, $z$, yields

$$\text{Pressure} : p(z) = p_0 \left[1 - \frac{\Gamma_d z}{T_0}\right]^{g/(\Gamma_d R)} \tag{1.24}$$

In Eq. (1.24), $p_0$ denotes the pressure on the surface ($z = 0$). The variation of density with height, that is, $\rho(z)$, can be easily obtained by applying the ideal gas law to Eqs. (1.23)–(1.24). We thus obtain:

$$\text{Density} : \rho(z) = \rho_0 \left[1 - \frac{\Gamma_d z}{T_0}\right]^{g/(\Gamma_d R)-1} \tag{1.25}$$

Note that $\Gamma_d$ has to be converted to (K per m) when used in Eqs. (1.24) or (1.25). While the actual pressure, $p$, does not affect wind turbine power production, it is important to understand that the air density, $\rho$, does affect the wind power density in Eq. (1.3). For example, at an elevation of $z = 3000$ m, the air density is only about 75% compared to that at sea level. Hence one might consider a rotor diameter must increase by a factor of $\sqrt{1/0.75} \approx 1.15$ to achieve the same wind power (W) passing through the rotor disk area as near sea level.

### 1.2.3 Statistical Description of Wind Data

The knowledge gained in analyzing measured wind data from a large number of wind sites and over large periods of time is very valuable in assessing new wind sites by projecting wind speed distributions from one location to another. Indeed, histograms of wind speed measurements have shown that wind occurrences do follow certain probability functions. Statistical methods can then be used to project measured data from one location to another. In general, a probability density function can be defined as:

$$\text{Probability Density Function} : \int_0^\infty p(U)\,dU = 1 \tag{1.26}$$

Changing the integration bounds in Eq. (1.26) for a known probability density function, $p(U)$, one can actually compute the probability of the wind speed to be within these integration bounds. The average wind speed, $\overline{U}$, is then obtained from

$$\text{Average Wind Speed} : \overline{U} = \int_0^\infty U p(U)\,dU \tag{1.27}$$

and, similarly, the standard deviation using the following relation:

$$\text{Standard Deviation} : \sigma_U = \sqrt{\int_0^\infty (U - \overline{U})^2 p(U)\,dU} \tag{1.28}$$

In a similar manner, a mean wind power density can be defined noting that according to Eq. (1.3), the "power in the wind" is proportional to the wind speed cubed.

$$\text{Mean Wind Power Density}: \frac{\overline{P}}{A} = \frac{1}{2}\rho \int_0^{\infty} U^3 p(U)\, dU = \frac{1}{2}\rho \overline{U^3} \tag{1.29}$$

Here we keep in mind that the average wind speed, $\overline{U}$, refers to the average wind speed at some height, $z$, above the ground. Hence the wind speed, $V_0$, for wind turbine design purposes is equal to the average wind speed at turbine hub height, that is, $\overline{U}(z_{hub})$. In the following, we briefly introduce the two wind speed distributions that are most used in the wind energy community.

### 1.2.3.1 Rayleigh Distribution

This wind-speed probability distribution is probably the simplest as it only requires knowledge of the average wind speed, $\overline{U}$. The probability density function for the Rayleigh distribution reads:

$$\text{Rayleigh}: p(U) = \frac{\pi}{2}\left(\frac{U}{\overline{U}^2}\right)\exp\left[-\frac{\pi}{4}\left(\frac{U}{\overline{U}}\right)^2\right] \tag{1.30}$$

Figure 1.15 shows some example wind-speed distributions using the Rayleigh probability density function of Eq. (1.30) for different values of the average wind speed, $\overline{U}$. It can be seen that as the mean wind speed, $\overline{U}$, increases, the probability of higher wind speeds to occur increases, while the peak probability decreases.

The variance of the Rayleigh distribution can be calculated using the following relation:

$$\text{Variance (Rayleigh)}: \sigma_U^2 = \int_0^{\infty} (U - \overline{U})^2\, p(U)dU \tag{1.31}$$

Note that it can be actually shown that $\sigma_U/\overline{U} = 0.523$ for all Rayleigh distributions, that is, independent of the average wind speed, $\overline{U}$. Here it is important to realize that, in

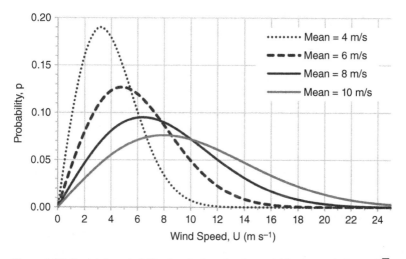

**Figure 1.15** Rayleigh probability density function for variable mean wind speed, $\overline{U}$.

general, the standard deviation of the wind distribution is quite high and that this poses some challenges to wind power forecasting.

### 1.2.3.2 Weibull Distribution

The Weibull probability density function is more general than the Rayleigh distribution and requires knowledge of both the average wind speed, $\overline{U}$, and the standard deviation, $\sigma_U$. The associated probability density function reads:

$$\text{Weibull} : p(U) = \frac{k}{c}\left(\frac{U}{c}\right)^{k-1} \exp\left[-\left(\frac{U}{c}\right)^k\right] \qquad (1.32)$$

The two unknown parameters $k$ and $c$ are functions of $\overline{U}$ and $\sigma_U$ and have to be determined separately, as shown next. The variance of the Weibull probability density function is governed by the following relation:

$$\text{Variance (Weibull)} : \sigma_U^2 = \overline{U}^2\left[\frac{\Gamma(1+2/k)}{\Gamma^2(1+2/k)} - 1\right] \qquad (1.33)$$

In Eq. (1.33), the "gamma function," $\Gamma(x)$, with $x = 1+2/k$ is defined as $\Gamma(x) = \int_0^\infty e^{-\tau}\tau^{x-1}d\tau$ and can be approximated using a series expansion according to Jamil (1994):

$$\Gamma(x) = x^{x-1}e^{-x}\sqrt{2\pi x}\left(1 + \frac{1}{12x} + \frac{1}{288x^2} - \frac{139}{51840x^3} + \dots\right) \qquad (1.34)$$

Furthermore, it can be shown that the average wind speed, $\overline{U}$, is related to $k$, $c$, and $\Gamma(x)$ as:

$$\text{Average Wind Speed} : \overline{U} = c\,\Gamma(1+1/k) \qquad (1.35)$$

In general, the average wind speed, $\overline{U}$, and variance, $\sigma_U^2$, are known from measurements at a given wind site. Equations (1.33) and (1.35) then define a system of two equations for the two unknown parameters $k$ and $c$; however, it is not a straightforward process to solve the respective equations. Fortunately, there are a number of approximations available, for example, a good approximation for $k$ according to Justus (1978):

$$\text{Approximate } k : k = \left(\frac{\sigma_U}{\overline{U}}\right)^{-1.086} \qquad (1.36)$$

Using Eq. (1.36) for $k$, subsequent solution of Eq. (1.35) for $c$ and using the series expansion in Eq. (1.34) yields all parameters needed to determine the Weibull probability density function given by Eq. (1.32). Note that the Rayleigh distribution is a special case of the Weibull distribution for $k = 2$. Alternatively, the parameter, $c$, can also be found from an approximation due to Lysen (1983):

$$\text{Approximate } c : \frac{c}{\overline{U}} = (0.568 + 0.433/k)^{-\frac{1}{k}} \qquad (1.37)$$

The primary advantage of the Weibull distribution is that it allows for variable standard deviation about the average wind speed, $\overline{U}$. Figure 1.16 shows some example Weibull distributions for an average wind speed of $\overline{U} = 6$ m s$^{-1}$. It can be seen that the Weibull parameter, $k$, controls the peak probability and the wind speed at which it occurs. Note from Eq. (1.26) that the integral under all $p(U)$ curves in Figure 1.16

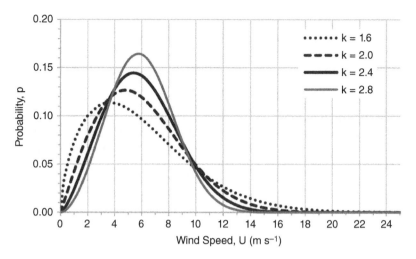

**Figure 1.16** Weibull probability density function for variable parameter, $k$ ($\overline{U} = 6\,\text{m}\,\text{s}^{-1}$).

is equal to 1. In addition, it can be understood from Eq. (1.36) in conjunction with Figure 1.16 that higher values for $k$ correspond to a smaller standard deviation, $\sigma_U$.

A more detailed discussion on properties of the Weibull distribution and statistical analysis methods of wind data in general can be found in Justus (1978), Johnson (1985), and Rohatgi and Nelson (1994).

### 1.2.4 Wind Energy Production Estimates

The energy production of a wind turbine can be estimated for a known wind turbine power curve, $P(U)$, and the probability density function, $p(U)$, of a given wind site. This can be defined in terms of an average wind turbine power, $\overline{P}$, as:

$$\text{Average Wind Turbine Power} : \overline{P} = \int_0^\infty P(U)\,p(U)\,dU \tag{1.38}$$

The average wind turbine power, $\overline{P}$, at a given wind site is always less than the rated power, $P_{Rated}$, of the wind turbine, simply due to the facts that the average wind speed is usually less than the rated wind speed, $V_{Rated}$, and that the wind turbine operates exclusively at wind speeds between $V_{In}$ and $V_{Out}$. The "capacity factor," $CF$, of a given wind turbine located at a particular wind site is defined as the ratio of $\overline{P}$ (actual produced wind power) to the rated wind power, $P_{Rated}$.

$$\text{Capacity Factor} : CF = \frac{\overline{P}}{P_{Rated}} \tag{1.39}$$

It is therefore desirable that the rated wind speed, $V_{Rated}$, be close to (if not less than) the average wind speed, $\overline{U}$, at a given wind site in order to maximize the capacity factor, $CF$. Note that the capacity factor is essentially an efficiency factor for the power production of a wind turbine at a given wind site, and does not provide any information about the actual aerodynamic efficiency of the wind turbine. Typical values for the capacity factor, $CF$, range between 0.3 and 0.4; higher values can be obtained for very well performing wind projects, while wind sites with $CF < 0.25$ are most probably struggling with projected revenue. The Annual Energy Production (AEP) is the total amount of energy (kWh) produced by a wind turbine or a given wind site with multiple turbines. It can be simply calculated from

$$\text{Annual Energy Production}: AEP = 8760 \cdot CF \cdot P_{Rated} \text{ [kWh]} \tag{1.40}$$

where the rated power is given in (kW) and we used that there are 8760 hours per year. As an example, a small wind project of four 1.5-MW rated wind turbines operating at a $CF = 0.4$ wind site has an AEP of $AEP \approx 21 \times 10^3$ kWh, enough energy to power 1200 homes using an average of 2 kW (or 48 kWh per day).

This chapter presented the essentials of the history of wind energy and a basic characterization of the wind resource. The remaining part of the book is devoted to teaching the reader an understanding of the aerodynamics of HAWTs of all sizes.

## References

Anderson, J.D. (1997). *A History of Aerodynamics and its Impact on Flying Machines*. Cambridge: Cambridge University Press.

Carson, R. (1962). *Silent Spring*. New York: Houghton Mifflin.

Counihan, J. (1975). Adiabatic atmospheric boundary layers: a review and analysis of data collected from the period 1880-1972. *Atmospheric Environment* 9: 871–905.

Dörner, H. (2002) *Drei Welten – Ein Leben: Prof. Dr. Ulrich Hütter*. Heilbronn.

Drachman, A.G. (1961). Heron's windmill. *Centaurus* 7 (2): 145–151.

Eggleston, D.M. and Stoddard, F.S. (1987). *Wind Turbine Engineering Design*. New York: Van Nostrand Reinhold.

Eldridge, F.R. (1980). *Wind Machines*, 2e. New York: Van Nostrand Reinhold.

Garratt, J.R. (1994). *The Atmospheric Boundary Layer*, Cambridge Atmospheric and Space Science Series, 290. New York: Cambridge University Press.

Golding, E.W. (1977). *The Generation of Electricity by Wind Power*. London: E. & F. N. Spon.

GWEC (2018) Annual Market Update 2017 – Global Wind Report. Available online at www.gwec.net (Last accessed: January 6, 2019).

Hiester, T. R. and Pennell, W. T. (1981) *The Meteorological Aspects of Siting Large Wind Turbines*. Pacific Northwest Laboratories Report PNL-2522, NTIS.

Jamil, M. (1994). Wind power statistics and evaluation of wind power density. *Wind Engineering* 18 (5): 227–240.

Jha, P.K., Duque, E.P.N., Bashioum, J.L., and Schmitz, S. (2015). Unraveling the mysteries of turbulence transport in a wind farm. *Energies* 8: 6468–6496.

Johnson, G.L. (1985). *Wind Energy Systems*. Englewood Cliffs, NJ: Prentice Hall.

Justus, C.G. (1978). *Winds and Wind System Performance*. Philadelphia, PA: Franklin Institute Press.

Logan, J., Marcy, C., McCall, J. et al. (2017) *Electricity Generation Baseline Report*. National Renewable Energy Laboratory Report No. NREL/TP-6A20–67645.

Lysen, E.H. (1983). *Introduction to Wind Energy*. Amersfoort, NL: SWD Publication SWD 8201.

Manwell, J.F., McGowan, J.G., and Rogers, A.L. (2009). *Wind Energy Explained*. Chichester: Wiley.

Meadows, D.H., Meadows, D.L., Randers, J., and Behrens, W.W. III, (1972). *The Limits to Growth*. New York: Universe Books.

NREL (2013) Life Cycle Greenhouse Gas Emissions from Electricity Generation (Fact Sheet). National Renewable Energy Laboratory, available online at www.nrel.gov/docs/fy13osti/57187.pdf. (Last accessed: January 7, 2019)

Putnam, P.C. (1948). *Power from the Wind*. New York: Van Nostrand Reinhold Company.

Rohatgi, J.S. and Nelson, V. (1994). *Wind Characteristics: An Analysis for the Generation of Wind Power*. Canyon, TX: Alternative Energy Institute.

Schlichting, H. (1968). *Boundary Layer Theory*, 6e. New York: McGraw-Hill.

Shepherd, D. G. (1990) The Historical Development of the Windmill, NASA Contractor Report 4337, DOE/NASA 52662.

Smeaton, J. (1759). An experimental enquiry concerning the natural powers of water and wind to turn mills, and other machines, depending on a circular motion. *Philosophical Transactions of the Royal Society* 100–174. https://doi.org/10.1098/rstl.1759.0019.

Spera, D.A. (ed.) (2009). *Wind Turbine Technology: Fundamental Concepts of Wind Turbine Engineering*. New York: ASME Press.

Vowles, H.P. (1932). Early evolution of power engineering. *Isis* 17 (2): 412–420.

Wortman, A.J. (1982). *Introduction to Wind Turbine Engineering*. Boston, MA: Butterworth.

Wyngaard, J.C. (2010). *Turbulence in the Atmosphere*. Chapter 10,. New York: Cambridge University Press.

## Further Reading

Eggleston, D.M. and Stoddard, F.S. (1987). *Wind Turbine Engineering Design*. New York: Van Nostrand Reinhold.

Manwell, J.F., McGowan, J.G., and Rogers, A.L. (2009). *Wind Energy Explained*. Chichester: Wiley.

Stull, R.B. (2000). *Meteorology for Scientists and Engineers*. Belmont, CA: Brooks/Cole/Cengage Learning.

Wyngaard, J.C. (2010). *Turbulence in the Atmosphere*. Chapter 10,. New York: Cambridge University Press.

# 2

# Momentum Theory

*Simplicity is the ultimate sophistication.*

– Leonardo da Vinci

## 2.1 Actuator Disk Model

Wind turbines operate in a complex flow environment that is time-varying, three-dimensional, and highly turbulent. This includes wind gusts, the interaction of turbine blades with turbulent eddying structures embedded within a sheared wind profile of the atmospheric boundary layer, and fluid-dynamic interactions with vortical wakes generated in a wind farm by upstream turbines. Today's utility-scale wind turbine blades are equipped with close-to-optimal airfoil sections along the blade radius that are designed specifically to local inflow conditions. Modern rotors are both pitch and speed controlled for optimal operation at wind speeds below rated (Region II) and controlled operation at rated power at wind speeds higher than the rated wind speed (Region III). Given the breath of complexity, it seems impractical to gain a basic understanding of rotor performance and blade loads without applying significant simplifications. Here, we revisit the theory as originally documented by Rankine (1865), W. Froude (1878), and R.E. Froude (1889).

### 2.1.1 Basic Streamtube Analysis

For this purpose, consider a wind turbine as a device that extracts momentum from the axial wind momentum flux passing through an actuator disk area, with the latter referring to the circular area swept by the rotor blades. Hence the rotor is reduced to an infinitesimally thin actuator disk of area, $A$:

$$\text{Actuator Disk Area:} \quad A = \frac{\pi}{4}D^2 \tag{2.1}$$

Here, $D$ is the diameter of the actuator disk. In fluid dynamics, a one-dimensional and axi-symmetric model is that of the streamtube, see Figure 2.1, that includes the actuator disk as a surface of pressure discontinuity.

It is further assumed that the velocity is uniform at each cross section of the streamtube, including the velocity at the rotor disk. The streamtube is bound by streamlines, thus preventing any mass, momentum, and energy flux across the bounding surface of

*Aerodynamics of Wind Turbines: A Physical Basis for Analysis and Design,* First Edition. Sven Schmitz.
© 2020 John Wiley & Sons Ltd. Published 2020 by John Wiley & Sons Ltd.
Companion website: www.wiley.com/go/schmitz/wind-turbines

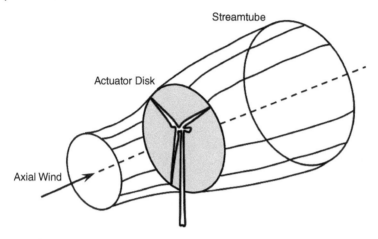

**Figure 2.1** Simplified axi-symmetric streamtube.

the streamtube. As the wind turbine (or actuator disk) extracts momentum and energy from the axial momentum flux, the velocity inside the streamtube gradually decreases and, consequently, the streamtube expands in order to satisfy mass conservation. The assumption of uniform cross-sectional flow inside the streamtube implies that there are no viscous shear effects within the streamtube. Therefore, viscous effects are neglected (inviscid), thus prohibiting the generation of vorticity (irrotational). In addition, the actuator disk is assumed to be stationary (no disk rotation), and a constant uniform wind enters the streamtube at all times (steady).

In summary, the assumptions associated with the actuator disk model are:

- One-dimensional
- Steady
- Incompressible
- Inviscid
- Zero vorticity (or Irrotational)
- No wake rotation

A cross-sectional view through the streamtube is shown in Figure 2.2. A uniform freestream wind profile of magnitude $V_0$ and ambient pressure $p_0$ enters the streamtube,

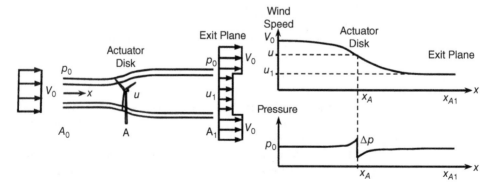

**Figure 2.2** Wind speed and pressure distribution inside axi-symmetric streamtube.

with an assumed uniform wind speed continuously decreasing inside the streamtube, resulting in its respective expansion. At the actuator disk, the uniform wind speed is $u$. At the exit plane of the streamtube, the wind speed has reduced to $u_1$. Note that the wind speed exterior to the streamtube remains unchanged at $V_0$.

As opposed to a continuous distribution of the wind speed inside the streamtube, the static pressure, $p$, is discontinuous at the actuator disk. The reason as to why there has to be a pressure discontinuity at the actuator disk can be explained as follows: The static pressure at both the entrance and exit of the streamtube is required to be equal to the ambient pressure, $p_0$. Nevertheless, the continuous flow deceleration inside the streamtube requires an increase in static pressure both upstream and downstream of the actuator disk according to Bernoulli's law under the assumptions of inviscid and irrotational flow. This can only be satisfied by means of a pressure discontinuity, $\Delta p$, at the actuator disk, which acts as an external force on the fluid inside the streamtube and is responsible for the extraction of axial momentum, energy, and power. As there is no mass flux across the bounding surface of the streamtube control volume, the mass flow rate, $\dot{m}$, is constant everywhere along the streamtube and equal to:

$$\text{Mass Flow Rate:} \quad \dot{m} = \rho V_0 A_0 = \rho u A = \rho u_1 A_1 \qquad (2.2)$$

In Eq. (2.2), $\rho$ is the fluid density and $A_0$, $A_1$ are the entrance and exit areas of the streamtube, respectively. Note that at this stage, all quantities except for $\rho$, $V_0$, and $A$ are unknown. The actuator disk produces a thrust force, $T$, in the streamwise direction by action of the pressure jump, $\Delta p$, across the actuator disk area, $A$, as:

$$\text{Thrust:} \quad T = \Delta p A \qquad (2.3)$$

This external thrust force in the streamwise direction, that is, being essentially a drag force, is the reaction of an equal-and-opposite force exerted on the decelerating fluid and is responsible for a change in axial momentum. Hence the thrust force, $T$, is the only external force acting on the streamtube control volume and is balanced by momentum fluxes entering and leaving the control volume. The thrust force, $T$, then becomes:

$$\text{Axial Momentum:} \quad T = \dot{m}(V_0 - u_1) \qquad (2.4)$$

Equating (2.3) and (2.4), one can find a first relation for the pressure jump, $\Delta p$, across the actuator disk after using $\dot{m} = \rho u A$ from Eq. (2.2):

$$\text{Pressure Jump:} \quad \Delta p = \rho u (V_0 - u_1) \qquad (2.5)$$

In addition, the actuator disk extracts kinetic energy from the wind at a power, $P$, equal to the kinetic energy fluxes entering and leaving the streamtube control volume as:

$$P = \frac{1}{2}\dot{m}(V_0^2 - u_1^2) \qquad (2.6)$$

In order to compute thrust and power by Eqs. (2.3) and (2.6), one needs to find wind speeds $u$ at the actuator disk and $u_1$ at the exit plane. This can be done by applying the incompressible Bernoulli equation upstream and downstream of the actuator disk. Note that Bernoulli's equation cannot be applied across discontinuities such as the pressure jump at the actuator disk as a pressure discontinuity necessitates different total pressure upstream and downstream of the actuator disk. Upstream of the actuator disk, the Bernoulli equation reads:

$$\text{Energy (Upstream:)} \quad p_0 + \frac{1}{2}\rho V_0^2 = p + \frac{1}{2}\rho u^2 \qquad (2.7)$$

Here, $p$ is the static pressure just upstream of the actuator disk. Applying the pressure jump, $\Delta p$, across the actuator disk, the Bernoulli equation downstream of the actuator disk becomes:

$$\text{Energy (Downstream:)} \quad (p - \Delta p) + \frac{1}{2}\rho u^2 = p_0 + \frac{1}{2}\rho u_1^2 \tag{2.8}$$

Note that $p - \Delta p$ is the static pressure just downstream of the actuator disk. Subtracting Eq. (2.8) from (2.7) results in a second expression for the pressure jump, $\Delta p$, as a function of unknowns $u$ and $u_1$ in the form:

$$\Delta p = \frac{1}{2}\rho(V_0^2 - u_1^2) \tag{2.9}$$

Using a binomial with $V_0^2 - u_1^2 = (V_0 - u_1)(V_0 + u_1)$ and equating (2.5) and (2.9) results in a relation for $u$ as:

$$u = \frac{1}{2}(V_0 + u_1) \tag{2.10}$$

It is interesting to note that the wind speed $u$ at the actuator disk becomes the simple arithmetic average of the respective speeds $V_0$ and $u_1$ entering and leaving the streamtube control volume. The relations obtained thus far are the result of applying the mass principle in (2.2), the momentum principle in (2.4), and the energy principle in (2.7) and (2.8) to the streamtube control volume in Figure 2.2.

### 2.1.2 Axial Induction Factor, *a*

In general, it is practical in engineering to introduce dimensionless coefficients. For this purpose, let us define an axial induction factor, $a$, using the wind speed, $u$, at the actuator disk as:

$$\text{Axial Induction Factor:} \quad a = 1 - \frac{u}{V_0} \tag{2.11}$$

Hence, $u/V_0 = 1 - a$ is a dimensionless wind speed at the actuator disk, with the factor of normalization being the freestream wind speed, $V_0$. Using Eq. (2.10), the wind speeds $u$ at the actuator disk and $u_1$ leaving the streamtube become:

$$u = (1 - a)V_0; \quad u_1 = (1 - 2a)V_0 \tag{2.12}$$

It is evident that for the special case of $a = 0$, both $u$ and $u_1$ are equal to $V_0$ and, consequently, both thrust, $T$, and power, $P$, are equal to zero according to Eqs. (2.4) and (2.6). Substituting Eqs. (2.11) and (2.12) into (2.4) and (2.6), respectively, rotor thrust and power are found to be

$$T = 2\rho V_0^2 a(1 - a)A \tag{2.13}$$

$$P = 2\rho V_0^3 a(1 - a)^2 A \tag{2.14}$$

after using Eq. (2.2) as $\dot{m} = \rho u A$. Note that the only remaining unknown in Eqs. (2.13) and (2.14) is the axial induction factor, $a$. Equations (2.13) and (2.14) describe rotor thrust, $T$, and power, $P$, as a function of the fluid density, $\rho$, the freestream wind speed, $V_0$, the actuator disk area, $A$, and the axial induction factor, $a$.

### 2.1.3 Rotor Thrust and Power

It is convenient writing both physical quantities in an appropriate dimensionless form. This can be done normalizing rotor thrust by the freestream dynamic-pressure force acting on the actuator disk, that is $\frac{1}{2}\rho V_0^2 A$, and rotor power by the available wind power flux passing through the actuator disk, that is $\frac{1}{2}\rho V_0^3 A$, simply equal to the product of the freestream dynamic-pressure force and the freestream wind speed itself (see Section 1.2.1). We can then define a thrust coefficient, $C_T$, and power coefficient, $C_P$, as:

$$\text{Thrust Coefficient:} \quad C_T = \frac{T}{\frac{1}{2}\rho V_0^2 A} = 4a(1-a) \tag{2.15}$$

$$\text{Power Coefficient:} \quad C_P = \frac{P}{\frac{1}{2}\rho V_0^3 A} = 4a(1-a)^2 \tag{2.16}$$

Both $C_T$ and $C_P$ are dimensionless quantities and exclusive functions of the axial induction factor, $a$. Rearranging Eqs. (2.15) and (2.16), and from Eq. (2.1) using the fact that $A \sim D^2$, rotor thrust and power can be written as:

$$\text{Thrust:} \quad T = T(V_0^2, D^2, C_T) \tag{2.17}$$

$$\text{Power:} \quad P = P(V_0^3, D^2, C_P) \tag{2.18}$$

These relations emphasize some important and universal dependencies for rotor thrust and power:

- Rotor thrust, $T$, is proportional to the freestream wind speed squared, the rotor (or disk) diameter squared, and the thrust coefficient.

In general, one aims to keep rotor thrust as low as possible, given all other constraints, as it is of no use to power production of a wind turbine rotor but merely a large drag-type force that has to be ultimately balanced by a turbine tower and foundation.

- Rotor power, $P$, is proportional to the freestream wind speed cubed, the rotor (or disk) diameter squared, and the power coefficient.

This means, in particular, that doubling the rotor diameter, $D$, results in a four-time increase of both rotor thrust and power, while doubling the freestream wind speed, $V_0$, increases rotor thrust four times but results in a rotor power eight times higher compared to the baseline. In other words, freestream wind speed, $V_0$, is the primary driver of rotor power due to the cubic relationship, followed by a quadratic dependence on rotor size, $D$, and a linear dependence on a dimensionless performance parameter, $C_P$. In the following, we refer to universal wind power dependences in Eqs. (2.17) and (2.18) as the "2-2-1 Law of Rotor Thrust" and the "3-2-1 Law of Rotor Power." Given the constraints of a wind resource, $V_0$, and rotor size, $D$, a basic design problem typically focuses on maximizing power coefficient, $C_P$. For a consistent comparison of blade designs, the rotor thrust coefficient, $C_T$, should be considered as an additional constraint to the design problem, as discussed in detail in Chapter 8.

### 2.1.4 Optimum Rotor Performance – The Betz Limit

Here, we proceed by plotting rotor thrust and power coefficients, $C_T$ and $C_P$, as a function of the axial induction factor, $a$, with the intent to finding some basic limitations on rotor thrust and power, see Figure 2.3.

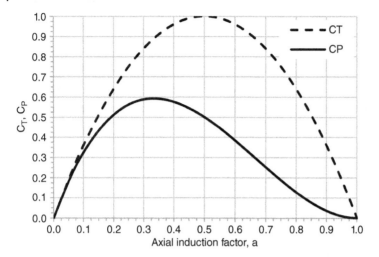

**Figure 2.3** Thrust and power coefficients $C_T$ and $C_P$ versus axial induction factor $a$ (actuator disk theory).

The thrust coefficient, $C_T$, in Figure 2.3, is of parabolic shape and symmetric about an axial induction factor of $a = 1/2$ at its maximum value of $C_T = 1$. This can be shown by setting the first derivative of Eq. (2.15), that is $dC_T/da = 4(1 - 2a)$, equal to zero. The existence of a maximum at $a = 1/2$ is given as $d^2C_T/da^2 = -8 < 0$. In that special case, according to Eq. (2.15), the rotor thrust is equal to the freestream dynamic-pressure force exerted on the actuator disk area, $A$. The thrust coefficient is equal to zero at $a = 0$ and $a = 1$, with the former referring to a streamtube of constant diameter and the latter to the vortex-ring state, see Section 2.1.6.

As for the power coefficient, $C_P$, its cubic dependence on the axial induction factor, $a$, results in one maximum, minimum, and inflection point. Also, $C_P = 0$ in the limits of $a = 0$ and $a = 1$ as is the case for the thrust coefficient, $C_T$. The maximum power coefficient, $C_{P,max}$, according to actuator disk theory under the idealized assumptions of one-dimensional, steady, incompressible, inviscid, and irrotational flow represents an upper performance limit for all subsequent analyses. We can determine $C_{P,max}$ by first finding the extrema of $C_P(a)$ in Eq. (2.16) by setting

$$\frac{dC_p}{da} = 4(1 - a)(1 - 3a) = 0 \tag{2.19}$$

and determining the roots to be $a = 1$ and $a = 1/3$. Finding the second derivative to be

$$\frac{d^2C_p}{da^2} = -8(2 - 3a) \tag{2.20}$$

it becomes clear that an inflection point exists for $a = 2/3$ such that $d^2C_P/da^2 = 0$ and, in particular, $d^2C_P/da^2 > 0$ for $a = 1$ (minimum) and $d^2C_P/da^2 < 0$ for $a = 1/3$ (maximum), see Figure 2.3. Hence the maximum power coefficient, $C_{P,max}$, occurs at an axial induction factor of $a = 1/3$ and becomes

$$C_{P,max} = \frac{16}{27} \approx 0.5926 \tag{2.21}$$

according to Eq. (2.16). This means in particular that we can at most capture 59.3% of the total wind power flux, $1/2\rho V_0^3 A$, passing through the actuator disk area, $A$. Equation (2.21) is called the "Betz limit" after the German aerodynamicist Albert Betz (1926). Burton et al. (2011) also partially attributes Eq. (2.21) to the British aeronautical pioneer Lanchester, while Okulov and van Kuik (2012) propose to also acknowledge the contribution by Joukowsky. The interested reader is referred to the accompanying early propeller theory in Lanchester (1915), Betz (1919), Glauert (1926), and Goldstein (1929). Equation (2.15) can be used to determine the corresponding thrust coefficient as:

$$C_{T,a=\frac{1}{3}} = \frac{8}{9} \approx 0.889 \qquad (2.22)$$

It should be mentioned that while the value for the thrust coefficient in Eq. (2.22) is associated with the "Betz limit" in actuator disk theory, it cannot be generalized that a thrust coefficient of $C_T = 8/9$ always applies to maximum power, as it is a mere result of the assumptions inherent to actuator disk theory.

### 2.1.5 Wake Expansion and Wake Shear

Figure 2.4 illustrates wake expansion and wake shear associated with the streamtube model as a function of thrust coefficient, $C_T$, and axial induction factor, $a$, respectively. As $C_T$ increases, the wind speed, $u_1$, leaving the streamtube decreases linearly with $a$ according to Eq. (2.12). The streamtube exit area, $A_1$, consequently increases to satisfy mass conservation inside the streamtube through Eq. (2.2). Remember that the streamtube is indeed a simplified concept with an impermeable fictitious surface that allowed us to apply mass, momentum, and energy balances within the assumptions of the actuator disk model. Note that $u_1/V_0 \to 1$ for $a \to 0$ in the limiting case of zero induction when both rotor thrust and power are zero. However, as $a$ increases, the discontinuity between the freestream wind speed outside and wind speeds inside the streamtube generates a viscous shear layer downstream of the actuator disk, which is in fact not compliant with the assumption of inviscid flow.

Furthermore, Eq. (2.12) reveals that $u_1/V_0 < 0$ for $a > 0.5$. In this case, there is reverse flow at the exit of the streamtube. This flow state is called the "turbulent wake state" and violates the assumptions of one-dimensional, inviscid, and irrotational flow of the actuator disk model. Therefore, we can already conclude that the actuator disk model is valid only for axial induction factors $a < 0.5$. Moreover, wake shear already becomes strong for $u_1/V_0 < 0.2$ or $a > 0.4$, respectively, and hence prior to occurrence of the turbulent wake state. This further limits the validity of the actuator disk model to axial induction factors $a < 0.4$ or $C_T < 0.96$, respectively. Note that wake expansion is governed by Eq. (2.2) and

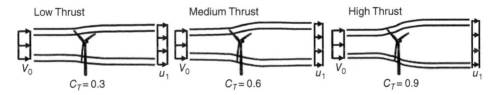

**Figure 2.4** Streamtube expansion for increasing thrust coefficient, $C_T$.

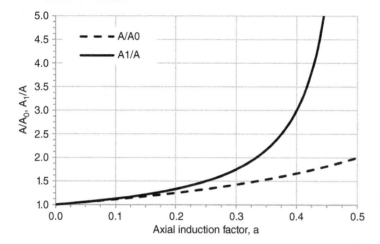

**Figure 2.5** Streamtube area ratios $A/A_0$ and $A_1/A$ versus axial induction factor $a$.

that the ratio of wind speeds at different streamtube cross sections is the inverse of the respective velocity ratio such that

$$\frac{A}{A_0} = \frac{1}{1-a} \quad ; \quad \frac{A_1}{A_0} = \frac{1-a}{1-2a}, \tag{2.23}$$

with the ratio of disk diameters scaling as the square root of the respective expression. The behavior of the area ratios is illustrated in Figure 2.5. It can be seen that $A_1/A$ becomes singular when approaching $a = 0.5$. It is interesting to note that $A_1/A = 3$ for $a = 0.4$, which has been experimentally shown to be a feasible limit of streamtube expansion according to Hansen (2008).

### 2.1.6 Validity of the Actuator Disk Model

Figure 2.6 presents a more detailed account on the validity of actuator disk theory, showing the thrust coefficient, $C_T$, for an extended range of the axial induction factor, $a$. It is worth noting that $a < 0$ corresponds to the "propeller state," with $u, u_1 > V_0$ according to Eqs. (2.11) and (2.12) and a resultant thrust force $T < 0$ that is directed upstream, see Eq. (2.4). In this case, the actuator disk acts as a true propulsion device, adding momentum and energy to the flow inside a contracting streamtube with $A_1 < A$ according to Eq. (2.2). In contrast, the actuator disk operates in the "windmill state" for $0 < a \leq 0.5$, see previous analysis, where momentum and energy are extracted at the actuator disk. In this case, the streamtube expands and the thrust force is directed downstream and acts as a drag force of the rotor system. The maximum power coefficient, $C_{P,max}$, occurs in this flow regime, see Section 2.1.4.

Momentum theory becomes invalid in the "turbulent wake state" for $a > 0.5$, plotted as a dashed line in Figure 2.6, as a result of $u_1$ becoming negative with reverse flow downstream of the actuator disk. As $a$ is further increased, the reverse flow region, which violates assumptions of actuator disk theory, progresses upstream to the actuator disk, reaching the actuator disk for $a = 1.0$ when $u$ becomes zero. Both thrust and power coefficients, $C_T$ and $C_P$, have now also reduced to zero and the rotor enters the "vortex-ring

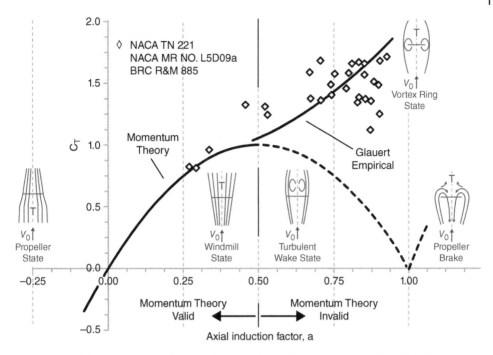

**Figure 2.6** Validity of actuator disk theory. Source: Adapted from Eggleston and Stoddard (1987).

state." While this flow state is irrelevant for wind turbine rotors, it is a dangerous flight state for helicopters.

In reality, however, the flow does not follow the (dashed) second half of the $C_T$ parabola for $0.5 < a < 1.0$. Starting at approximately $a = 0.4$, the $C_T$ curve begins to deviate from momentum theory, indicated by the solid line "Glauert Empirical," with the limit case of $C_T = 2.0$ at $a = 1.0$. The symbols surrounding the empirical curve are experimental data obtained by Glauert and others on propeller-type devices. The steep increase in $C_T$ for $a > 0.4$ is attributed to flow separation and stall. In general, one would aim at operating a wind turbine rotor with $0 < a \leq 0.4$ and as close as possible to $C_{P, max}$ at $a = 1/3$.

### 2.1.7 Summary – Actuator Disk Model

The following gives a brief summary of the actuator disk model:

- Assumptions: One-dimensional, steady, incompressible, inviscid, irrotational, no wake rotation
- Rotor Power: $P = P(V_0^3, D^2, C_P)$
- Wind turbine power production driven by …
  - Wind resource, $V_0$
  - Rotor size, $D$
  - Blade/rotor design, $C_P$ and $C_T$
  - "Betz limit": $C_{P, max} \approx 0.5926$

## 2.2 Rotor Disk Model

The preceding analysis of actuator disk theory resulted in a maximum theoretical power coefficient, that is, the Betz limit, with both rotor thrust and power coefficients, $C_T$ and $C_P$, being exclusive functions of an axial induction factor, $a$. In rotor disk theory, on the other hand, angular momentum and kinetic energy are added to the wake of a rotor disk rotating at an angular speed, $\Omega$. The addition of wake rotation is a reaction to the torque generated by the rotor disk in the circumferential (or angular) direction. We will see that this results in a loss in the maximum theoretical power coefficient, which is a direct result of sustaining the added wake rotation. Here, an angular induction factor and a tip speed ratio are introduced that relate the angular momentum in the wake to the rotor torque and thus power.

### 2.2.1 Extended Streamtube Analysis

The rotor disk model remains to being a greatly simplified model of a wind turbine rotor, though it is one step forward toward a full physical model. Here a wake angular velocity component, $\omega$, is added as a "swirl" to the wake flow downstream of the rotor. In rotor disk theory, the following assumptions apply to the flow:

- One-dimensional
- Steady
- Incompressible
- Inviscid
- Zero vorticity (or irrotational)

Figure 2.7 shows a schematic of the rotor disk concept in which wake rotation is imposed to an axial streamline downstream of the rotor disk.

Added wake rotation amounts to angular kinetic energy in the wake, resulting in an efficiency loss and lower maximum power coefficient, $C_{P,\,max}$, as expected from actuator disk theory. As the angular kinetic energy is a reaction to the torque generated

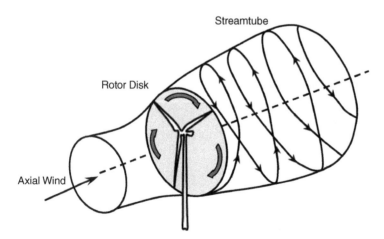

**Figure 2.7** Simplified streamtube including wake rotation.

by the rotor disk, wake rotation appears in opposite sense (see Figure 2.7), and we expect the associated loss to be higher at higher torque, that is, for a slow-rotating disk. On the other hand, high-speed rotors with low torque are expected to have lower losses associated with wake rotation. We recall from classical mechanics that rotor torque is equivalent to a change in angular momentum, being a change in wake angular speed, $\omega$, in the present case of a rotor disk. For an annulus at location $r$ on the rotor disk and of width $dr$, an incremental torque and power can now be defined as:

$$\text{Incremental Torque:} \quad dQ = (\dot{m}\, r^2)\,\omega \tag{2.24}$$

$$\text{Incremental Power:} \quad dP = (\dot{m}\, r^2)\,\omega\,\Omega \tag{2.25}$$

Here, the term in parentheses is the mass moment of inertia flowing through the annulus of the rotor disk at radial location $r$. The incremental mass flow rate is defined according to Eq. (2.2) as $\dot{m} = \rho u\, dA$, with $dA = 2\pi r\, dr$ being the area of the rotor disk annulus. In Eq. (2.25), $\Omega$ is the disk/rotor angular speed. Figure 2.8 illustrates an annulus of the rotor disk.

The integrated rotor torque and power can now be obtained by integrating over all annuli of the rotor disk:

$$\text{Rotor Torque:} \quad Q = 2\pi\rho \int_0^R u\omega r^3 dr \tag{2.26}$$

$$\text{Rotor Power:} \quad P = 2\pi\rho\Omega \int_0^R u\omega r^3 dr \tag{2.27}$$

As a next step, we need to find a relation between the added wake rotation, $\omega$, and the pressure jump, $\Delta p$, across the actuator/rotor disk. In rotor disk theory, we consider a control volume rotating at the angular speed, $\Omega$, with the rotor disk and apply the energy relation just upstream (+) and downstream (−) of the rotor disk. In doing so we assume that the pressure drop (or jump), $\Delta p$, as described in actuator disk theory is used exclusively by the rotor disk to generate torque (and power) by adding wake rotation, $\omega$, see also Glauert (1935), Schmitz (1955), or Wilson et al. (1976). Upstream of the rotor disk (+), the rotating energy reads:

$$\text{Rotating Energy, Disk (+):} \quad p + \frac{1}{2}\rho(\Omega r)^2 \tag{2.28}$$

Here, $\Omega$ is again the angular speed of the rotor disk and $r \leq R$ is the radial location of a rotor disk annulus, as depicted in Figure 2.8. Applying the pressure jump, $\Delta p$, across

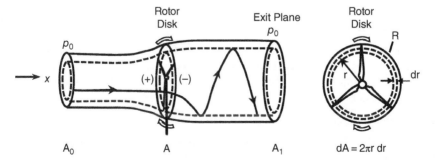

**Figure 2.8** Illustration of a rotor disk annulus.

the rotor disk, the energy relation downstream (−) of the rotor disk becomes:

$$\text{Rotating Energy, Disk }(-): \quad (p - \Delta p) + \frac{1}{2}\rho[(\Omega + \omega)r]^2 \tag{2.29}$$

In Eq. (2.29), $p - \Delta p$ is the static pressure just downstream of the rotor disk, similar to Eq. (2.8) in actuator disk theory, and $\omega$ is an added angular speed at the rotor disk, that is, added wake rotation. Equating (2.28) from (2.29) results in an expression for $\Delta p$ based on rotor disk theory:

$$\text{Rotor Disk:} \quad \Delta p = \rho\left(\Omega + \frac{1}{2}\omega\right)\omega r^2 = 2\rho\left(1 + \frac{\omega}{2\Omega}\right)\frac{\omega}{2\Omega}\Omega^2 r^2 \tag{2.30}$$

Equation (2.30) is a rotational equivalent to actuator disk theory in Eq. (2.9). Note that $\Delta p$ becomes zero in the absence of added wake rotation, $\omega$. We further assume that the added wake rotation, $\omega$, is small compared to the angular speed, $\Omega$, of the rotor disk. In this case, Wilson et al. (1976) argue that the pressure in the far wake downstream of the rotor disk is still equal to the ambient pressure, $p_0$, as considered in actuator disk theory. The following section will now combine actuator disk and rotor concepts into a combined theory.

## 2.2.2 Angular Induction Factor, $a'$

It is again practical to introduce a dimensionless coefficient for further analysis. Equation (2.30) suggests defining an angular induction factor $a'$ as:

$$\text{Angular Induction Factor:} \quad a' = \frac{\omega}{2\Omega} \tag{2.31}$$

As the rotor disk of radius $R$ has an angular velocity $\Omega$, it further makes sense to introduce a dimensionless tip speed ratio as:

$$\text{Tip Speed Ratio:} \quad \lambda = \frac{\Omega R}{V_0} \tag{2.32}$$

Equation (2.32) simply describes the ratio of the angular velocity at the edge of the rotor disk to the incoming wind speed, $V_0$. A local speed ratio can then be defined as the respective fraction of $\lambda$ seen at a local disk annulus as:

$$\text{Local Speed Ratio:} \quad \lambda_r = \frac{r}{R}\lambda \tag{2.33}$$

Note that $\lambda_r$ ranges between 0 at the axis of rotation and $\lambda$ at the edge of the rotor disk. Using the definition for $a'$, $\lambda$, and $\lambda_r$, the expression for $\Delta p$ in Eq. (2.30) can now be written as:

$$\text{Pressure Jump (Rotor Disk:)} \quad \Delta p = 2\rho V_0^2 a'(1 + a')\lambda_r^2 \tag{2.34}$$

Note that the limits of rotor disk theory occur for $\Delta p \to 0$ for either $a' = 0$ or $a' = -1$, with the former referring to zero thrust and the latter to canceling the angular velocity of the rotor disk. Combining Eqs. (2.3) and (2.13) from Section 2.1, the pressure jump across the actuator disk becomes:

$$\text{Pressure Jump (Actuator Disk:)} \quad \Delta p = 2\rho V_0^2 a(1 - a) \tag{2.35}$$

Equating (2.34) and (2.35) results in a combined actuator-/rotor disk concept and a first relation between $a$ and $a'$ as:

$$\text{1st Relation between }a\text{ and }a': \quad \frac{a(1 - a)}{a'(1 + a')} = \lambda_r^2 \tag{2.36}$$

Equation (2.36) will prove itself to be essential when deriving optimum rotor performance including wake rotation, see Section 2.2.4. As a next step, let us first revisit rotor torque and power in Eqs. (2.26) and (2.27), this time using definitions of the dimensionless axial and angular induction factors in Eqs. (2.11) and (2.31) to substitute for $u$ and $\omega$. We thus obtain the following relations:

$$\text{Rotor Torque:} \quad Q = 4\pi\rho V_0\Omega \int_0^R a'(1-a)r^3 dr \tag{2.37}$$

$$\text{Rotor Power:} \quad P = 4\pi\rho V_0\Omega^2 \int_0^R a'(1-a)r^3 dr \tag{2.38}$$

Note again the limiting cases of zero torque and power for no added wake rotation ($a'=0$), for which also $a=0$ to satisfy Eq. (2.36), or the special case of complete induction with $a=1$ (see also Figure 2.3) in conjunction with the validity of the actuator disk concept as discussed in Section 2.1.6.

### 2.2.3  Rotor Torque and Power

Next, it is convenient writing rotor torque and power in their respective dimensionless forms. Here we normalize rotor torque by the freestream dynamic-pressure force acting on the actuator disk times the disk radius, that is, $\frac{1}{2}\rho V_0^2 AR$, and rotor power by the available wind power flux passing through the actuator disk, that is, $\frac{1}{2}\rho V_0^3 A$, as was done in Section 2.1.3. We can then define a torque coefficient, $C_Q$, and power coefficient, $C_P$, according to rotor disk theory as:

$$C_Q = \frac{Q}{\frac{1}{2}\rho V_0^2 AR} = \frac{8}{\lambda^3}\int_0^\lambda a'(1-a)\lambda_r^3\, d\lambda_r \tag{2.39}$$

$$C_P = \frac{P}{\frac{1}{2}\rho V_0^3 A} = \frac{8}{\lambda^2}\int_0^\lambda a'(1-a)\lambda_r^3\, d\lambda_r \tag{2.40}$$

Both $C_Q$ and $C_P$ are dimensionless quantities and functions of the tip speed ratio $\lambda$, with the axial and angular induction factors satisfying Eq. (2.36) and requiring one additional relation between $a$ and $a'$ for closure. Note the stronger convergence of $C_Q$ with $\lambda$ when compared to $C_P$. After using $A\sim D^2$ from Eq. (2.1), we can now write rotor torque and power as:

$$\text{Torque:} \quad Q = Q(V_0^2, D^3, C_Q(\lambda)) \tag{2.41}$$

$$\text{Power:} \quad P = P(V_0^3, D^2, C_P(\lambda)) \tag{2.42}$$

Let us emphasize some important and universal dependencies for rotor torque and power in rotor disk theory:

• Rotor torque, $Q$, is proportional to the freestream wind speed squared, the rotor (or disk) diameter cubed, and the torque coefficient as a function of tip speed ratio.

For given power, rotor torque behaves with the inverse of rotor (or disk) angular velocity; that is, $\sim 1/\Omega$. This means in particular that rotor torque decreases with increasing tip speed ratio, $\lambda$, but increases with the third power of rotor size as $D^3$. This is an important result as it affects sizing of the low-speed shaft of horizontal-axis wind turbines, though not necessarily in an adverse sense as the area moment of inertia scales up with the fourth power of a circular shaft diameter. The dependency in Eq. (2.41) describes a universal "2-3-1 Law of Rotor Torque."

- Rotor power, $P$, is proportional to the freestream wind speed cubed, the rotor (or disk) diameter squared, and the power coefficient as a function of tip speed ratio.

This has been described in Section 2.1.3 as the "3-2-1 Law of Rotor Power." As discussed previously, we expect to achieve maximum attainable rotor power for high-speed rotors as both torque and angular induction behave as $\sim 1/\Omega$, thus having rotor disk theory becoming actuator disk theory in the asymptotic limit, as derived in the following section.

### 2.2.4 Optimum Rotor Performance Including Wake Rotation

It is expected that adding wake rotation to the flow as a reaction to rotor torque leads to an efficiency loss when compared to the Betz limit of $C_{P,\,max} \approx 0.5926$ as derived in Section 2.1.4. Furthermore, we concluded with basic reasoning that we expect to approach the Betz limit for high tip speed ratios, $\lambda$, (or high angular velocities, $\Omega$) as, for a given rotor power, rotor torque and thus wake rotation tend to zero for increasing $\Omega$. A question of practical importance is, though, how high $\lambda$ has to be, that is, $\lambda = 5, 10, \ldots$, in order to achieve power coefficients within, for example, 1% of the Betz limit? To answer this question, we proceed by further inspecting Eq. (2.40) and note that the power coefficient, $C_P$, becomes a maximum, if the integrand factor $a'(1-a)$ is maximized for a given $\lambda_r$ distribution. Hence the problem reduces to the following:

$$Maximize \quad f(a,a') = a'(1-a) \tag{2.43}$$

As a next step, we evaluate first and second derivatives of Eq. (2.43). Note that $a$ and $a'$ are not independent variables as they are coupled through Eq. (2.36). We therefore consider $a' = a'(a)$ for a given $\lambda_r$ distribution and determine the derivatives as:

$$\frac{df}{da} = \frac{da'}{da}(1-a) - a' \tag{2.44}$$

$$\frac{d^2f}{da^2} = \frac{d^2a'}{da^2}(1-a) - 2\frac{da'}{da} \tag{2.45}$$

The function $f(a, a')$ has a maximum for $\frac{df}{da} = 0$ and $\frac{d^2f}{da^2} < 0$ for the respective value of $a$. Setting Eq. (2.44) equal to zero yields the following:

$$\text{Required Condition for Maximum:} \quad \frac{da'}{da} = \frac{a'}{1-a} \tag{2.46}$$

Equation (2.46) is, in theory, an ordinary differential equation for $a'(a)$, though with, at least thus far, unknown boundary conditions along each annulus of the rotor disk in terms of the local speed ratio, $\lambda_r$. Nevertheless, Eq. (2.46) can be used for substitution, if we write the first relation between $a$ and $a'$, that is, Eq. (2.36) in the following form

$$a(1-a) = a'(1+a')\lambda_r^2 \tag{2.47}$$

and take the derivative with respect to $a$ on both sides of Eq. (2.47) to obtain

$$1 - 2a = \frac{da'}{da}(1+2a')\lambda_r^2 \tag{2.48}$$

which, after substituting Eq. (2.46) for $\frac{da'}{da}$ and Eq. (2.36) for $\lambda_r^2$, can be written as

$$\frac{a}{1-2a} = \frac{1+a'}{1+2a'} \tag{2.49}$$

and solved easily for $a'$ to obtain:

$$\text{2nd Relation between } a \text{ and } a' : \quad a' = \frac{1 - 3a}{4a - 1} \tag{2.50}$$

Note that Eq. (2.50) describes an extremum of $f(a, a')$ as Eq. (2.46) was used to obtain the former relation. It is interesting to note that Eq. (2.50) reduces to optimum actuator disk theory for $a = 1/3$ and has a singularity for $a = 1/4$, both of which will be discussed further next. In general, wake rotation is added $(a' \geq 0)$ for $a \in [1/4; 1/3]$. Equations (2.36) and (2.50) constitute a system of two equations for the two unknowns $a$ and $a'$ at each annulus of the rotor disk and the local speed ratio, $\lambda_r$, as an additional parameter. Before solving the system of equations and integrating for the maximum power coefficient, $C_{P, max}$, in Eq. (2.40), we first have to prove, however, that Eq. (2.50) does indeed represent a maximum. This can be done by evaluating the sufficient condition for a maximum, that is, $d^2 f / da^2 < 0$, as:

$$\text{Sufficient Condition for Maximum:} \quad \frac{d^2 a'}{da^2}(1 - a) - 2\frac{da'}{da} < 0 \tag{2.51}$$

For added wake rotation $(a' > 0)$ and $a \in [1/4; 1/3]$, it becomes clear from Eq. (2.46) that $da'/da \geq 0$ so that Eq. (2.51) reduces to proving simply that $d^2 a'/da^2 < 0$. Taking the derivative with respect to $a$ of Eq. (2.48) yields

$$-2 = \lambda_r^2 \left( \frac{d^2 a'}{da^2}(1 + 2a') + 2\left( \frac{da'}{da} \right)^2 \right) \tag{2.52}$$

which can be easily rearranged to show that indeed

$$\frac{d^2 a'}{da^2} = -\frac{2}{1 + 2a'} \left( \frac{1}{\lambda_r^2} + \left( \frac{da'}{da} \right)^2 \right) < 0 \tag{2.53}$$

and that hence Eq. (2.50) describes a maximum for the power integrand $a'(1 - a)$ in Eq. (2.40). Before evaluating the integral, however, we note that the second relation between $a$ and $a'$, that is, Eq. (2.50), is a sole function of $a$, and substitution into the first relation between $a$ and $a'$, that is, Eq. (2.36), results in

$$\lambda_r^2 = \frac{(1 - a)(1 - 4a)^2}{(1 - 3a)} \tag{2.54}$$

which can be written as a third order polynomial for the optimal axial induction factor $a(\lambda_r)$ across the rotor disk as:

$$16a^3 - 24a^2 + (9 - 3\lambda_r^2)a + (\lambda_r^2 - 1) = 0 \tag{2.55}$$

It can be shown following Bronstein et al. (1997) that the discriminant is always less than zero and hence that Eq. (2.55) has three real roots. Upon further investigation, it can be shown that one real root does indeed exist for $a \in [1/4; 1/3]$, which can be found very efficiently using Newton's method. Note that Eq. (2.55) describes a universal relation for the optimum axial induction factor, $a$, as a function of the local speed ratio, $\lambda_r$. The corresponding optimum angular induction factor is $a'(a(\lambda_r))$ determined from Eq. (2.50). The optimum induction factors are plotted in Figure 2.9 versus the local speed ratio, $\lambda_r$.

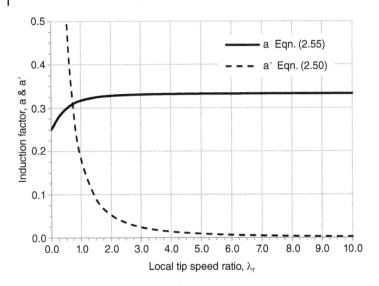

**Figure 2.9** Universal solutions for optimum induction factors *a* and *a'* as a function of local speed ratio, $\lambda_r$

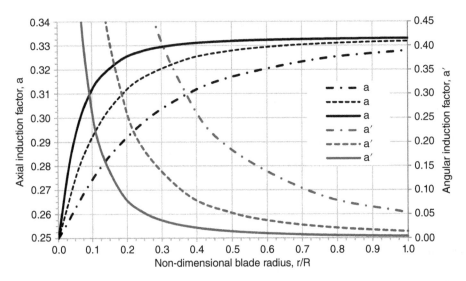

**Figure 2.10** Spanwise distributions of optimum induction factors *a* and *a'* for varying tip speed ratio ($\lambda = 2$ (dash-dot), $\lambda = 4$ (dash), and $\lambda = 8$ (solid) lines).

The universal solutions for the optimum induction factors *a* and *a'* of Eqs. (2.55) and (2.50), respectively, can then be scaled to a given actual tip speed ratio, $\lambda$, using Eq. (2.33). Figure 2.10 shows spanwise distributions of optimum induction factors, *a* and *a'*, for tip speed ratios $\lambda = 2, 4, 8$. It can be seen in Figure 2.10 that the optimum axial induction factor, *a*, increases along the blade radius toward $a = 1/3$, that is, the optimum according to actuator disk theory as derived in Section 2.1.4, though faster for increasing tip speed ratio, $\lambda$. A similar behavior is seen for the optimum angular induction factor, *a'*, which approaches slowly the limit of actuator disk theory (no wake rotation) toward $a' = 0$. It is not surprising that both $a$ and $a'$ approach their respective asymptotic limits toward the outer edge of the rotor disk as this behavior scales with increasing local speed ratio, $\lambda_r = \lambda \cdot r/R$. What is interesting to note, though, is that for an operating tip speed ratio of $\lambda = 4$, the inner one-third of the rotor disk are notably affected by angular induction, which has an effect on optimum blade design and other effects such as rotational augmentation.

We can now proceed by evaluating the integral in Eq. (2.40) to obtain $C_{P,max}(\lambda)$, using results from the preceding analysis. However, we need an expression for $d\lambda_r$, which can be found by taking the derivative of Eq. (2.54) with respect to $a$ as:

$$2\lambda_r d\lambda_r = \frac{6(4a-1)(1-2a)^2}{(1-3a)^2} da \qquad (2.56)$$

Substituting Eqs. (2.50), (2.54), and (2.56) into Eq. (2.40) yields

$$C_{P,max} = \frac{24}{\lambda^2} \int_{a_1}^{a_2} \left[ \frac{(1-a)(1-2a)(1-4a)}{(1-3a)} \right]^2 da \qquad (2.57)$$

where the integration bounds, $a_1$ and $a_2$, are solutions to Eq. (2.55) with $a_1 = a(0) = 1/4$ and $a_2 = a(\lambda)$. Using the substitution $x = 1 - 3a$ and its derivative $dx/da = -3$ along with polynomial division, Eq. (2.57) becomes

$$C_{P,max} = \frac{1}{\lambda^2}\left(\frac{2}{9}\right)^3 \int_{x_2}^{x_1} \left[ 8x^2 + 18x + 3 - \frac{2}{x} \right]^2 dx \qquad (2.58)$$

with $x_1 = 1 - 3a_1 = 1/4$ and $x_2 = 1 - 3a_2$. Note that the negative sign due to the substitution has been removed by switching the integration bounds. Expanding Eq. (2.58) and subsequent integration results in

$$C_{P,max} = \frac{1}{\lambda^2}\left(\frac{2}{9}\right)^3 \left\{ \frac{64}{5}x^5 + 72x^4 + 124x^3 + 38x^2 - 63x - 12ln(x) - 4x^{-1} \right\}_{x_2=(1-3a_2)}^{x_1=1/4} \qquad (2.59)$$

for the maximum power coefficient, $C_{P,max}(\lambda)$, according to rotor disk theory (or ideal rotor with rotation). Table 2.1 shows sample values for $a_2$ from solutions of Eq. (2.55) and the resultant maximum power coefficient, $C_{P,max}$, from Eq. (2.59) as a function of tip speed ratio, $\lambda$. Note that $a_2$ has to be computed to high accuracy due to the dominance of the last two terms in Eq. (2.59) for small $x$.

The corresponding plot of maximum power coefficient, $C_{P,max}$, versus tip speed ratio, $\lambda$, is shown in Figure 2.11. It can be seen that a relatively small tip speed ratio of $\lambda = 5$ gets to within 4% of the Betz limit from actuator disk theory and to within 1.5% at $\lambda = 10$. These are encouraging results as wind turbine power extraction appears to be close-to-optimal for relatively slow-rotating machines, a good result considering that dynamic loads scale with the squared of the rotor angular velocity; that is, $\Omega^2$ (see Chapter 8). Figure 2.11 also shows that for tip speed ratios $\lambda \leq 2$, a significant loss in $C_{P,max}$ is associated with adding wake rotation. One can also think of this as being a consequence of a slow-rotating disk not being able to work on much of the mass flow passing through the disk area.

**Table 2.1** Example values of maximum power coefficient, $C_{P,max}$, as a function of tip speed ratio, $\lambda$, and integration bound, $a_2$.

| $\lambda$ | $a_2$ | $C_{P,max}$ |
|---|---|---|
| 0.5 | 0.298 346 | 0.289 4 |
| 1.0 | 0.316 987 | 0.415 5 |
| 1.5 | 0.324 456 | 0.477 2 |
| 2.0 | 0.327 896 | 0.511 2 |
| 2.5 | 0.329 700 | 0.531 9 |
| 5.0 | 0.332 367 | 0.570 4 |
| 7.5 | 0.332 899 | 0.580 8 |
| 10.0 | 0.333 088 | 0.585 2 |

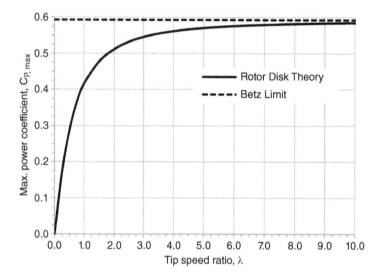

**Figure 2.11** Maximum power coefficient, $C_{P,max}$, versus tip speed ratio, $\lambda$ (Rotor Disk Theory).

## 2.2.5 Validity of the Rotor Disk Model

Both actuator and rotor disk models are basic concepts intended to define performance limits of horizontal-axis wind turbines and must always be seen from this point of view. It is interesting that trying to bring both theories together results in some challenging conceptual difficulties. First, the assumption of one-dimensional flow is questionable when adding wake rotation; however, it appears that angular induction remains low for $\lambda \geq 4$ and remains low in general at outboard disk annuli that produce most of the torque and power. Another interesting question may arise as to how to evaluate the thrust coefficient, $C_T$, if the pressure jump, $\Delta p$, across the disk is assumed to be constant in actuator disk theory but not so in rotor disk theory. In particular, is $C_T$ governed by Eq. (2.15) or by integrating $a(1-a)$ over the disk area? Again, one can argue that $\Delta p$,

$a \approx const.$ over the outer disk area and for $\lambda \geq 4$. As far as the irrotationality assumption is concerned, it could be easily argued that this is a direct consequence of one-dimensional flow; however, we have seen that this is strictly not the case. Nevertheless, irrotational flow would remain in the presence of wake rotation, if $a' \sim 1/r$. In reality, though, the solution for $a'$, Eq. (2.36), behaves more as $a' \sim 1/r^{1.8}$, thus leading to a small violation of the irrotationality assumption. In addition, the assumption that $\Delta p$ is used exclusively to add rotation to the wake is a simple, maybe the simplest, way of accounting for wake rotation, but is likely not fully physically correct either, particularly for low tip speed ratios. Here, a more in-depth analysis on variants of rotor disk theory is discussed in Sørensen (2016). Differences exist but, as we shall see in Chapter 3, some type of tip loss modeling has to be applied anyhow in blade-element momentum theory (BEMT) to account for a finite number of blades and associated three-dimensional effects. In that sense, the particular way of accounting for wake rotation from a disk model perspective may not be as important as originally thought because, again, modern designs for horizontal-axis wind turbines operate at design tip speed ratios, $\lambda$, between four and six in Region II and at smaller $\lambda$ under controlled conditions in Region III. Nevertheless, low tip speed ratios, $\lambda$, do occur during startup (practically in Region I.5) and although the wind turbine produces very low power, it does affect the torque coefficient $C_Q(\lambda) = C_P(\lambda)/\lambda$, see Eq. (2.39), which is an important parameter with respect to accelerating the rotor against its resisting inertia and bearing friction.

### 2.2.6 Summary – Rotor Disk Model

The following gives a brief summary of the rotor disk model:

- Assumptions: One-dimensional, Steady, Incompressible, Inviscid, Zero vorticity (or Irrotational)
- Includes effect of Tip Speed Ratio: $\lambda = (\Omega R)/V_0$
- Rotor Torque: $Q = Q(V_0^2, D^3, C_Q(\lambda))$
- Rotor Power: $P = P(V_0^3, D^2, C_P(\lambda))$
- Wind Turbine Power Production driven by …
  - Wind Resource, $V_0$
  - Rotor Size, $D$
  - Blade/Rotor Design, $C_Q$ & $C_P$
  - Approaches "Betz limit" of $C_{P,max} \approx 0.5926$ for high $\lambda$

## References

Betz, A. (1919). *Schraubenpropeller mit geringstem Energieverlust*. Delft: Göttinger Nachrichten.

Betz, A. (1926). *Windenergie und ihre Ausnützung durch Windmühlen*. Göttingen: Vandenhoeck und Ruprecht.

Bronstein, I.N., Semendjajew, K.A., Musiol, G., and Mühlig, H. (1997). *Taschenbuch der Mathematik*, 57–58. Frankfurt am Main: Verlag Harri Deutsch.

Burton, T., Jenkins, N., Sharpe, D., and Bossanyi, E. (2011). *Wind Energy Handbook* (Chapter 3.10). Chichester: Wiley.

Eggleston, D.M. and Stoddard, F.S. (1987). *Wind Turbine Engineering Design*. New York: Van Nostrand Reinhold Co.

Froude, W. (1878). On the Elementary Relation Between Pitch, Slip and Propulsive Efficiency. *Transactions of the Institution of Naval Architect* 19: 47.

Froude, R.E. (1889). *On the part played in propulsion by difference in fluid pressure.* *Transactions of the Institution of Naval Architect* 30: 390–405.

Glauert, H. (1926) *The Analysis of Experimental Results in the Windmill Brake and Vortex Ring States of an Airscrew.*, ARCR R&M No. 1026.

Glauert, H. (1935). Airplane Propellers. In: *Aerodynamic Theory*, vol. 4, Division L (ed. W.F. Durand), 169–360. Berlin: Julius Springer.

Goldstein, S. (1929). *On the vortex theory of screw propeller. Royal Society Proceedings (A)* 123: 440.

Hansen, M.O.L. (2008). *Aerodynamics of Wind Turbines*. London: Earthscan.

Lanchester, F.W. (1915). A contribution to the theory of propulsion and the screw propeller. *Transactions of the Institution of Naval Architects* 57: 98.

Okulov, V.L. and van Kuik, G.A.M. (2012). *The Betz–Joukowsky limit: on the contribution to rotor aerodynamics by the British, German and Russian scientific schools. Wind Energy* 15: 335–344.

Rankine, W.J.M. (1865). *On the mechanical principles of the action of propellers. Transactions of the Institution of Naval Architect* 6: 13.

Schmitz, G. (1955) Theorie und Entwurf von Windrädern optimaler Leistung (Theory and design of windwheels with an optimum performance). *Wiss. Zeitschrift der Universität Rostock*, 5. Jahrgang 1955/56.

Sørensen, J.N. (2016). *General Momentum Theory for Horizontal Axis Wind Turbines*. London: Springer.

Wilson, R.E., Lissaman, P.B.S., and Walker, S.N. (1976). *Aerodynamic Performance of Wind Turbines*. Energy Research and Development Administration, ERDA/NSF/04014-76/1.

## Further Reading

Branlard, E. (2017). *Wind Turbine Aerodynamics and Vorticity-Based Methods – Fundamentals and Recent Applications*. London: Springer.

Burton, T., Jenkins, N., Sharpe, D., and Bossanyi, E. (2011). *Wind Energy Handbook* (Chapter 3.10). Chichester: Wiley.

Eggleston, D.M. and Stoddard, F.S. (1987). *Wind Turbine Engineering Design*. New York: Van Nostrand Reinhold Co.

Hansen, M.O.L. (2008). *Aerodynamics of Wind Turbines*. London: Earthscan.

Leishman, J.G. (2000). *Principles of Helicopter Aerodynamics*. Cambridge: Cambridge University Press.

Manwell, J.F., McGowan, J.G., and Rogers, A.L. (2002). *Wind Energy Explained*. Chichester: Wiley.

Sørensen, J.N. (2016). *General Momentum Theory for Horizontal Axis Wind Turbines*. London: Springer.

Stepniewski, W.Z. and Keys, C.N. (1984). *Rotary-Wing Aerodynamics*. New York: Dover.

# 3

# Blade Element Momentum Theory (BEMT)

> *It is some sort of tragedy that many get caught up in the idea of generating power from the wind and attempt to build machines before they have mastered the disciplines required.*
>
> — D. M. Eggleston and F. S. Stoddard (1987)

## 3.1 The Blade Element – Incremental Torque and Thrust

Blade element theory is the next conceptual step toward a full physical representation of wind turbine aerodynamics. While actuator-/rotor disk theories represent the rotor as an infinitesimally thin surface, blade element theory considers individual blades that sweep over the disk area and subdivides each blade into discrete blade elements. Each blade element consists of an airfoil section that generates both lift and drag forces (two-dimensional). The upstream wind speed, $V_0$, is assumed to be uniform at all times (steady). Blade elements are further assumed to operate independent of each other and in the absence of radial flow gradients (no radial flow). Cumulative three-dimensional effects are accounted for by associated root-/tip-loss factors applied to each blade element (root-/tip losses).

The assumptions associated with blade element theory are:

- Two-dimensional
- Steady
- No radial flow
- Root-/tip losses

An illustration is given in Figure 3.1. Note that the lower airfoil surface faces the wind speed, $V_0$. This is necessary in order to have sectional airfoil lift and drag forces project favorably to generate rotor torque and thus power. Here the angle of projection is the blade flow angle, $\Phi(r)$, shown here as the angle between the directions of lift and thrust that, as shall be seen later, is also equal to the angle between the local blade element flow and the direction of torque. It is worth noting that the projected sectional airfoil (or profile) drag force reduces the torque generating component of the sectional airfoil lift force. We therefore expect an associated loss in maximum attainable power coefficient, $C_{P,max}$, as a result of including airfoil (or profile) drag, a property that we will find to depend on rotor size and blade design. We further note that lift and drag forces project

*Aerodynamics of Wind Turbines: A Physical Basis for Analysis and Design,* First Edition. Sven Schmitz.
© 2020 John Wiley & Sons Ltd. Published 2020 by John Wiley & Sons Ltd.
Companion website: www.wiley.com/go/schmitz/wind-turbines

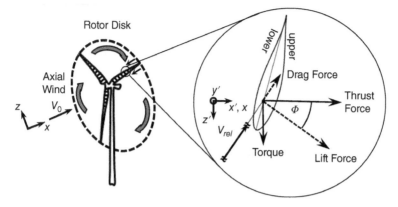

**Figure 3.1** Principle of blade element theory.

onto a streamwise thrust force in the direction of the incoming wind speed, therefore essentially being a "rotor drag force."

It is imperative to understand these basic force projections of a wind turbine rotor and their fundamental difference compared to a propeller where the upper airfoil surface faces the incoming wind speed, thus generating a true (propelling) thrust force in the upstream direction, which is realized at the expense of opposite torque (and power) that has to be provided by an engine. In this case, momentum and energy are added to the flow and the actuator-/rotor disk contracts rather than expands, see also Figure 2.6 for an axial induction factor, $a$, less than zero.

### 3.1.1 Airfoil Nomenclature

As stated before, each blade element consists of an airfoil section. Figure 3.2 shows a cambered airfoil with the following geometric specifications:

- *Leading edge.* Upstream end point of mean camber line
- *Trailing edge.* Downstream end point of mean camber line
- *Chord, c.* Distance between leading and trailing edges
- *Mean camber line.* Line halfway between upper and lower airfoil surfaces
- *Angle of attack, α.* Angle between the chord line and the incoming velocity vector.

Also shown in Figure 3.2 are the airfoil lift and drag forces along with the pitching moment, which are all integrated values of the respective pressure and skin friction

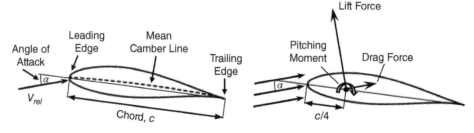

**Figure 3.2** Airfoil nomenclature.

distributions along the airfoil surface contour and defined in the airfoil frame of reference as:

- *Lift force, L.* Resultant force perpendicular to local incoming flow
- *Drag force, D.* Resultant force in direction of local incoming flow
- *Pitching moment, $M_p$.* Resultant moment about 1/4-chord point

Note that the lift force is perpendicular to the airfoil chord line only for an angle of attack, $\alpha$, being equal to zero. Hence for a wind turbine rotor, the directions of blade element lift and drag forces are essentially decoupled from the blade geometry but depend on the local inflow angle to a blade element. For a symmetric airfoil, that is, collinear camber and chord lines, the zero-lift angle of attack is $\alpha_0 = 0°$; however, this is not the case for an airfoil with positive camber where $\alpha_0 < 0°$. More details are given in Chapter 4. Airfoil profile drag does have an adverse effect on torque and therefore rotor power as seen in Figure 3.1; this will be further quantified in the following sections. The airfoil pitching moment typically acts nose-down (negative) over angles of attack, $\alpha$, for attached flow and contributes to a torsional moment about the blade pitch axis. For subsequent analyses, it is once more convenient to introduce dimensionless coefficients.

$$\text{Lift coefficient:} \quad c_l = \frac{dL}{\frac{1}{2}\rho V_{rel}^2 dA_c} = \frac{Lift\ force}{Dynamic\ pressure\ force\ on\ blade\ element}$$

$$\text{Drag coefficient:} \quad c_d = \frac{dD}{\frac{1}{2}\rho V_{rel}^2 dA_c} = \frac{Drag\ force}{Dynamic\ pressure\ force\ on\ blade\ element}$$

$$(3.1)$$

$$\text{Moment coefficient:} \quad c_m = \frac{dM_p}{\frac{1}{2}\rho V_{rel}^2 dA_c c} = \frac{Pitching\ moment}{Dynamic\ moment\ on\ blade\ element}$$

In Eq. (3.1), $dA_c = c\, dr$ is the "wetted" area of an infinitesimal blade element. We further note that the local inflow velocity, $V_{rel}$, at radial station $r$ is used as a reference velocity in Eq. (3.1) as opposed to the incoming wind speed, $V_0$, which is used for integrated thrust, torque, and power coefficients. In general, coefficients of airfoil lift, drag, and moment are documented as functions of the angle of attack, $\alpha$, for a given Reynolds number, $Re$. Here, a chord-based Reynolds number is defined local to the blade element as:

$$\text{Reynolds number (chord-based)}: \quad Re_c = \frac{\rho\, V_{rel}\, c}{\mu} \quad\quad (3.2)$$

Here, $\rho$ is the fluid density and $\mu$ is the fluid dynamic viscosity. For high-Reynolds number attached flow, the magnitude of the drag coefficient is typically just a few percent of the airfoil lift coefficient. For the following analyses, we simply assume that airfoil data are available from lookup tables as a function of the local angle of attack, $\alpha$, and local blade element Reynolds number, $Re_c$. A more detailed account on airfoil theory and families of airfoils for wind turbine applications is given in Chapter 4.

### 3.1.2 Blade Element Velocity and Force/Torque Triangles

Figure 3.3 illustrates the basic transformations for blade element thrust, $dT_B$, and torque, $dQ_B$, along with the respective velocity relations. The angle of transformation is the local

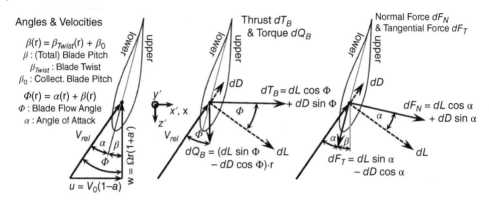

**Figure 3.3** Velocity and force/torque triangles at blade element.

blade flow angle, $\phi(r)$, which is both the angle between the thrust direction and the local lift vector and also the angle between the direction of the local relative velocity, $V_{rel}$, and the rotor plane. Note that the directions of sectional thrust and torque are defined by a fixed reference frame, while the directions of lift, $dL$, and drag, $dD$, as well as normal and tangential forces, $dF_N$ and $dF_T$, vary from one blade element to another. In Figure 3.3, blade element normal and tangential forces, $dF_N$ and $dF_T$, are defined perpendicular to and in the direction of, though pointing upstream, the airfoil chord line at a particular blade element. Note that normal and tangential forces are useful for quantitative comparisons against integrated surface pressure data, with examples to follow in subsequent chapters.

Given sectional lift and drag forces, $dL$ and $dD$, one can then compute the incremental thrust and torque for a given blade element at station $r$ of one blade as:

$$\text{Incremental thrust (one blade)}: \quad dT_B = dL \cos \phi + dD \sin \phi \quad (3.3)$$

$$\text{Incremental torque (one blade)}: \quad dQ_B = (dL \sin \phi - dD \cos \phi)\, r \quad (3.4)$$

In general, both incremental thrust and torque are generated primarily by the sectional lift force, $dL$. It becomes once more apparent that profile drag, $dD$, is a two-way loss in the sense that it adds to the rotor thrust force, which is of no use to the generation of rotor power, and reduces the power-producing sectional torque. One might argue that it would be advantageous to design rotor blades such that the resultant local blade flow angle, $\phi(r)$, along the blades is as large as possible in order for the lift force to generate maximum torque and the drag force to minimally reduce torque. However, as we shall see later, this argument is constrained by the angular velocity governing the direction of $V_{rel}$ for the outer half of the blade. One has also to keep in mind that resultant forces scale with $V_{rel}^2 \sim r^2$ such that the primary contributor to the incremental torque scales as $dQ_B \sim r^3 \sin \Phi$, typically resulting in the maximum incremental torque occurring around $r/R = 0.80 \mp 0.10$.

In addition, it is important to recognize in Figure 3.3 that the local blade flow angle, $\phi(r)$, is the sum of the local angle of attack, $\alpha(r)$, and local blade pitch angle, $\beta(r)$, as:

$$\text{Angle Relation}: \quad \phi(r) = \alpha(r) + \beta(r) \quad (3.5)$$

Here, the local angle of attack, $\alpha(r)$, is defined as the angle between the directions of the relative velocity vector and the local chord line. On the other hand, the local blade pitch angle, $\beta(r)$, is the angle between the local chord line and the rotor plane and describes how much a local blade section is pitched (nose-down) into the wind. In the wind energy community, the "blade tip pitch angle," $\beta_{Tip} = \beta(R)$, is typically used as a reference angle to describe the collective blade pitch over Regions II/III of the power curve of a pitch-controlled turbine and in parked conditions. From the perspective of a blade element airfoil, a positive blade pitch, that is, $\beta(r) > 0$, is defined as "nose-down." (This is an important fundamental difference between a wind turbine and a propeller, with the latter having the upper surface facing the incoming wind/airspeed and positive pitch being defined as "nose-up." An author's rule-of-thumb that encompasses both rotating devices states that "positive pitch equals 'nose into the wind' – always.") Note that in general, $\beta(r)$ is the sum of a "built-in" (or manufactured) twist, $\beta_{Twist}(r)$, plus a collective blade pitch angle, $\beta_0$, equal to $\beta_{Tip} - \beta_{Twist}(R)$, such that the blade tip is pitched at the desired $\beta_{Tip}$. Of further note is that Eq. (3.5) is essential to both blade design and pitch control. For example, considering small perturbations around a given blade flow angle distribution, $\phi(r)$, one can design the local blade pitch distribution, $\beta(r)$, for an advantageous angle-of-attack distribution, $\alpha(r)$, with optimal lift-to-drag ratio; alternatively, one could prevent blade stall or simply control torque (and thus rotor power) by pitching a local blade element further into the wind, that is, increasing the local blade pitch angle, $\beta(r)$, with the effect of lowering $\alpha(r)$ to angles below the airfoils' stall angles or to angles of attack with respective lift forces that satisfy a power constraint.

The velocity triangle in Figure 3.3 further allows us to write the relative velocity, $V_{rel}$, as a function of the axial wind speed, $V_0$, axial-/angular induction factors, $a$ and $a'$, and the local speed ratio, $\lambda_r = \Omega r / V_0$, as:

$$\text{Relative Velocity:} \quad V_{rel}(r) = V_0 \sqrt{(1 - a(r))^2 + (1 + a'(r))^2 \, \lambda_r^2} \tag{3.6}$$

In addition, the following trigonometric relations on the velocity triangle will be of use in subsequent analyses:

$$\text{Velocity Triangle:} \quad \tan \phi(r) = \frac{1}{\lambda_r} \frac{1 - a(r)}{1 + a'(r)} \tag{3.7}$$

$$\sin \phi(r) = \frac{V_0}{V_{rel}(r)} [1 - a(r)] \tag{3.8}$$

$$\cos \phi(r) = \frac{\Omega r}{V_{rel}(r)} [1 + a'(r)] \tag{3.9}$$

Having derived the basics of blade element torque and thrust generation, the next step is now to relate blade element theory to the fundamental relations obtained from actuator-/rotor disk theories.

## 3.2 Combining Momentum Theory and Blade Element Theory through $a$, $a'$, and $\Phi$

To this end, the flow state at a given blade element is essentially given by the axial-/angular induction factors, $a(r)$ and $a'(r)$. The relative velocity, $V_{rel}(r)$, at a blade element can then be found using Eq. (3.6) and the local blade flow angle, $\phi(r)$, using any of the trigonometric relations in Eqs. (3.7)–(3.9).

### 3.2.1 Sectional Thrust and Torque in Momentum and Blade Element Theory

In the following, we combine the thus far independent relations of actuator-/rotor disk theories, that is, momentum theory, with relations for a sectional blade element at radial station $r$. As for momentum theory, we remind ourselves that those relations apply uniformly over a disk annulus $dA = 2\pi r \, dr$. Writing Eq. (2.13) in differential form for an incremental disk-annulus thrust and similarly Eq. (2.37) for an incremental disk-annulus torque using Eq. (2.31), we obtain:

**Momentum Theory (Disk annulus)**

$$\text{Incremental thrust:} \quad dT = 4a(1-a)\,\rho\, V_0^2\, \pi\, r\, dr \tag{3.10}$$

$$\text{Incremental torque:} \quad dQ = 4a'(1-a)\,\rho\, V_0\, \pi\, r^3\, \Omega\, dr \tag{3.11}$$

In order to proceed with the determination of sectional thrust and torque in blade element theory, one must first define the respective incremental lift and drag forces at each blade element using Eq. (3.1) in differential form as:

$$\text{Incremental lift (one blade):} \quad dL = c_l \frac{1}{2}\rho\, V_{rel}^2\, c\, dr \tag{3.12}$$

$$\text{Incremental drag (one blade):} \quad dD = c_d \frac{1}{2}\rho\, V_{rel}^2\, c\, dr \tag{3.13}$$

Next, we multiply the relations for incremental thrust and torque for one blade, that is, Eqs. (3.3) and (3.4), by the blade number, $B$, and use Eqs. (3.12) and (3.13) to obtain incremental thrust and torque of all blades for a given blade element at station $r$ as:

**Blade Element Theory ($B$ blades):**

$$\text{Incremental thrust:} \quad dT = B\frac{1}{2}\rho V_{rel}^2 (c_l \cos\phi + c_d \sin\phi)\, c\, dr \tag{3.14}$$

$$\text{Incremental torque:} \quad dQ = B\frac{1}{2}\rho V_{rel}^2 (c_l \sin\phi - c_d \cos\phi)\, c\, r\, dr \tag{3.15}$$

The preceding relations form the basis for the design of ideal rotors as presented in Section 3.3. A general BEM solution method, on the other hand, consists of equating (3.10), (3.14) and (3.11), (3.15), respectively, and is presented in Section 3.5.

### 3.2.2 Rotor Thrust and Power in Blade Element Theory

In the case of all quantities in Eqs. (3.10), (3.11), and (3.14), (3.15) being known from analysis and/or design, one can integrate for the respective rotor thrust and power as:

$$\text{Thrust:} \quad T = B\frac{1}{2}\rho \int_{r_{Root}}^{R} V_{rel}^2 (c_l \cos\phi + c_d \sin\phi)\, c\, dr \tag{3.16}$$

$$\text{Power:} \quad P = B\frac{1}{2}\rho\Omega \int_{r_{Root}}^{R} V_{rel}^2 (c_l \sin\phi - c_d \cos\phi)\, c\, r\, dr \tag{3.17}$$

Here, $r_{Root}$ is the so-called root cut-out, typically defined as the radial station of the blade-hub juncture. As for rotor thrust and power coefficients, it is convenient to introduce a local blade solidity, $\sigma'$, as:

$$\text{Local Solidity:} \quad \sigma' = \frac{Bc}{2\pi r} \tag{3.18}$$

Note that $\sigma'$ is an important design parameter and describes the fraction of a given rotor annulus occupied by the total blade surface of $B$ blades with chord $c$ at radial station $r$. The rotor performance coefficients can then be written as:

Thrust Coefficient:

$$C_T = \frac{T}{\frac{1}{2}\rho V_0^2 A} = 2 \int_{r_{Root}/R}^{1} \sigma' \left(\frac{V_{rel}}{V_0}\right)^2 (c_l \cos\phi + c_d \sin\phi)\frac{r}{R}\,d\frac{r}{R} \qquad (3.19)$$

Power Coefficient:

$$C_P = \frac{P}{\frac{1}{2}\rho V_0^3 A} =$$
$$2\lambda \int_{r_{Root}/R}^{1} \sigma' \left(\frac{V_{rel}}{V_0}\right)^2 (c_l \sin\phi - c_d \cos\phi)\left(\frac{r}{R}\right)^2 d\frac{r}{R} \qquad (3.20)$$

All quantities in Eqs. (3.19) and (3.20) are dimensionless, and the velocity ratio, $V_{rel}/V_0$, can be written in terms of the blade flow angle, $\phi(r)$, and the respective induction factors using trigonometric relations on the velocity triangle in Eqs. (3.8) and (3.9).

## 3.3 Aerodynamic Design and Performance of an Ideal Rotor

The basic relations for momentum theory and blade element theory in Section 3.2.1 can now be used for a first design exercise applied to ideal rotors. We will first consider an ideal rotor without wake rotation, also called a "Betz Optimum Rotor," and then perform the design of an ideal rotor with rotation using radial distributions of optimal axial- and angular induction factors from Section 2.2. We choose the following design specifications as an example:

**Given: Example Design Parameters for Ideal Rotor**

- 3-Bladed Rotor, $B = 3$
- Tip Speed Ratio, $\lambda = 6$
- Root Cut-Out, $r_{Root}/R = 0.10$
- Lift Coefficient (Blade Element), $c_l = 0.69$
- Angle of Attack (Blade Element), $\alpha = 6.3°$

Note that in this particular case, the lift coefficient $c_l(\alpha) = 2\pi\alpha$ follows the relation for a symmetric thin airfoil in inviscid flow, see for example, Anderson (2001) or Chapter 4, with $\alpha$ in *rad*. Given these design parameters, the task is to design an ideal rotor blade planform:

**Find: Blade Planform**

- Blade Pitch, $\beta = \beta(r/R)$
- Non-Dimensional Blade Chord, $\frac{c}{R} = \frac{c}{R}(r/R)$

In the following, we will design ideal rotor planforms and introduce an optimum design parameter in conjunction with the combined effects of root cut-out and wake rotation on rotor performance.

### 3.3.1 The Ideal Rotor Without Wake Rotation

The ideal rotor without rotation is also referred to as the Betz Optimum Rotor due to its relation to the "Betz limit" derived in Section 2.1.4. As such the optimum axial and angular induction factors are:

$$\text{Betz limit:} \quad a = \frac{1}{3}; a' = 0 \tag{3.21}$$

**Blade Pitch, $\beta = \beta(r/R)$**

Substituting Eq. (3.21) into a trigonometric relation on the velocity triangle, that is, here Eq. (3.7), and replacing the blade flow angle using Eq. (3.5) yields the following:

$$\text{Blade Pitch:} \quad \beta(r/R) = \tan^{-1}\left(\frac{2}{3\lambda} \cdot \frac{1}{r/R}\right) - \alpha \tag{3.22}$$

Here, we used the definition of the local speed ratio in Eq. (2.33). Note that Eq. (3.22) is in *rad* and, if desired, has to be converted to *deg* (or °) by multiplying with $180°/\pi$.

**Non-Dimensional Blade Chord, $\frac{c}{R} = \frac{c}{R}(r/R)$**

For the ideal rotor without wake rotation, the incremental thrust relations in Eqs. (3.10) and (3.14) become:

$$\text{Momentum Theory:} \quad dT = \frac{8}{9}\rho V_0^2 \pi r \, dr \tag{3.23}$$

$$\text{Blade Element Theory:} \quad dT = B\frac{1}{2}\rho V_{rel}^2 \, c_l \cos\phi \, c \, dr \tag{3.24}$$

Substituting $V_{rel}$ in Eq. (3.24) by another trigonometric relation on the velocity triangle, that is, here Eq. (3.8), and equating (3.23) and (3.24) results in an optimum blade parameter, $\sigma'c_l$:

$$\text{Optimum Blade Parameter:} \quad \sigma'c_l = 2\sin(\alpha + \beta)\tan(\alpha + \beta) \tag{3.25}$$

In Eq. (3.25), $\sigma'$ is the local solidity as defined in Eq. (3.18). It becomes apparent that the product of blade number, local chord, and section lift coefficient, that is, $Bcc_l$, does have an optimal value at each radial station, $r$. Hence, the actual blade chord can be, at least to some degree, controlled by the blade number and the section lift coefficient, which is an important relationship with respect to both blade weight and blade structural properties. Using once more relations on the velocity triangle in Figure 3.3, specifically Eqs. (3.5)–(3.9), and induction factors from Eq. (3.21) we find

$$\sin(\alpha + \beta) = \frac{1}{\sqrt{1 + \left(\frac{3}{2}\lambda\right)^2 (r/R)^2}} \; ; \quad \tan(\alpha + \beta) = \frac{2}{3\lambda}\frac{1}{r/R} \tag{3.26}$$

for the *sin* and *tan* in Eq. (3.25). Subsequent substitution and solving for the non-dimensional blade chord, $c/R$, results in:

$$\text{Non-Dimensional Blade Chord:} \quad \frac{c}{R}(r/R) = \frac{8\pi}{3\lambda Bc_l}\frac{1}{\sqrt{1 + \left(\frac{3}{2}\lambda\right)^2 (r/R)^2}} \tag{3.27}$$

Figure 3.4 shows the ideal rotor blade planform consisting of blade pitch, $\beta$, and non-dimensional blade chord, $c/R$, as functions of the non-dimensional blade radius, $r/R$. It

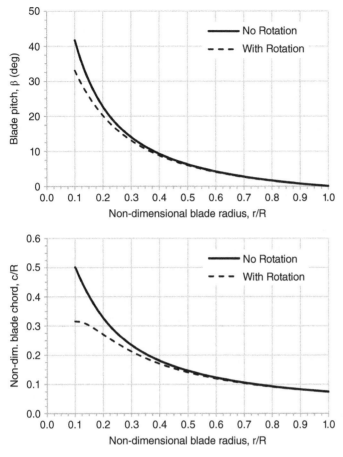

**Figure 3.4** Example – ideal rotor blade planforms with/without wake rotation ($B = 3$, $\lambda = 6$, $c_l = 0.69$, $\alpha = 6.3°$).

can be seen that both decrease toward the blade tip, with relatively high values for blade pitch and non-dimensional blade chord at the innermost blade stations close to the root cut-out. Such high values are quite unrealistic given the assumption of no radial flow in both actuator-/rotor and blade element theories. It is worth mentioning at this stage that this is, at least in part, a result of ignoring wake rotation and three-dimensional root-/tip effects. Nevertheless, we realize that the non-dimensional blade chord could be reduced by operating the innermost blade stations at a higher section lift coefficient, a strategy for selecting appropriate wind turbine airfoils that we will explore in more detail in Chapter 4.

### 3.3.2 The Ideal Rotor with Wake Rotation

The induced flow field at the rotor disk for the ideal rotor with wake rotation is derived in Section 2.2.4. The associated optimum axial and angular induction factors are:

Optimum (Wake Rotation):    $a(\lambda_r)$ from Eq. (2.55);

$$a' = \frac{1 - 3a}{4a - 1} \tag{3.28}$$

**Blade Pitch, $\beta = \beta(r/R)$:**

Using Eqs. (3.5) and (3.7) on the velocity triangle for the optimal solution of $a$ and $a'$, the optimum blade pitch, $\beta(r/R)$, becomes:

$$\text{Blade Pitch:} \quad \beta(r/R) = \tan^{-1}\left( \frac{(4a - 1)(1 - a)}{\lambda a} \cdot \frac{1}{r/R} \right) - \alpha \tag{3.29}$$

Here, Eq. (3.28) determines the radial distribution of $a$ and $a'$, and we note that Eq. (3.29) is in *rad*.

**Non-Dimensional Blade Chord, $\frac{c}{R} = \frac{c}{R}(r\,/\,R)$**

For the ideal rotor with wake rotation, the incremental thrust relations in Eqs. (3.10) and (3.14) now appear in simplified form as:

$$\text{Momentum Theory:} \quad dT = 4a(1 - a)\, \rho\, V_0^2\, \pi\, r\, dr \tag{3.30}$$

$$\text{Blade Element Theory:} \quad dT = B\frac{1}{2}\rho V_{rel}^2\, c_l \cos\phi\, c\, dr \tag{3.31}$$

Again, substituting $V_{rel}$ in Eq. (3.31) by the *sin* relation on the velocity triangle, that is, Eq. (3.8), and equating (3.30) and (3.31) results in the same optimum blade parameter, $\sigma' c_l$, as for the ideal rotor without rotation:

$$\text{Optimum Blade Parameter:} \quad \sigma' c_l = 2 \sin(\alpha + \beta) \tan(\alpha + \beta) \tag{3.32}$$

However, it has to be kept in mind that $\beta$ and hence the blade flow angle $\phi = \alpha + \beta$ differ between ideal rotor considerations with and without wake rotation. This can be readily seen by evaluating the *sin* and *tan* relations on the velocity triangle with

$$\sin(\alpha + \beta) = \frac{(4a - 1)(1 - a)}{\sqrt{(4a - 1)^2(1 - a)^2 + (\lambda a)^2(r/R)^2}};$$

$$\tan(\alpha + \beta) = \frac{(4a - 1)(1 - a)}{\lambda a} \cdot \frac{1}{r/R} \tag{3.33}$$

for the *sin* and *tan* in Eq. (3.32). Subsequent substitution and solving for the non-dimensional blade chord, $c/R$, results in:

Non-Dimensional Blade Chord:

$$\frac{c}{R}(r/R) = \frac{4\pi}{\lambda B c_l} \frac{(4a - 1)^2(1 - a)^2}{a\sqrt{(4a - 1)^2(1 - a)^2 + (\lambda a)^2(r/R)^2}} \tag{3.34}$$

Figure 3.4 also shows the ideal rotor blade planform including wake rotation consisting of blade pitch, $\beta$, and non-dimensional blade chord, $c/R$, as functions of the non-dimensional blade radius, $r/R$. In comparison to the ideal rotor without wake rotation in Section 3.3.1, it can be seen here that both blade pitch and non-dimensional blade chord exhibit more realistic values along the inboard blade half and, in particular, at the innermost blade stations close to the root cut-out. Note that the blade root chord could be further reduced for the same induced flow, that is, the same $a$ and $a'$, by increasing the blade section $c_l$, while keeping a constant value for the optimum blade parameter, $\sigma' c_l$. Again, operating inboard blade stations at higher $c_l$ using suitable airfoils is indeed a

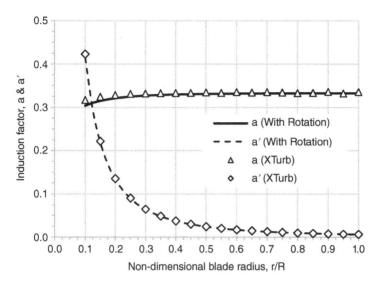

**Figure 3.5** Radial distributions of optimum induction factors *a* and *a′* for $\lambda = 6$ (lines: Eq. (2.55); symbols: *XTurb* analysis).

common design methodology that is further described in Chapters 4 and 8. For the outboard blade half, blade pitch and non-dimensional blade chord are practically the same for ideal rotors with and without wake rotation. This is an important result, which is related to the optimal solution of *a* and *a′* in Eq. (3.28). Figure 3.5 shows exact solutions for the induction factors, *a* and *a′*, and it can be seen that, as discussed in Chapter 2, the respective values approach the Betz limit of Eq. (3.21) toward the blade tip.

Also shown in Figure 3.5 are results obtained by *XTurb* (Schmitz 2012) for the inductions factors, *a* and *a′*, for the ideal rotor with wake rotation. Here the *XTurb* code was run in ANALYSIS mode at a tip speed ratio of $\lambda = 6$, with blade pitch, $\beta$, and normalized blade chord, $c/R$, specified by Eqs. (3.29) and (3.34), respectively. An inviscid airfoil polar for a thin symmetric airfoil (*ThinSymmAirf-Inv.polar*) was used for the analysis. Note that the settings in *IdealRotorWithRotation1.inp* are such that no root-/tip corrections, which are discussed in the following section, were used. The *XTurb* results plotted as symbols in Figure 3.5 serve as a good verification of the ideal rotor design. Integrated rotor performance parameters are given next as:

---

**XTurb Example 3.1 Ideal Rotor with Rotation (IdealRotorWithRotation1.inp)**

```
3.3.2 - Ideal Rot. w/ Rotat. 1        ***** XTurb V1.9 -  OUTPUT *****
  Blade Number        BN =            3


                                 +    BLADE  ELEMENT  MOMENTUM  THEORY  (BEMT)    +

  Number    TSR      PITCH [deg]      CT        CP       CPV       CB       CBV
       1   6.0000    0.0080        0.8802   -0.5726    0.0000   -0.1976    0.0000
```

---

It is important to note that in *XTurb*, the force/moment convention is such that values $C_P < 0$ refer to power extraction from the fluid and hence *positive* power production.

Consequently, $C_p$ values obtained from *XTurb* have to be plotted as $-C_p$ to be consistent with documented power coefficients. Note that the *XTurb* computed power coefficient is less than the theoretical optimum of $C_{P,max} = 0.5759$ as shown in Figure 2.11. One contributing cause for this discrepancy is that the theoretical optimum assumes no root cut-out. Hence one can either adjust the integration bounds in Eq. (2.59) or (3.20), respectively, to account for $r_{Root}/R$ and obtain an adjusted value for the theoretical optimum of $C_{P,max} = 0.5711$. Nevertheless, *XTurb* computed $C_p$ is within 0.26% (absolute) of the adjusted $C_{P,max}$ and can be considered as good verification of the blade design for the ideal rotor with wake rotation. In the following section, we will introduce tip and root loss factors to account for three-dimensional effects at both blade ends.

## 3.4  Tip and Root Loss Factors

To this end, we have assumed that all radial blade elements operate as airfoil sections (two-dimensional). However, this assumption breaks down at both blade tip and root, if the blade is aerodynamically loaded (pressure jump across actuator/rotor disk), as higher-pressure upstream flow (lower blade surface) slips around the blade tip and root to lower-pressure downstream flow (upper blade surface), thus generating distinct tip and root vortices that trail downstream into the wake. These edge vortices induce a downwash along the blade radius (or lifting line) that reduces the angle of attack, $\alpha$, of blade elements close to the tip and root sections by locally increasing the axial induction factor, $a$, and, consequently, also reduce aerodynamic lift and hence sectional thrust and torque/power. Combined tip and root loss factors, $F(r)$, along the blade radius do, at least to some extent, account for three-dimensional vortex-induced effects as a consequence of a finite number of blades.

In the next sections, a brief introduction to some classical work is given (Sections 3.4.1–3.4.3), followed by a description of more recent research approaches to tip modeling (Section 3.4.4) for the advanced reader. In Section 3.5.1, the sectional thrust and torque relations of momentum and blade element theory of Section 3.2 are adjusted to include a total tip-/root loss correction, $F(r)$.

### 3.4.1  Prandtl Blade Number Correction versus Glauert Tip Correction – Historical Perspective

The tip-loss factor most commonly used today in combined blade element momentum theory (BEMT) methods is often referred to as the "Prandtl Tip-Loss Factor." On a historical note, however, this is not fully correct as Prandtl's original work on the subject was not intended to be a tip-loss factor but an approximation for the circulation distribution on a rotating blade. Prandtl derived this approximation as an addendum to the dissertation of Betz (1919), which also means that the approximation is strictly valid only for an aerodynamically ideal rotor with a finite number of blades, see Section 3.3.1. According to the formulation derived by Betz (1919), an ideal rotor refers to one where trailing helicoidal vortex sheets move at constant helical pitch, albeit uniform axial induction across the rotor disk. Prandtl's ingenious approach was to approximate the helical wake by a stack of semi-infinite planar vortex sheets, with consecutive vortex

**Figure 3.6** Prandtl's approximation of helical vortex sheets using semi-infinite planar vortex sheets.

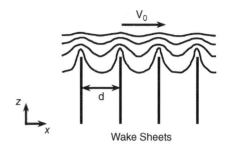

Wake Sheets

sheets separated by a distance, $d$, determined as:

$$\text{Vortex-Sheet Spacing (Prandtl):} \quad d = \frac{2\pi R}{B} \frac{1}{\sqrt{1+\lambda^2}} \quad (3.35)$$

Here, $B$ is the number of blades and $\lambda$ the tip speed ratio. An illustration is given in Figure 3.6.

The problem then reduces to finding the solution for a perturbation potential around the edges of the stack of semi-infinite planes. The perturbation potential is shown to be proportional to the ratio of circulations of a rotor with $B$ blades, $\Gamma_B$, and that of a full rotor disk, that is, infinite number of blades, $\Gamma_\infty$. This can be done using conformal mapping, see Betz (1919) or Chapter 8 of Sørensen (2016) for the mathematical details. Further ignoring induced velocities at the tip leads to Prandtl's approximation, $F_B$, to account for a finite number of blades:

$$\text{Prandtl:} \quad F_B(r) = \frac{B\Gamma_B}{\Gamma_\infty} = \frac{2}{\pi}\cos^{-1}\left[\exp\left(-\frac{B}{2}\sqrt{1+\lambda^2}(1-r/R)\right)\right] \quad (3.36)$$

It was Glauert, however, who used Prandtl's blade number correction of Eq. (3.36) to construct an explicit tip correction for BEM models, see Glauert (1935), by interpreting $F_B$ not as a ratio of circulations but as the ratio of a local and averaged induced velocity on the vortex sheets. In the end, the functional relationship is the same, but Glauert was able to make Eq. (3.36) be consistent with BEM models. For this reason, Glauert conducted a local blade flow angle approach such that:

$$\text{Vortex-Sheet Spacing (Glauert):} \quad d \cong \frac{2\pi r}{B}\sin\phi(r) \quad (3.37)$$

In conjunction with Eqs. (3.35) and (3.36), this results in Glauert's tip correction, $F_T$, expressed in the following form:

$$\text{Glauert (Tip Loss):} \quad F_T(r) = \frac{2}{\pi}\cos^{-1}\left[\exp\left(-\frac{B}{2}\frac{1-r/R}{r/R}\frac{1}{\sin\phi(r)}\right)\right] \quad (3.38)$$

Equation (3.38) describes the tip correction used in most BEMT codes today. A loss factor is typically associated with a fraction less than one. Indeed, it can be easily verified that $F_T \in [0;1]$ in Eq. (3.38). At the blade tip ($r/R = 1$), the Glauert tip correction equals zero as it is assumed that the pressure on lower (pressure) and upper (suction) blade surface equates. Consequently, there is no force being generated right at the blade tip, a condition consistent with lifting line theory (see Chapter 6). Note that toward the blade root ($r/R \to 0$), on the other hand, the Glauert tip correction, $F_T$, quickly approaches a value of one, that is, no loss associated as a result of the tip vortex; however, here a

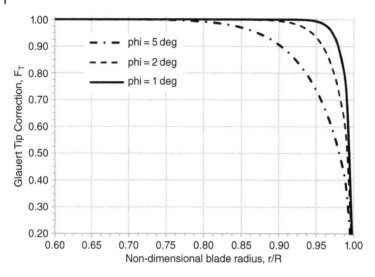

**Figure 3.7** Radial comparisons of Glauert tip correction (or Prandtl tip-loss factor) for different blade flow angle, $\phi$ ($\phi = 5°$ (dash-dot), $\phi = 2°$ (dash), and $\phi = 1°$ (solid)).

root loss model becomes active as described in Section 3.4.2. Furthermore, it is evident from Eq. (3.38) that as the blade number $B$ increases, the Glauert tip correction, $F_T$, approaches the limiting case of a full actuator/rotor disk with increasing blade solidity. Of further interest is the dependence of $F_T$ on the local blade flow angle, $\phi$, which is a direct result of Glauert's approximation for the vortex-sheet spacing generated by $B$ blades. It appears from Eq. (3.38) that less tip loss, that is, $F_T$ closer to one, is associated with a smaller blade flow angle, $\phi$. Figure 3.7 shows a radial distribution of the Glauert tip correction factor for three assumed constant values of $\phi$.

With reference to the angle relation in Eq. (3.5), a smaller $\phi$ results in a smaller angle of attack, $\alpha$, for a given blade pitch, $\beta$. Hence a particular blade section is more lightly loaded and therefore subject to less tip loss. Note, however, that this is a linearized *Gedankenexperiment* (German for "thought experiment"), assuming small perturbations to a non-linear problem. Nevertheless, this opens up an interesting question about whether or not rotor design with the same radial $c_l$ distribution achieves a different power coefficient, $C_P$, depending on whether or not a cambered as opposed to a symmetric airfoil is used in the tip region. With reference to the velocity triangle in Figure 3.3, there is no straightforward answer to this question as the incremental torque, $dQ_B$, in Eq. (3.4) is $\sim \sin \phi$; hence reducing $\phi$ decreases $dQ_B$, which may or may not outweigh the favorable effect of reducing $\phi$ in the tip correction (or tip loss) $F_T$.

### 3.4.2 A Total Tip-/Root Loss Correction

A simple adjustment can be made to the Glauert tip correction in Eq. (3.38) to obtain a corresponding root loss factor, $F_R$, as:

$$\text{Root Loss}: \quad F_R(r) = \frac{2}{\pi}\cos^{-1}\left[\exp\left(-\frac{B}{2}\frac{r/R - r_{root}/R}{r_{root}/R}\frac{1}{\sin \phi(r)}\right)\right] \tag{3.39}$$

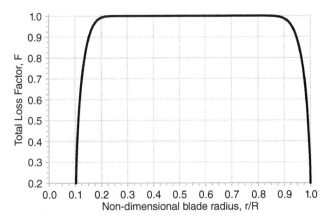

**Figure 3.8** Radial variation of total tip-/root loss correction for a three-bladed ideal rotor with rotation at $\lambda = 6$. Source: Data from *XTurb_Output_Method.dat*.

Note that $F_R = 0$ is satisfied at the blade root ($r/R = r_{root}/R$) and that $F_R \rightarrow 1$ at the blade tip ($r/R = 1$) for a small root cut-out, that is, $r_{root}/R \rightarrow 0$. It should be mentioned that though Eq. (3.39) is used in the present or very similar forms in most BEMT methods, it is only a simple manual adjustment of Eq. (3.38) with no direct physical reasoning. In fact, one could easily argue that Prandtl's approximation in Figure 3.6 is not intended for use as a root loss correction, also because some interaction between root vortices of adjacent blades can be expected, which is not accounted for in Eq. (3.39). Nevertheless, Eq. (3.39) has been proven to give acceptable results. A total tip-/root loss correction can then be defined as:

$$\text{Total Tip-/Root Loss Correction:} \quad F(r) = F_R(r) \cdot F_T(r) \tag{3.40}$$

In essence, the total tip-/root correction of Eq. (3.40) modifies the local induced flow from finite-number blade element theory and thus makes it possible to set it equal to the corresponding expression from momentum theory, as will be shown in Section 3.5. An example of a radial distribution of the total tip-/root loss correction, $F$, is shown in Figure 3.8 for the ideal rotor with wake rotation from Section 3.3.2. The corresponding *XTurb* input file is *IdealRotorWithRotation2.inp* (with $RLOSS = 1$, $TLOSS = 1$). Resulting *XTurb* integrated rotor performance parameters become:

---

**XTurb Example 3.2 Ideal Rotor with Rotation and Tip-/Root Loss (IdealRotorWithRotation2.inp)**

```
3.3.2 - Ideal Rot. w/ Rotat. 2      ***** XTurb V1.9 -  OUTPUT *****
   Blade Number        BN =  3

                              +     BLADE ELEMENT MOMENTUM THEORY (BEMT)    +

   Number   TSR      PITCH [deg]    CT       CP       CPV       CB       CBV
        1   6.0000   0.0080         0.8280  -0.5156   0.0000   -0.1827   0.0000
```

Note that the *XTurb* computed power coefficient of $C_P = 0.5156$ induced tip-/root loss is notably less than the reference case of $C_P = 0.5726$ without tip-/root loss, see Section 3.3.2. Here the induced tip-/root loss accounts for a 10% reduction in power coefficient, still assuming an inviscid airfoil polar. It is important to keep in mind, however, that this particular rotor as designed in Section 3.3.2 was optimized ignoring the combined effects of tip-/root loss. In fact, it can be easily shown by modifying *Ideal-RotorWithRotation2.inp* to analyze the blade over a range of tip speed ratios that the actual $C_{P,max}$ for this rotor is obtained at a higher tip speed ratio, $\lambda$, and/or blade tip pitch angle, $\beta_{Tip}$.

### 3.4.3 Limitations of Classical Tip-/Root Corrections

Prandtl's approximation using semi-infinite planar vortex sheets to represent the helical wake downstream of a $B$-bladed rotor is very elegant (as is everything that Prandtl did) because it leads to a closed-form solution for a perturbation potential around the vortex sheets. In reality, however, the approximation is better suited for propellers than it is for turbines. The fact that the assumed vortex sheets are planar implies that the rotor is lightly loaded so that the vortex sheets do not deform and/or roll up. For wind turbines, on the other hand, the nature of the velocity triangle, see Figure 3.3, is such that the rotor is practically always highly loaded if the objective is to maximize $C_P$. This is primarily a result of $dT_B \sim \cos\phi$ while $dQ_B \sim \sin\phi$ in Eqs. (3.3) and (3.4), with $\phi < 30°$ particularly in the outboard one-third of the blade radius. Furthermore, Prandtl assumed that planar vortex sheets advect downstream at a constant axial velocity, an assumption that is also not strictly valid if the rotor loading is not as ideal as assumed by Betz (1919). In the end, one can also expect that Prandtl's approximation, $F_B$, performs better for blade numbers $B = 4 \dots 8$ (propellers) rather than $B = 2 \dots 3$ for modern wind turbines, simply because the relation is scaled to a full rotor disk with an infinite number of blades. In addition, Glauert's interpretation of $F_B$ being a ratio of induced velocities rather than circulations adds another difficulty with respect to the fact that Glauert assumes the mass flow rate to be unaffected. As for the root loss correction, it is quite obvious that Prandtl's model of semi-infinite vortex sheets is not directly constructed to be used as a root correction, the major difficulty being that root ends of $B$ blades are in their respective vicinities, which implies at least some interaction. Though the classical tip-/root corrections do have all these limitations, their mathematical elegance and simplicity of use have proven themselves in many engineering applications. The following section provides some introduction to research efforts made to remedy one or more of these inherent limitations. In general, alternative approaches to tip modeling can be categorized as those that modify $F_T$ in Eq. (3.38) and those that approach the problem from a different perspective.

### 3.4.4 Modern Approaches to Tip Modeling

It was Wilson and Lissaman (1974) and Lissaman et al. (1976) who were the first to look again at the tip-loss problem and suggested that both the induced velocities and the mass flow rate be affected by $F_T$. The downside of this approach is, however, that the combined equations become more complex and, as de Vries (1979) showed, the orthogonality condition between the induced velocity and relative blade velocity (tangential velocity) is not satisfied. De Vries (1979) then added a further correction to satisfy the orthogonality condition, showing in the end practically the same results when compared to the baseline correction proposed by Wilson and Lissaman. Hence, it can be argued whether the mass flow rate needs to be corrected at all if results obtained do not show adequate and consistent improvement. In general, the observation has been that the classical Glauert tip correction, $F_T$, overpredicts blade tip loads for highly loaded wind turbine blades, see for example, theses by Micallef (2012) and Ramdin (2017) for an in-depth discussion. This had led researchers to different ideas of finding alternative concise, yet consistent, ways of better predicting blade tip loads, a few of which are briefly described next. The following subsections are meant for the advanced reader with some prior knowledge of BEMT solution algorithms and vortex methods. A first-time reader is advised to proceed directly to Section 3.5 and revisit Sections 3.4.4.1–3.4.4.4 at a later stage.

### 3.4.4.1 Correction of Normal-/Tangential Force Coefficients (Shen et al.)

The basic motivation for Shen et al. (2005a,b) is a small inconsistency that exists at the blade tip where $F_T \to 0$, but not necessarily in the corresponding angle of attack at the tip, $\alpha_T$, though the tip force should tend to zero due to pressure equalization from pressure to suction sides of the blade. Shen et al. proposed to correct the airfoil data close to the tip in the form of their respective resulting normal and tangential force coefficients, $C_n^r$ and $C_t^r$, as:

$$C_n^r = F_1 \cdot C_n, \quad C_t^r = F_1 \cdot C_t \tag{3.41}$$

Shen et al. further argue that the correction factor, $F_1$, in Eq. (3.41) is of the same form as the original Glauert tip correction, $F_T$, in Eq. (3.38), though with an embedded coefficient, $g$, to obtain the following form:

$$\text{F1 Correction:} \quad F_1(r) = \frac{2}{\pi} \cos^{-1} \left[ \exp \left( -g \frac{B}{2} \frac{1 - r/R}{r/R} \frac{1}{\sin \phi(r)} \right) \right] \tag{3.42}$$

In general, the $g$ coefficient depends on the blade number, $B$, tip speed ratio, $\lambda$, as well as the blade pitch and chord distributions in the tip region; however, Shen et al. propose a simpler form as:

$$g \text{ Coefficient:} \quad g = \exp(-c_1 (B\lambda - c_2)) + 0.1 \tag{3.43}$$

The parameters $c_1$ and $c_2$ in Eq. (3.43) are found empirically by comparison to measured normal forces in the tip regions of the NREL Phase VI and Swedish WG 500 rotors at low and high tip speed ratio, specifically $c_1 \approx 0.125$ and $c_2 = 21$.

### 3.4.4.2 Helical Model for Tip Loss (Branlard et al.)

The classical tip correction due to Glauert is based on Prandtl's approximation of a lightly loaded rotor wake by a stack of planar (non-deforming) semi-infinite vortex sheets. On the other hand, vortex theory as described in Chapter 6, uses either a prescribed or free "helical" wake structure where the resulting induced flow at the velocity triangle mutually effects all radial trailing wake vortex filaments, whose strengths is determined by radial gradients in the blade bound circulation (also called trailing vorticity). Hence, radial variations in the helical pitch as well as wake expansion and distortion effects are included that better account for highly loaded wind turbine blades. Branlard et al. (2013)

and Branlard and Gaunaa (2014) consider the tip-loss factor as the ratio of induced velocities from a helical rotor wake with an infinite number of blades to one with $B$ blades as:

$$\text{Branlard (Tip Loss):} \quad F_T(r) = \frac{a_\infty(r)}{a_B(r)} = \frac{u_\infty(r)}{u_B(r)} \tag{3.44}$$

In Eq. (3.44), $u_B(r)$ is computed as a superposition of semi-infinite helical vortex filaments, $r'_{B,j}$, whose individual pitch angles depend on the local velocity triangle at the respective radial station they originate from, specifically:

$$\text{Axial induced velocity (Helix):} \quad u_B(r) = \sum_j u_{helix}(r, r'_{B,j}, B, \Gamma_{B,j}) \tag{3.45}$$

Here, $\Gamma_{B,j}$ denotes the radial ($j$) circulation distribution along $B$ blades. Equation (3.45) can be computed analytically using the Biot–Savart induction law, see also Chapter 6, within a BEMT algorithm, or determined from a table lookup of pre-computed load cases. The induced velocity of a rotor with an infinite number of blades, $u_\infty(r)$, is simply obtained from the limit of a vortex cylinder as the ratio of the local circulation and the overall helix pitch. More details can be found in Branlard (2017) and Branlard and Gaunaa (2016). A related approach of incorporating elements of vortex theory into blade element theory is also suggested by Wood et al. (2016).

### 3.4.4.3 Decambering Effect at Blade Tip (Sørensen et al.)

The conceptual idea of the decambering effect goes back to Montgomerie (1995) who demonstrated that the induced flow from a helical wake whose vortex filaments originate from a blade lifting line is subject to a small error associated with the actual blade planform in the tip region. The error is caused by the fact that the sectional angle of attack, $\alpha(r)$, is typically determined at the airfoil quarter-chord point, $c/4$, while the helical wake (or trailing vorticity) originates at the blade trailing edge. At the trailing edge, the total flow induction is higher than at the quarter-chord point due to both the helical wake itself and the action of the bound vorticity along the blade quarter-chord line. Consequently, the angle of attack determined at the airfoil quarter-chord point, $c/4$, is a bit higher than the actual induced flow, which can be interpreted as the actual flow applying a finite amount of "decambering" (i.e. reducing the effective angle of attack) on the sectional airfoil. From thin-airfoil theory, the camber line slope, $\eta'_c$, is equal to the ratio of normal and tangential velocities, $u_n$ and $u_t$, relative to the local chord line. The additional camber line slope (or decambering correction), $\Delta\eta'_c$, is determined by assuming that the center of pressure is located at the quarter-chord point, $c/4$, and by subtracting the respective $c/4$ induction from the local induction:

$$\text{Decambering Correction:} \quad \Delta\eta'_c(r) = \frac{u_t(r) - u_{t,c/4}(r)}{V_{rel}(r)\cos\alpha(r) + [u_n(r) - u_{n,c/4}(r)]} \tag{3.46}$$

Here, $u_t(r) = \sum_j u_{t,helix}(r, r'_{B,j}, B, \Gamma_{B,j})$ and $u_n(r) = \sum_j u_{n,helix}(r, r'_{B,j}, B, \Gamma_{B,j})$ are determined from the Biot–Savart induction law. Sørensen et al. (2014, 2015) then transform Eq. (3.46) into a respective correction for the sectional lift coefficient, that is, $\Delta c_l(r)$, and employ the Kutta–Joukowski theorem to compute an equivalent correction for the radial circulation, $\Delta\Gamma(r)$, which iteratively corrects the radial circulation, $\Gamma(r)$, along the blades.

In the end, the decambering correction is blade planform dependent, that is, the correction has to be either computed a priori by a free-vortex method for a given blade planform or iteratively as part of the BEMT solution algorithm. More information and promising validation examples can be found in Sørensen et al. (2014, 2015) and Sørensen (2016).

### 3.4.4.4 Extended Glauert Tip Correction Using a $g$ Function (Schmitz and Maniaci 2016)

The challenge associated with improving the prediction of blade tip loads for highly loaded wind turbine blades is that, ideally, a generalized tip correction is consistent with variations in tip speed ratio, blade pitch angle, characteristics of tip airfoils, and tip shapes. Furthermore, it is desirable that a generalized correction does not add computational expense to a given BEMT solution algorithm. The $g$ function approach by Schmitz and Maniaci (2016) was inspired by works presented in Sections 3.4.4.1–3.4.4.3, in that free-wake vortex theory is utilized to take into account the combined effects of wake expansion, wake distortion (vortex and vortex-sheet rollup), and planform effects; however, the goal here is to develop a simple, though general, $g$ function that is embedded in the classical Glauert tip correction in Eq. (3.38) as:

$$\text{Extended Glauert:} \quad F_T^*(r) = \frac{2}{\pi}\cos^{-1}\left[\exp\left(-g(r)\frac{B}{2}\frac{1-r/R}{r/R}\frac{1}{\sin\phi(r)}\right)\right] \quad (3.47)$$

Schmitz and Maniaci (2016, 2017) developed an inverse BEMT-type approach in which the resulting system of two combined BEMT equations is solved for a total loss factor, $F^*(r)$, and the relative velocity ratio, $V_{rel}(r)/V_0$, rather than axial and angular induction factors, $a$ and $a'$, as presented in Section 3.5.1. Given the local blade flow angle, $\phi(r)$, as computed by a free-wake vortex method that uses the same airfoil data and blade planform specifications, the corresponding BEMT quantities $F^*(r)$ and $V_{rel}(r)/V_0$ can be determined analytically. For the total loss factor, $F^*(r)$, the exact solution becomes:

$$\text{Schmitz and Maniaci (Exact):} \quad F^*(r) = \frac{\sigma'(c_l\cos\phi(r) - c_d\sin\phi(r))}{4\sin\phi(r)\left(\frac{V_{rel}(r)}{V_0}\cos\phi(r) - \lambda_r\right)}\frac{V_{rel}(r)}{V_0}$$

$$(3.48)$$

Here, $\sigma'$ is the local solidity as defined in Eq. (3.18), $\lambda_r$ is the local speed ratio per Eq. (2.33), and $c_l$, $c_d$ are sectional airfoil lift and drag coefficients, all of which are functions of $r$, but not shown so in Eq. (3.48) for the sake of brevity. The exact expression for the relative velocity ratio, $V_{rel}(r)/V_0$, depends on whether or not a local blade element is operating in the windmill or turbulent wake state, see Maniaci and Schmitz (2016). Using a free-wake vortex method based on higher-order distributed vorticity elements, *WindDVE*, see method details in Bramesfeld and Maughmer (2008), Basom (2010), and Maniaci (2013), a number of $g$ functions can be determined, see Schmitz and Maniaci (2017), by solving Eq. (3.47) for $g(r)$ as:

$$\text{Extended Glauert,} g : \quad g(r) = -\frac{2}{B}\frac{r/R}{1-r/R}\sin\phi(r)\ln\left[\cos\left(\frac{\pi}{2}\frac{F^*(r)}{F_R(r)}\right)\right] \quad (3.49)$$

Note that $\phi(r)$ is computed by the *WindDVE* code and Eq. (3.40) was used in the form $F^*(r) = F_R(r)\cdot F_T^*(r)$, with $F_R(r)$ from Eq. (3.39) and $F^*(r)$ from the analytical expression

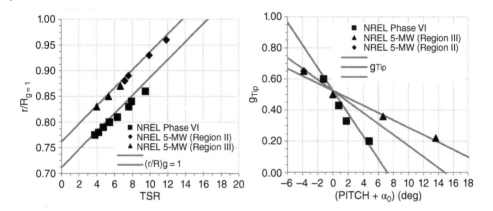

**Figure 3.9** General $g$ function parameters (Schmitz and Maniaci 2016, 2017).

in Eq. (3.48). This inverse approach can be applied to any free-wake vortex method, although one has to be aware that the argument of the cos function in Eq. (3.49) is well defined, that is, $|F^*(r)/F_R(r)| \in [0; 1]$, which was found to not necessarily be the case anywhere else than in the tip region. A parameter study for different rotors results in a very compact expression for the $g$ function as:

$$g \text{ function:} \quad g(r) = \begin{cases} 1, r/R \le (r/R)_{g=1} \\ 1 - \dfrac{(r/R) - (r/R)_{g=1}}{1 - (r/R)_{g=1}} \cdot (1 - g_{Tip}), r/R > (r/R)_{g=1} \end{cases} \quad (3.50)$$

Hence, according to Eq. (3.50), the $g$ function is active only for $r/R > (r/R)_{g=1}$, while the classical Glauert tip correction applies unaltered with $g = 1$ inboard of $(r/R)_{g=1}$. The parameters $(r/R)_{g=1}$ and $g_{Tip}$ themselves show some simple trends with tip speed ratio, $\lambda$, tip pitch angle, $\beta_{Tip}$, the airfoil zero-lift angle at the 97.5% radial station, $(\alpha_0)_{97.5\%}$, and the respective local chord ratio, $(c/R)_{97.5\%}$, for $r/R > (r/R)_{g=1}$. Here the 97.5% radial station is assumed to represent average conditions in the tip region. Empirical functional relationships for $(r/R)_{g=1}$ and $g_{Tip}$ are given as:

$$g \text{ parameters:} \quad \begin{cases} (r/R)_{g=1} = \dfrac{\pi}{180}\lambda + \left(\dfrac{\pi}{4} - \left(\dfrac{c}{R}\right)_{97.5\%}\right) \\ g_{Tip} = -(\beta_{Tip} + (\alpha_0)_{97.5\%})\dfrac{180}{\pi}\left(\dfrac{c}{R}\right)_{97.5\%} + \dfrac{\pi}{6} \end{cases} \quad (3.51)$$

Equations (3.50), (3.51) substituted into (3.47) constitute the tip correction ($TLOSS = 2$) in the *XTurb* code, with $(r/R)_{g=1} \in [\pi/4; 1]$ and $g_{Tip} \in [0.1; 1]$. Some quantitative example comparisons of the extended Glauert tip correction against measured data are shown for example, in Chapter 6. Figure 3.9 shows *WindDVE* computed $g$ parameters for the NREL Phase VI and NREL 5-MW turbines and approximated lines following Eq. (3.51).

Note that while the concept of the $g$ function is general, the validity of empirical Eq. (3.51) is, as are tip corrections in the preceding sections, based on a finite number of analyzed turbine operating cases. Hence, Eq. (3.51) may be subject to empirical improvements, while the true challenge to the advanced student lies in elucidating a theoretical connection to general vortex rollup.

*Epilogue* on *Tip Modeling*: Continued active research on tip modeling in the wind energy community is very important to ensure current and future viability of wind energy enabled by horizontal-axis wind turbines. We are now at a stage where CPU speed is nearly stagnant and so is the support for optimized compilers for various computing platforms. It is therefore that efficient and robust BEMT solution algorithms are likely not to be replaced in the near future by high-resolution free-wake vortex methods or even fully resolved computational fluid dynamics (CFD) analyses, at least not at the design stage. On the contrary, increased competitiveness of wind produced energy and healthy competition within the wind turbine research and development industry requires accurate, robust, and efficient design tools for further optimized wind turbine blades and increased annual energy production (AEP). The nature of a close-to-optimal wind turbine rotor is being highly loaded, and it is here where a fundamental gap still exists between classical theory and state-of-the-art wind turbines in use today. Some attempts to closing this knowledge gap have been presented but continued research in the area of tip modeling is paramount.

## 3.5 BEM Solution Method

In the following, the sectional thrust and torque relations of momentum and blade element theory of Section 3.2 are adjusted to include a total tip-/root loss correction, $F(r)$, as outlined in the previous section. These relations form the basis of a combined BEMT. In particular, the present section derives explicit relations for the axial and angular induction factors, $a(r)$ and $a'(r)$, along with some examples of iterative solution methodologies, including various models that apply in case of a turbulent wake state (see Section 2.1.6). A simplified BEMT is presented in Section 3.6, resulting in some additional theoretical limits for the maximum power coefficient, $C_{P,\,max}(\lambda)$, as a function of an airfoil performance parameter. The effect of various blade design parameters on the rotor power coefficient are discussed in detail in Section 3.7 by means of example analyses using the *XTurb* code.

### 3.5.1 A System of Two Equations for Two Unknowns, *a* and *a'*

In order to combine momentum and blade element theories, the former is adjusted to account for three-dimensional vortex-induced effects associated with a finite number of blades by multiplying Eqs. (3.10) and (3.11) with a total tip-/root loss correction as:

**Momentum Theory (Disk annulus):**

Incremental thrust: $\quad dT = 4a(1-a)\,\rho\,V_0^2\,\pi\,r\,F\,dr$ $\qquad\qquad\qquad$ (3.52)

Incremental torque: $\quad dQ = 4a'(1-a)\,\rho\,V_0^2\,\pi\,r^3\,\Omega\,F\,dr$ $\qquad\qquad$ (3.53)

The relations for blade element theory, see Eqs. (3.14) and (3.15), are unaltered and simply restated here for completeness:

**BEM (*B* blades), Eqs. (3.14) and (3.15):**

Incremental thrust: $\quad dT = B\dfrac{1}{2}\rho V_{rel}^2(c_l\cos\phi + c_d\sin\phi)\,c\,dr$

Incremental torque: $\quad dQ = B\dfrac{1}{2}\rho V_{rel}^2(c_l\sin\phi - c_d\cos\phi)\,c\,r\,dr$

A combined BEMT is obtained by equating (3.52) and (3.14) and, respectively, (3.53) and (3.15). We thus obtain the following:

**BEMT: 2 Equations for $a$ and $a'$**

$$4a(1-a) \rho V_0^2 \pi r F = B\frac{1}{2}\rho V_{rel}^2 (c_l \cos\phi + c_d \sin\phi) c \tag{3.54}$$

$$4a'(1-a) \rho V_0^2 \pi r^3 \Omega F = B\frac{1}{2}\rho V_{rel}^2 (c_l \sin\phi - c_d \cos\phi) c r \tag{3.55}$$

Specifically, Eqs. (3.54) and (3.55) constitute a system of two equations for the two unknown axial- and angular induction factors, $a$ and $a'$. Note that both Eqs. (3.54) and (3.55) are non-linear in $a$ and $a'$, preventing a direct solution for the respective variables. However, using the trigonometric relations on the velocity triangle in Figure 3.3, that is, Eqs. (3.7)–(3.9), Eq. (3.18) for the local solidity, $\sigma'$, Eq. (2.33) for the local speed ratio, $\lambda_r$, and some algebra, we obtain:

Thrust:  $\quad 4aF = \sigma'\,(1-a)(c_l \cos\phi + c_d \sin\phi)/\sin^2\phi \tag{3.56}$

Torque:  $\quad 4a'F = \sigma'\,(1+a')(c_l \sin\phi - c_d \cos\phi)/(\sin\phi \cos\phi) \tag{3.57}$

At this stage, we can solve explicitly for the axial and angular induction factors, $a$ and $a'$, as

$a(r):\quad a = \{1 + 4 F \sin^2\phi/[\sigma'\,(c_l \cos\phi + c_d \sin\phi)]\}^{-1} \tag{3.58}$

$a'(r):\quad a' = \{-1 + 4 F \sin\phi \cos\phi/[\sigma'\,(c_l \sin\phi - c_d \cos\phi)]\}^{-1} \tag{3.59}$

Note that all quantities in Eqs. (3.58) and (3.59) are functions of the radial station, $r$, by means of an inner dependence on the blade flow angle, $\phi(r)$, with $F = F(r, \phi(r))$ and $c_l, c_d = c_l(\alpha(r)), c_d(\alpha(r))$, where $\alpha(r) = \phi(r) - \beta(r)$ through Eq. (3.5). In other words, knowledge of the radial distribution of the blade flow angle, $\phi(r)$, is a sufficient condition to determine the flow state at a blade element.

### 3.5.2 Iterative BEM Solution Methodologies – Analyzing a Given Blade Design

In the following, a small selection of BEM iterative solution methodologies is presented. It is worth mentioning that there are probably as many algorithms available as there are BEM researchers. While each algorithm has its own peculiarities and specifics to achieve robust convergence, we focus here primarily on communalities and note only major differences in the respective methodologies. At first, the *blade analysis* problem is defined as:

**Given: Rotor Operating Conditions and Blade Planform**

- Tip Speed Ratio, $\lambda$
- Blade Tip Pitch Angle, $\beta_{Tip} = \beta(r/R = 1)$
- Blade Number, $B$
- Root Cut-Out, $r_{Root}/R = 0.10$
- Blade Pitch, $\beta = \beta(r/R)$
- Non-Dimensional Blade Chord, $\frac{c}{R} = \frac{c}{R}(r/R)$
- Airfoil Tables, $c_l(\alpha)$ and $c_d(\alpha)$

For a given blade design and rotor operating conditions, the task is then to compute integrated rotor thrust and power coefficients, that is, $C_T$ and $C_P$. In BEMT, a solution strategy consists of an outer loop over all blade elements, with an inner iterative solution scheme aiming to converge the flow state at each blade element. Following, rotor thrust and power coefficients, $C_T$ and $C_P$, are determined by integrating the converged flow states over all radial blade elements, see next.

**Find: Rotor Performance Coefficients, $C_T$ and $C_P$**

- **For each blade element, ...**
- Iterative Solution, $\begin{cases} \text{Blade Flow Angle, } \phi \text{ Eq.(3.7)} \\ \text{Airfoil Data, } c_l \text{ and } c_d \text{ Tip-/Root Corr., } F \\ \text{Induction Factors, } a \text{ and } a' \end{cases}$
- Velocity Ratio, $V_{rel}/V_0$ Eq. (3.6)
- **... until all blade elements are converged!**
- **Integrate over blade radius, ...**
- Thrust Coefficient, $C_T$ Eq. (3.19)
- Power Coefficient, $C_P$ Eq. (3.20)

In the subsequent sections, two solution strategies are presented. While particulars with respect to the exact implementation might be different than noted, the objective here is to provide the reader with representative solution methodologies used in the wind energy community. Figure 3.10 shows a flowchart of the basics of both methodologies for one blade element.

### 3.5.2.1 Simultaneous Solution of $a$ and $a'$
Most BEM solution algorithms in use today are rooted in the two-equation approach described before and illustrated in Figure 3.10 (left). Differences among the various

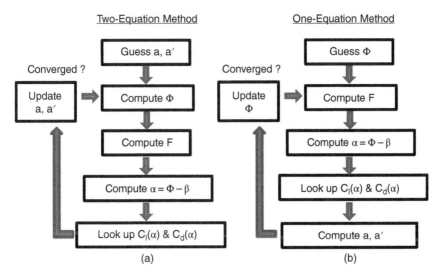

**Figure 3.10** Flowchart of BEM solution algorithms: (a) two-equation method and (b) one-equation method.

approaches are primarily in the specific sequence of iterative equations, treatment of the "turbulent wake state," see Figure 2.6 and Section 3.5.3, and initial conditions. Some examples of initial conditions are given next:

$$\text{Initial Conditions:} \quad a = 0 \; ; \; a' = 0$$

$$a = \frac{1}{3} \; ; \; a' = 0 \text{ or } a' = \frac{2}{9\lambda_r^2}$$

$$a = \frac{1}{4}[2 + \pi \lambda_r \sigma' - \sqrt{4 - 4\pi \lambda_r \sigma' + \pi \lambda_r^2 \sigma'[8\beta + \pi \sigma']}] \; ; \; a' = 0 \tag{3.60}$$

The first initial condition listed in Eq. (3.60) appears to be trivial as it describes a zero-induction flow state; however, it is the most general initial condition with respect to arbitrary analysis settings for the rotor tip speed ratio, $\lambda$, and blade pitch setting, $\beta_{Tip}$. Indeed, this is the initial condition implemented in *XTurb* where solution robustness is ensured by applying suitable under-relaxation factors (*AXRELAX, ATRELAX*) in the BEMT input list. The second choice for initial conditions in Eq. (3.60) is based on an ideal rotor in the absence of tip-/root losses; hence this is known to be advantageous only if the rotor is operated in the vicinity of optimal conditions for $\lambda$ and $\beta_{Tip}$, but may actually lead to convergence instability otherwise. The third initial condition listed in Eq. (3.60) is given as implemented in the *AeroDyn* code, see Moriarty and Hansen (2004), and is essentially an approximation of Eq. (3.58) under the assumption of small blade flow angles, $\phi$, which may not be a good starting point for $\lambda$ and $\beta_{Tip}$ being such that the rotor is operating at relatively high $\phi$ along the blade radius. As mentioned previously, the actual choice of initial condition is, as other peculiarities of a given solution algorithm, quite subjective. Some BEM solution algorithms actually have an inner loop where the angular induction factor, $a'$, is converged before updating the axial induction factor, $a$. Further variations apply to whether a fixed-point iteration or Newton's method is used to solve Eqs. (3.58) and (3.59) to convergence within at least single-precision accuracy, see McWilliam and Crawford (2011) for a more detailed discussion.

### 3.5.2.2 Root-Finding Method of Single Equation for Φ

The basic idea for the one-equation method proposed by Ning (2014) is that at a given blade element, the blade flow angle, $\phi$, determines the flow state. This has the advantage that both induction factors, $a$ and $a'$, and sectional lift and drag coefficients, $c_l$ and $c_d$, can be determined directly from the local blade flow angle, $\phi$, instead of an inner dependence of $c_l$ and $c_d$ on $a$ and $a'$, as is done in two-equation methods, see Figure 3.10. A single equation relating the three variables $\phi$, $a$, and $a'$ has been introduced as Eq. (3.7) from the velocity triangle in Figure 3.3, which is written here in the form of a residual equation as:

$$\text{Single Residual Equation:} \quad f(\phi) = \frac{\sin \phi}{1 - a} - \frac{\cos \phi}{\lambda_r(1 + a')} = 0 \tag{3.61}$$

Equation (3.61) can be initialized with any initial condition listed in Eq. (3.60). Reducing the set of two non-linear equations for the axial-/angular induction factors, $a$ and $a'$, to a one-equation method has the additional advantage of being able to draw from a suite of one-dimensional root-finding algorithms. In his work, Ning (2014) investigates several approaches and suggests the use of Brent's method, see Brent (1971), as a robust method for one-dimensional non-linear functions when $\phi$ is properly bracketed.

The particular equations for $a$ and $a'$ depend on the specific flow regime in Figure 2.6, including Eqs. (3.58) and (3.59), with the initial propeller brake state marking one of the singularities of Eq. (3.61).

*Epilogue on BEM Solution Methodologies*: The author wishes to emphasize the importance of solution robustness of BEM algorithms. It is known from experience that optimization methods relying on a particular BEM algorithm may exploit actual weaknesses of the BEM analysis method at specific blade sections and operating conditions, rather than finding the true optimal solution. It is therefore that all optimal solutions have to be assessed critically to the extent possible, with a clear and complete understanding of the flow state and the physical suitability of a solution. Here, the efficiency of BEM algorithms with respect to the number of iterations required is actually not as important as, for a fair comparison, one would also have to consider the total number of algorithmic operations within one iteration, the total clock time required for conditional "If, then …" clauses, and optimal compiler settings of a given programming language. Given today's processing power of laptop and desktop computers, all BEM solution algorithms in use today analyze a given blade design within a fraction of a second. A "real" time benefit is gained by an experienced blade aerodynamicist and designer who critically assesses the solutions obtained.

### 3.5.3 Thrust Coefficient in the Turbulent Wake State, $a > 0.4$

The validity of actuator disk theory has been discussed in Section 2.1.6, see also Figure 2.6. In particular, classical momentum theory for turbine rotors is strictly valid only in the "windmill state" and for an axial induction factor, $a < 0.4$, or respective thrust coefficient, $C_T < 0.96$, that is, prior to entering the turbulent wake state. Note that in a combined BEM solution algorithm, a given radial blade section may indeed operate in either the windmill state or turbulent wake state, and even switch between the two during the iterative solution process. As Eq. (3.58) was derived under the assumption of a given blade section operating in the windmill state, an alternate relation for $a$ in the turbulent wake state is mandatory, also simply because the solution would otherwise be non-unique and alternate between two possible solutions on the classical $C_T$ parabola. In the following, a representative selection of available models is presented, with emphasis on their respective limitations. First, though, let us develop a specific criterion for entering the turbulent wake state at a given blade element by defining an incremental thrust coefficient, $dC_T$, from blade element theory as:

$$\text{Incremental thrust coefficient:} \quad dC_T = \frac{\sigma'(1-a)^2(c_l \cos\phi + c_d \sin\phi)}{\sin^2\phi} \tag{3.62}$$

Here Eq. (3.62) is obtained by normalizing Eq. (3.14) by $\frac{1}{2}\rho V_0^2(2\pi r)dr$ and using further Eqs. (3.8) and (3.18). With reference to combined BEMT, Eq. (3.56), and the condition $a > 0.4$ from momentum theory, the criterion for entering the turbulent wake state during the BEM solution process becomes:

$$\text{Turbulent Wake State:} \quad dC_T > 0.96\,F \tag{3.63}$$

Note that Eq. (3.63) includes a total tip-/root loss factor, $F$, indicating that tip sections in particular are prone to operate in the turbulent wake state.

### 3.5.3.1 Glauert Empirical Relation

The first mathematical relation for the turbulent wake state was derived by Glauert (1926) as an empirical approximation to measured data by Lock et al. (1926). Glauert suggested that the thrust coefficient, $C_T$, in the turbulent wake state behaves as:

$$\text{Glauert Empirical Relation:} \quad C_T = 0.889 - \frac{0.0203 - (a - 0.143)^2}{0.6427} \qquad (3.64)$$

Consequently, Eq. (3.64) behaves such that $C_T = 0.96$ for $a = 0.4$ at the beginning of the turbulent wake state and $C_T = 2.0$ for $a = 1.0$ as a limit case of flow stagnation at the rotor disk (beginning of "vortex ring state"). Solving Eq. (3.64) for the physically relevant solution for $a$ yields the following:

$$a(r): \quad \text{Glauert } a = 0.143 + \sqrt{0.0203 - 0.6427(0.889 - dC_T)} \qquad (3.65)$$

Here, $dC_T$ is used at a given blade element according to Eq. (3.62). Note that Glauert's empirical relation does not account for tip-/root losses by means of a suitable correction factor, $F$. Indeed, as shown in Figure 3.11, a gap exists between a sectional $dC_T$ curve (assuming a typical value for $F$ in the tip-/root region) and Eq. (3.65). This can lead to discontinuities in $a$ along the blade radius or incomplete convergence at a given blade element. Symbols in Figure 3.11 are measured data extracted from Figure 2.6.

This problem can be remedied, see Eggleston and Stoddard (1987), by adjusting Eq. (3.65) such that:

$$a(r): \quad \text{Glauert } a = \frac{1}{F}[0.143 + \sqrt{0.0203 - 0.6427[0.889 - dC_T]}] \qquad (3.66)$$

It becomes clear, though, from Figure 3.11 that the "gap" has not been eliminated as this simple adjustment is indeed not conforming with the turbulent wake state criterion in Eq. (3.63).

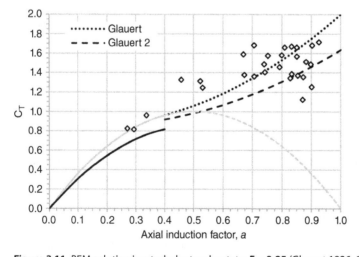

**Figure 3.11** BEM solution in a turbulent wake state, $F = 0.85$ (Glauert 1926, 1935).

### 3.5.3.2  1st-Order Approximation (Wilson, Burton)

A different approach was suggested by Wilson (1994) and a later variation by Burton et al. (2011). In this case, a line tangential to the classical $C_T$ curve is used in the turbulent wake state that reaches a value of an assumed $C_{T1}$ at $a = 1$ (as opposed to the limiting case of $C_T = 2$ as assumed by Glauert). The objective here is to better fit measured data, see Figure 2.6, or even that of a flat circular plate, see Hoerner (1965), in the deep turbulent wake state. The associated relation for $C_T$ in the turbulent wake state becomes:

$$\text{Empirical (Wilson, Burton)}: \quad C_T = C_{T1} - 4(\sqrt{C_{T1}} - 1)(1 - a) \tag{3.67}$$

Here the intercept with the classical $C_T$ curve from momentum theory now occurs at $a_T = 1 - \frac{1}{2}\sqrt{C_{T1}}$. This means in particular that the criterion for entering the turbulent wake state becomes a function of $C_{T1}$ as:

$$\text{Turbulent Wake State}: \quad dC_T > 2\left(1 - \frac{1}{2}\sqrt{dC_{T1}}\right)\sqrt{dC_{T1}}\,F \tag{3.68}$$

While Wilson (1994) suggests $dC_{T1} = 1.6$ ($a_T = 0.368$), Burton et al. (2011) assume a higher value of $dC_{T1} = 1.816$ ($a_T = 0.326$). The actual equation for the axial induction factor, $a$, in the turbulent wake state at a given blade element is obtained from Eq. (3.67) as:

$$a(r): \quad \text{Wilson, Burton } a = 1 - \frac{dC_{T1} - dC_T}{4(\sqrt{dC_{T1}} - 1)} \tag{3.69}$$

Figure 3.12 illustrates, however, that the aforementioned gap issue for $F < 1$ in the tip-/root region is not resolved as it has not been properly accounted for. Note that symbols in Figure 3.12 are measured data extracted from Figure 2.6.

### 3.5.3.3  2nd-Order Approximation (Buhl)

The approach suggested by Buhl (2005) is specifically designed to resolve the gap issue, while retaining the primary behavior of the Glauert empirical relation. Here, a general

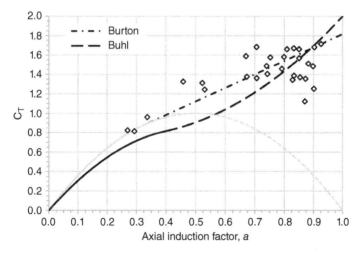

**Figure 3.12** BEM solution in a turbulent wake state, $F = 0.85$ (Burton et al. 2011; Buhl 2005).

second-order model for the thrust coefficient, $C_T$, is proposed as:

$$\text{2nd-Order Model (Buhl):} \quad C_T = b_0 + b_1 a + b_2 a^2 \tag{3.70}$$

Equation (3.70) requires a total of three model constraints for the unknown coefficients $b_0$, $b_1$, and $b_2$. In reference to the Glauert empirical relation, these are:

$$C_T - \text{Model Constraints (Buhl):} \quad \begin{cases} C_T = 4a(1-a)F, \ a = 0.4 \\ \dfrac{dC_T}{da} = 4F(1-2a), \ a = 0.4 \\ C_T = 2, \ a = 1.0 \end{cases} \tag{3.71}$$

Substituting the left-hand-side of Eq. (3.71) with the appropriate form of Eq. (3.70) results in a system of three linear equations for the unknown coefficients $b_0$, $b_1$, and $b_2$. The thrust coefficient, $C_T$, in the turbulent wake state then becomes:

$$\text{General Glauert (Buhl):} \quad C_T = \frac{8}{9} + \left(4F - \frac{40}{9}\right)a + \left(\frac{50}{9} - 4F\right)a^2 \tag{3.72}$$

We note in Figure 3.12 that $C_T$ from Eq. (3.72) is always tangent to the relation in the windmill state, thus eliminating the gap issue. Subsequent solution of Eq. (3.72) for the physically relevant solution of the axial induction factor, $a$, in the turbulent wake state results in the following:

$$a(r): \quad \text{Buhl } a = \frac{18F - 20 - 3\sqrt{dC_T(50 - 36F) + 12F(3F - 4)}}{36F - 50} \tag{3.73}$$

This relation can be readily implemented into BEM solution algorithms. Note that Eq. (3.73) is the turbulent wake state relation for $a$ used in *XTurb*.

*Epilogue on BEM and Turbulent Wake State*: One has to be aware that all models are approximations of the behavior in the turbulent wake state where the fundamental assumptions of combined BEMT break down. Nevertheless, these simple models have proven themselves to be useful and of reasonable accuracy. The large scatter in measured data in the turbulent wake state itself leaves room for physical model interpretation. However, $a < 0.6$ for most cases, except for some particular tip loadings. It is worth mentioning that the particular choice of the turbulent wake state model within a BEM solution algorithm has to be considered in conjunction with the tip-/root loss model used. Therefore, one has to always consider both when conducting verification and validation (*V&V*) studies between different BEM solution algorithms and data comparisons.

## 3.6 Simplified BEMT (Wilson and Lissaman 1974)

To this end, combined BEMT as derived in previous sections, including tip-/root loss and airfoil profile drag, does not provide additional analytical insight into the respective combined effects on the rotor power coefficient, $C_P$. The work of Wilson and Lissaman (1974) is pivotal in that some simplifications were applied to the combined BEM Eqs. (3.56) and (3.57) that let us approximate, at least to some extent, the effect of airfoil profile drag on the power coefficient, $C_P$, with reference to rotor disk theory in Chapter 2.

We begin by simplifying Eq. (3.56) with $F = 1$ and $c_d = 0$ such that the following is obtained:

Simplified Thrust Equation:    $4a = \sigma'(1-a)c_l \cos\phi/\sin^2\phi$ (3.74)

Similarly, simplifying Eq. (3.57) with $F = 1$ and $c_d = 0$ yields the following expression:

Simplified Torque Equation:    $4a' = \sigma'(1+a')c_l/\cos\phi$ (3.75)

Next, we use Eqs. (3.74) and (3.75) on the velocity triangle and can write Eq. (3.7) in a simpler form as:

Approximate Relation on Velocity Triangle:    $\tan\phi = \dfrac{a'}{a}\lambda_r$ (3.76)

Note that Eq. (3.76) is indeed approximate when seen on the velocity triangle in Figure 3.3 as it is practically the inverse relation on the perturbations $a$ and $a'$. The resulting Eqs. (3.74)–(3.76) are essential to a simplified BEMT. We proceed by using Eqs. (2.33) and (3.8) and writing the general relation for the rotor power coefficient, $C_P$, from blade element theory, see Eq. (3.20), in the following form:

Power Coefficient:    $C_P = \dfrac{2}{\lambda^2}\displaystyle\int_{r_{Root}/R}^{1} \sigma' c_l \dfrac{(1-a)^2}{\sin\phi}[1-(c_d/c_l)\cot\phi]\lambda_r^2\, d\lambda_r$ (3.77)

It is important to understand that the simplifications, that is, $F = 1$ and $c_d = 0$, have been added to the system of two equations for the two unknowns $a$ and $a'$; however, airfoil profile drag cannot be neglected when integrating for the rotor power coefficient, $C_P$. Next, we substitute Eq. (3.76) back into Eq. (3.74) and solve for $\sigma' c_l(1-a)/\sin\phi$ to obtain:

$$\sigma' c_l(1-a)/\sin\phi = 4a'\lambda_r$$ (3.78)

Subsequent substitution of Eq. (3.78) into (3.77) yields an approximate relation for the rotor power coefficient, $C_P$, with:

Power Coefficient (Simplified BEMT):

$$C_P = \dfrac{8}{\lambda^2}\int_{r_{Root}/R}^{1} a'(1-a)[1-(c_d/c_l)\cot\phi]\lambda_r^3\, d\lambda_r$$ (3.79)

Equation (3.79) constitutes a relation for the rotor power coefficient, $C_P$, in simplified BEMT. Furthermore, it is written in a form that we know from rotor disk theory in Chapter 2, that is, Eq. (2.40).

It becomes apparent that whether or not considering optimum radial distributions for $a$ and $a'$, airfoil profile drag, $c_d$, reduces $C_P$, with all other parameters held constant. In reference to Figure 3.3, this is a result of airfoil profile drag generating a torque opposite to the direction of blade rotation. Equation (3.79) is often used to argue that an optimum blade design is achieved when all blade sections operate at the maximum "finesse," or $(c_l/c_d)_{max}$, of the respective section airfoils. While this is a very good guideline for blade design, it is not fully correct because one has to be aware of the assumptions that went into the derivation of Eq. (3.79). In addition, Eq. (3.79) actually suggests minimum losses for $(c_d/c_l)\cot\phi$ becoming a minimum. We will revisit this discussion in later chapters on optimized blade design but acknowledge that practical blade designs perform close to their best when $(c_l/c_d)_{max}$ occurs close to the actual design $c_l$.

## 3.7 Effect of Design Parameters on Power Coefficient

The preceding sections have introduced combined BEMT and iterative solution methodologies to compute integrated rotor performance coefficients for given rotor operating conditions and blade planform parameters, see in particular Section 3.5.2. As a blade designer, however, one would hope to have some additional understanding about how the various blade design parameters do affect the rotor power coefficient, $C_P$. In particular, we could ask ourselves:

- What is the effect of blade number and solidity on $C_P$ as a function of tip speed ratio?
- How is $C_P$ affected for a blade being collectively pitched into or out of the wind?
- What is the effect of airfoil properties, such as $c_d$ itself and $c_l/c_d$, on the rotor power coefficient, $C_P$?
- What is the effect of rotor speed in conjunction with airfoil $c_l/c_d$?
- What are the advantages/disadvantages of a two-bladed versus a three-bladed rotor?

The goal of this section is to provide a comprehensive set of analyses to answer these questions at least in part, with the understanding that a sole focus on the rotor power coefficient, $C_P$, does not include constraints on rotor thrust, root-flap bending moment, blade solidity, and hence weight and so on. The reader is referred to Chapter 8 on optimization for more in-depth analyses and examples that include some design constraints. Here, we proceed by giving further attention to Eq. (3.79), that is, the power coefficient according to simplified BEMT. In particular, optimum solutions for the axial-/angular induction factors, $a$ and $a'$, can be obtained from Chapter 2, and corresponding ideal rotors can be designed according to Section 3.3.2. The result is a family of optimum rotor configurations where the blade number and solidity, airfoil profile drag, and tip speed ratio can be varied for assumed conditions. A single approximate equation for the rotor power coefficient, $C_P$, can be found as:

$$\text{Maximum Power Coefficient (Simplified BEMT):} \quad C_{P,max} =$$

$$\frac{16}{27}\lambda\left[\lambda + \frac{1.32 + \left(\frac{\lambda-8}{20}\right)^2}{B^{2/3}}\right]^{-1} - \frac{0.57\lambda^2}{\frac{c_l}{c_d}\left(\lambda + \frac{1}{2B}\right)} \qquad (3.80)$$

Equation (3.80) is presented here in the form as given in Manwell (2009). The results date back to Wilson and Lissaman (1976) who actually refer to an earlier paper by Rohrbach and Worobel (1975) and Rohrbach (1976), the former being a very compact and comprehensive piece of work that is rarely credited. Equation (3.80) recovers the documented data to within 0.5% for tip speed ratios, $\lambda$, between 4 and 20, airfoil $c_l/c_d > 25$, and the blade number $B \leq 3$. The beauty of Eq. (3.80) is that principal design parameters can be varied independently, thus giving us a practical understanding of the effects of the respective design parameters. As we discuss results obtained by Eq. (3.80), we acknowledge that Eq. (3.80) is an approximation in itself obtained from results of a simplified BEMT.

### 3.7.1 Effect of Blade Number and Solidity

The effect of blade number, $B$, on the maximum rotor power coefficient, $C_{P,\,max}$, is easiest investigated by means of the rotor solidity, $\sigma$, defined as:

$$\text{Rotor Solidity:} \quad \sigma = \int_{r_{Root}/R}^{1} \sigma' d\frac{r}{R} \tag{3.81}$$

Here, $\sigma' = Bc/(2\pi r)$ is the local solidity as defined in Eq. (3.18). In words, the rotor solidity, $\sigma$, is defined as the ratio of total blade area to rotor disk area, or the fraction of the rotor disk area occupied by blade area. For large wind turbines, the rotor solidity, $\sigma$, hovers around 0.05, depending on the design $c_l$ distribution, but can be substantially higher for smaller wind turbines. Here, we assume respective optimum conditions from rotor disk theory in Chapter 2 and focus on the sole effect of blade number, $B$, for zero airfoil profile drag, that is, $c_d = 0$. Results obtained by Eq. (3.80) are shown in Figure 3.13, including the limit cases from actuator-/rotor disk theories from Chapter 2 for comparison.

It becomes clear from Figure 3.13 that a one-bladed rotor is associated with a notable loss in maximum attainable power coefficient when compared to the rotor disk limit of an ideal rotor with wake rotation (and infinite number of blades). It is further interesting to see that while a two-bladed rotor performs notably better than its one-bladed counterpart, increasing the blade number to three results in a much smaller change. In other words, the solution converges slowly with blade number to the limit of an infinite number of blades. This means in particular that it probably does not make much sense to consider a rotor with more than three blades as the gain in $C_{P,\,max}$ is small compared to the added complexity of rotor weight, and so on; however, one has to keep in mind that a higher blade number results in a proportionally higher start-up torque. We also note that assuming $c_d = 0$, a two-bladed rotor can achieve the same rotor power coefficient

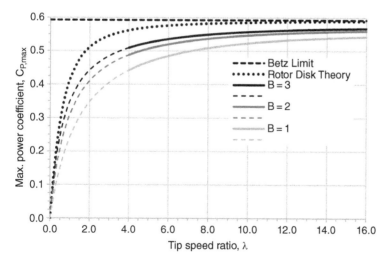

**Figure 3.13** Maximum power coefficient, $C_{P,max}$, versus tip speed ratio, $\lambda$ – effect of blade number, $B$ ($c_d = 0$).

at a bit higher tip speed ratio than a three-bladed rotor, a feasible option if tip speed is not constrained as may be the case in offshore versus onshore applications.

### 3.7.2 Effect of Profile Drag

The behavior changes to some extent when considering the effect of airfoil profile drag, $c_d$. Here, we focus exclusively on a three-bladed rotor and apply Eq. (3.80) to various lift-to-drag ratios, $c_l/c_d$. Results obtained are shown in Figure 3.14.

The Betz limit along with the rotor disk limit of an ideal rotor with wake rotation (and infinite number of blades) is shown again for comparison. The actual reference case is that of $c_d = 0$ for a three-bladed rotor ($B = 3$). We can see in Figure 3.14 that the lift-to-drag ratio, $c_l/c_d$, does have an appreciable effect on the maximum attainable power coefficient, $C_{P,max}$, where values of $C_{P,max} > 0.50$ can be obtained for $c_l/c_d > 100$, while $C_{P,max} < 0.40$ for $c_l/c_d < 25$. These are basic, though important findings with respect to scaled wind turbine aerodynamics as the section airfoil lift-to-drag ratio, $c_l/c_d$, is primarily a function of section Reynolds number, $Re_c$. A more detailed discussion is presented in Chapter 4. Here we note an approximate rule-of-thumb that (i) $c_l/c_d > 100$ for MW-size wind turbines, (ii) $c_l/c_d = 25 - 100$ for the broad range of kW-size wind turbines, and (iii) $c_l/c_d < 25$ for W-size wind turbines used in many small-scale wind-tunnel experiments. We will learn in later chapters that this has implications on scaling wind turbine aerodynamics from full-scale to model-scale conditions.

### 3.7.3 Combined Effects of Blade Number, Solidity, and Profile Drag

In reality, the lift-to-drag ratio, $c_l/c_d$, is not constant along the blade radius as it depends on the blade section Reynolds number, $Re_c$, and the airfoil itself that changes

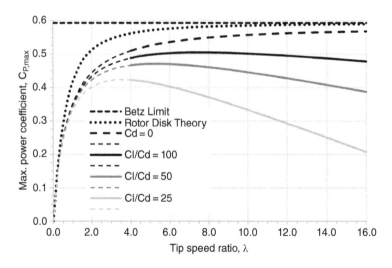

**Figure 3.14** Maximum power coefficient, $C_{P,max}$, versus tip speed ratio, $\lambda$ – effect of lift-to-drag ratio, $c_l/c_d$ ($B = 3$).

along the blade radius. To investigate the combined effects of section $c_l/c_d$ and blade number/solidity, let us analyze the generic PSU 1.5-MW turbine using the *XTurb* code in ANALYSIS mode for a variable blade number, $B$ (*BN* in the *XTurb* input file) and including root-/tip losses as default settings. The PSU 1.5-MW turbine is a generic wind turbine designed to represent a utility-scale wind turbine, see also Schmitz (2015) for more details. Results obtained for a three-bladed rotor are shown next as listed in *XTurb_Output.dat*:

---

**XTurb Example 3.3  PSU 1.5-MW Turbine – XTurb_Output.dat (PSU-Ch3-7-3.inp)**

```
3.7.3 - PSU 1.5-MW               ***** XTurb V1.9    -   OUTPUT *****
   Blade Number        BN =   3


                           +     BLADE  ELEMENT  MOMENTUM  THEORY  (BEMT)     +

  Number    TSR      PITCH [deg]     CT         CP        CPV        CB        CBV
     1     1.0000     0.0000        0.0709    -0.0090    0.0132    -0.0122    -0.0217
     2     2.0000     0.0000        0.1249    -0.0584    0.0291    -0.0248    -0.0126
     3     3.0000     0.0000        0.2362    -0.1594    0.0340    -0.0502    -0.0069
     4     4.0000     0.0000        0.3698    -0.2685    0.0333    -0.0799    -0.0039
     5     5.0000     0.0000        0.5155    -0.3734    0.0268    -0.1109    -0.0020
     6     6.0000     0.0000        0.6516    -0.4456    0.0251    -0.1408    -0.0012
     7     7.0000     0.0000        0.7519    -0.4799    0.0267    -0.1640    -0.0009
     8     8.0000     0.0000        0.8263    -0.4857    0.0340    -0.1820    -0.0008
     9     9.0000     0.0000        0.8860    -0.4725    0.0454    -0.1969    -0.0007
    10    10.0000     0.0000        0.9397    -0.4507    0.0597    -0.2109    -0.0007
    11    11.0000     0.0000        0.9908    -0.4227    0.0772    -0.2245    -0.0007
    12    12.0000     0.0000        1.0402    -0.3889    0.0985    -0.2380    -0.0007
    13    13.0000     0.0000        1.0888    -0.3492    0.1236    -0.2514    -0.0007
    14    14.0000     0.0000        1.1370    -0.3032    0.1530    -0.2649    -0.0007
    15    15.0000     0.0000        1.1847    -0.2501    0.1868    -0.2784    -0.0007
    16    16.0000     0.0000        1.2320    -0.1894    0.2257    -0.2920    -0.0007
```

---

Note again that in the force/moment convention used in *XTurb*, values $C_P < 0$ refer to power extraction from the fluid and hence *positive* power production. Consequently, $C_P$ values obtained from *XTurb* have to be plotted as $-C_P$ to be consistent with documented power coefficients. On the contrary, the viscous contribution to the power coefficient, $C_{PV} > 0$, refers to a power loss, or added power to the fluid to sustain blade motion, and is consistent with the force/torque triangle in Figure 3.3. It can be further seen that $C_{PV}$ is only a small fraction of the total $C_P$ close to the design tip speed ratio, $\lambda$, but can become substantial for both low and high $\lambda$ due to either blade stall or unfavorable section force/torque projection in the direction of thrust, that is, $C_T$. Figure 3.15 shows results obtained using the *XTurb* input file *PSU-Ch3-7-3.inp* for different blade numbers. It can be seen that the maximum power coefficient, $C_{P,max}$, increases with increasing blade number, $B$, and occurs at a progressively lower tip speed ratio, $\lambda$. In addition, the $C_{P,max}$ versus $\lambda$ curve becomes narrower around the optimal tip speed ratio. While a lower optimal $\lambda$ is advantageous for blade loads and noise, increasing the blade number increases rotor weight, and the "narrowing" effect of the $C_{P,max}$ versus $\lambda$ curve makes the rotor operate less efficiently off its design point.

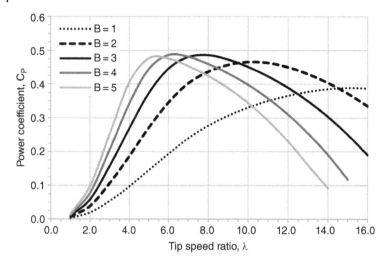

**Figure 3.15** Maximum power coefficient, $C_{P,max}$, versus tip speed ratio, $\lambda$ – effect of blade number, **B** (PSU 1.5-MW turbine, $\beta_{Tip}$, pitch $= 0°$).

The combined effects of rotor solidity for a variable section lift-to-drag ratio can be summarized as:

**Low Solidity**

- Broad and flat $C_{P,max}$ versus $\lambda$ curve
- $C_{P,max}$ changes little over a wide range of tip speed ratio, $\lambda$
- $C_{P,max}$ occurs at a higher $\lambda$ and is lower than for a higher blade number, $B$

**High Solidity**

- Narrower $C_{P,max}$ versus $\lambda$ curve with sharper peak
- $C_{P,max}$ becomes more sensitive to changes in tip speed ratio, $\lambda$
- $C_{P,max}$ occurs at a lower $\lambda$ and is higher $=>$ Higher rotor torque

Overall, Figure 3.15 suggest once more that a blade number of $B = 2 \dots 3$ may be the best choice for a wind turbine rotor. Note that the rotor solidity ranges from $\sigma = 0.0159$ ($B = 1$) to $\sigma = 0.0795$ ($B = 5$), with $\sigma = 0.0477$ ($B = 3$) being a typical rotor solidity for a utility-scale wind turbine.

### 3.7.4 Effects of Rotor Speed and Blade Pitch

The effects of rotor speed, rpm, and blade tip pitch, $\beta_{Tip}$ (or PITCH), are studied next. Here, we consider the generic PSU 1.5-MW turbine in PREDICTION mode, with rotor specifications and performance metrics given in dimensional form. The associated *XTurb* input file is *PSU-Ch3-7-4.inp*, including root-/tip losses as default settings. A representative output sample from *XTurb_Output.dat* is given here:

# XTurb Example 3.4 PSU 1.5-MW Turbine – XTurb_Output.dat (PSU-Ch3-7-4.inp)

```
3.7.4 - PSU 1.5-MW              ***** XTurb V1.9      -  OUTPUT *****
  Blade Number      BN =   3

                     +      BLADE ELEMENT MOMENTUM THEORY (BEMT)    +

PREDICTION
Blade Radius      BRADIUS  =   38.5 [m]
Air Density       RHOAIR   =    1.225 [kg/m**3]
Air Dyn. Visc.    MUAIR    =    0.000018 [kg/(m*s)]
Number of Cases   NPRE     =   25

Thrust           T = 0.5*RHOAIR*VWIND**2.*(pi*BRADIUS**2.)*CT
Power            P = 0.5*RHOAIR*VWIND**3.*(pi*BRADIUS**2.)*CP
Torque           TO = P / RPM
Bending Moment   BE = 0.5*RHOAIR*VWIND**2.*(pi*BRADIUS**2.)*BRADIUS*CB

Number  VWIND[m/s] RPM[1/min] TSR    PITCH[deg]  CT       CP      CB        T[N]          P[W]          TO[Nm]         BE[Nm]
...
  8      8.0000    14.0000    7.0555  0.0000     0.7567  -0.4809  -0.1652   138126.0116   -702276.8069   -479017.8091   -1160877.2063
...
```

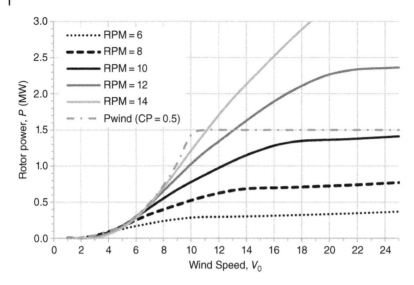

**Figure 3.16** Rotor power, *P*, versus wind speed, $V_0$ – effect of rotor speed, ***rpm*** (PSU 1.5-MW turbine, $\beta_{Tip}$, pitch = 0°).

Note that the rotor speed, rpm, can be computed from the tip speed ratio, $\lambda$, and wind speed, $V_0$, according to:

$$\text{Rotor Speed:} \quad \text{rpm} = \frac{\lambda\, V_0}{R}\frac{60}{2\pi} \tag{3.82}$$

Here, $R$ is the blade radius. Results obtained for rotor power, $P$, versus wind speed, $V_0$, are shown in Figure 3.16 for a constant blade tip pitch angle of $\beta_{Tip} = 0°$. It becomes apparent that rotor power, $P$, increases with rotor speed, rpm, as a direct consequence of Eq. (3.17) with $P \sim \Omega V_{rel}^2$. For comparison, a wind turbine power curve for the same rotor diameter is shown assuming a power coefficient of $C_P = 0.50$ in Region II, with the corresponding rotor power capped at $P = 1.5$ MW. It seems that additional power capture might be possible in Region II of the ($C_P = 0.50$) power curve; however, one has to keep in mind that the blade tip speed, $V_{Tip}$, behaves as:

$$\text{Blade Tip Speed:} \quad V_{Tip} = \lambda\, V_0 = \Omega R = \left(\text{rpm}\frac{2\pi}{60}\right) R \tag{3.83}$$

Indeed, the generic PSU 1.5-MW turbine has a maximum rotor speed of rpm = 13.75, resulting in a maximum tip speed of only $V_{Tip} = 55.4\frac{m}{s}$, which is a conservative value with respect to typical onshore noise constraints for $V_{Tip}$ in the range of $65\frac{m}{s} - 75\frac{m}{s}$.

It is further interesting to see in Figure 3.16 that the generic PSU 1.5-MW turbine behaves as a stall-controlled turbine for constant rpm = 10 and $\beta_{Tip}$, PITCH=0°. Consequently, the wind turbine would still be able to operate purely speed controlled, if the blade pitch mechanism would not work for any reason. Nevertheless, the power curves shown in Figure 3.16 illustrate that this would be associated with a substantial power production loss in Region II of the power curve. On the other hand, Figure 3.17 shows the effect of blade tip pitch $\beta_{Tip}$ (or PITCH) for a constant rotor speed, *rpm* = 14. It can be seen that rotor power, *P*, can be controlled in Region III by means of blade tip pitch, $\beta_{Tip}$ (or PITCH), at a constant maximum rotor speed. In reference to the velocity triangle in Figure 3.3, this is realized by collectively decreasing the local angle of attack, $\alpha$, along the blade radius and thus section lift and torque.

In fact, the rotor cut-in speed increases substantially with increasing blade tip pitch. This is because part of the rotor is operating in the propeller state, with axial induction factors, $a < 0$. The interested reader can further analyze computed induction factors in *XTurb_Output_Method.dat*.

### 3.7.5  Aerodynamic Considerations – Two Blades versus Three Blades

A common discussion is always that of a two-bladed versus a three-bladed rotor. It seems indeed from Figure 3.15 that a two-bladed rotor is at a disadvantage due to both lower $C_{P,\,max}$ and the former occurring at a higher tip speed ratio, $\lambda$, which may impact noise constraints with respect to maximum tip speed, $V_{Tip}$, and also structural constraints, for example, centrifugal forces $\sim\Omega^2$. The question is, though, whether or not such a comparison is complete as one might argue that the two-bladed rotor had not been given a fair chance in comparison to a three-bladed design. Let us explore this a bit further by noting that a two-bladed rotor could in fact obtain the same rotor solidity, $\sigma$, compared to its three-bladed counterpart, if the radial distribution of local blade chord, $c/R$, were increased by 50% all along. The results obtained with a corresponding *XTurb* input file *PSU-Ch3-7-5.inp* are shown here (including root-/tip losses as default settings):

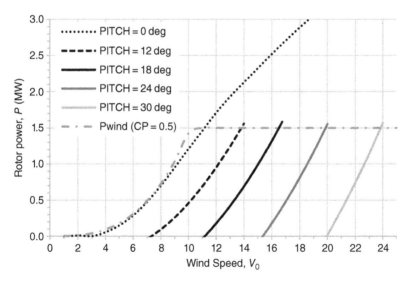

**Figure 3.17** Rotor power, *P*, versus wind speed, $V_0$ – effect of blade tip pitch, $\beta_{Tip}$, pitch (PSU 1.5-MW turbine, rpm = 14).

---

**XTurb Example 3.5  PSU 1.5-MW Turbine – XTurb_Output.dat (PSU-Ch3-7-5.inp)**

```
3.7.5 - PSU 1.5-MW              ***** XTurb V1.9     -   OUTPUT *****
    Blade Number        BN =  2

                              +     BLADE ELEMENT MOMENTUM THEORY (BEMT)    +

    Number    TSR     PITCH [deg]      CT        CP       CPV       CB        CBV
       1    1.0000    0.0000        0.0665   -0.0089    0.0123   -0.0173   -0.0201
       2    2.0000    0.0000        0.1226   -0.0584    0.0278   -0.0368   -0.0119
       3    3.0000    0.0000        0.2350   -0.1575    0.0326   -0.0751   -0.0065
       4    4.0000    0.0000        0.3674   -0.2629    0.0317   -0.1190   -0.0037
       5    5.0000    0.0000        0.5106   -0.3627    0.0257   -0.1644   -0.0019
       6    6.0000    0.0000        0.6444   -0.4311    0.0240   -0.2083   -0.0012
       7    7.0000    0.0000        0.7422   -0.4632    0.0260   -0.2420   -0.0008
       8    8.0000    0.0000        0.8154   -0.4692    0.0337   -0.2683   -0.0007
       9    9.0000    0.0000        0.8764   -0.4584    0.0453   -0.2912   -0.0007
      10   10.0000    0.0000        0.9317   -0.4393    0.0595   -0.3128   -0.0007
      11   11.0000    0.0000        0.9845   -0.4139    0.0771   -0.3339   -0.0007
      12   12.0000    0.0000        1.0355   -0.3825    0.0984   -0.3548   -0.0007
      13   13.0000    0.0000        1.0854   -0.3447    0.1236   -0.3755   -0.0007
      14   14.0000    0.0000        1.1349   -0.3003    0.1529   -0.3963   -0.0007
      15   15.0000    0.0000        1.1834   -0.2484    0.1867   -0.4170   -0.0007
      16   16.0000    0.0000        1.2314   -0.1885    0.2257   -0.4377   -0.0007
```

Figure 3.18 shows the corresponding $C_p$ versus $\lambda$ curves, with the $B = 2$, 3 cases from Figure 3.15 shown for comparison. It is quite interesting to note that a two-bladed rotor scaled to the same rotor solidity, $\sigma$, does indeed achieve a very similar $C_p$ versus $\lambda$

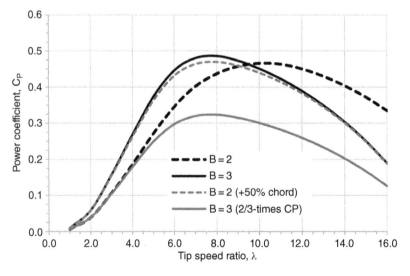

**Figure 3.18** Power coefficient, $C_p$, versus tip speed ratio, $\lambda$ – 2 blades versus three blades (PSU 1.5-MW turbine, $\beta_{Tip}$, pitch $= 0°$).

distribution compared to the three-bladed rotor. Note that the remaining discrepancy, or $C_p$ deficit of the solidity-scaled two-bladed rotor, is a result of higher tip loss for a lower blade number, see the Glauert tip correction in Eq. (3.38). In the end, we find that a solidity-scaled two-bladed rotor does practically perform very similar to a three-bladed rotor, a result that stands in relation to an optimum blade parameter, $\sigma' c_l$, as introduced in Section 3.3. Of further interest is a 2/3-scaled $C_p$ curve ($B = 3$) in Figure 3.18 that signifies difficulties in rotor start-up torque (or $P/\Omega$) at low $\lambda$ for lower-solidity rotors.

It should also be mentioned, nonetheless, that we have neglected the effect of increased blade section Reynolds numbers, $Re_c$ (see *XTurb_Output_1.dat*), on the respective airfoil data of the solidity-scaled two-bladed rotor. Though blade section Reynolds numbers are on the order of several millions for a 1.5-MW rated wind turbine, some benefit in the maximum lift-to-drag ratio, $(c_l/c_d)_{max}$, can be expected at 1.5 $Re$ that may actually close part of the $C_p$ deficit of the solidity-scaled two-bladed rotor in Figure 3.18. Furthermore, all analyses have been performed at the same blade pitch angle, $\beta_{Tip} = 0°$, and the reader is encouraged to further explore the performance space using the *XTurb* code.

### 3.7.6 Analysis of a MW-Scale Pitch-/Speed-Controlled Wind Turbine

In the following, let us foster our current knowledge and understanding of wind turbine aerodynamics by analyzing in detail a three-bladed utility-scale wind turbine. Here, we take the example of the generic PSU 1.5-MW turbine, see Schmitz (2015). This turbine is a "by hand" student-designed 1.5-MW rated research wind turbine. It has been iteratively adjusted within a graduate course taught by the author at the Pennsylvania State University. There are no claims that this turbine is indeed an absolute optimum in any respect. On the contrary, the generic PSU 1.5-MW turbine is viewed as a representative baseline with some room for improvement, both from the standpoint of optimal aerodynamic efficiency and AEP.

The airfoil distribution along the PSU 1.5-MW turbine blade is summarized in Table 3.1. Here, Delft University (DU) airfoils are used exclusively along the blade, except in the cylindrical root region. Airfoil data tables ($Re = 3 \cdot 10^6$) are taken from

**Table 3.1** Airfoil distribution (PSU 1.5-MW turbine).

| r/R | Airfoil |
|---|---|
| <0.15 | Cylinder |
| 0.15–0.18 | DU 00-W2-401 |
| 0.18–0.40 | DU 00-W2-350 |
| 0.40–0.50 | DU 97-W-300 |
| 0.50–0.60 | DU 91-W2-250 |
| 0.60–0.82 | DU 93-W-210 |
| 0.82–1.00 | DU 95-W-180 |

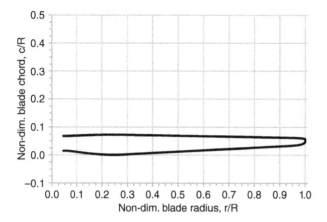

**Figure 3.19** Radial distribution of non-dimensional blade chord, *c/R* (PSU 1.5-MW turbine).

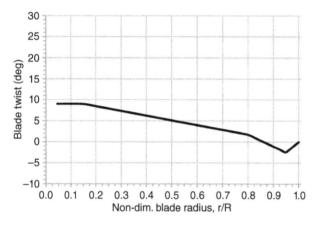

**Figure 3.20** Radial distribution of blade twist, $\beta_{Twist}$ (PSU 1.5-MW turbine).

Jonkman et al. (2009) who applied three-dimensional corrections to airfoil data documented in appendix A of Kooijman et al. (2003).

The radial distribution of non-dimensional blade chord, $c/R$, is shown in Figure 3.19, with a maximum of $c/R = 0.0733$ occurring at $r/R = 0.25$. The resulting rotor solidity for the three-bladed rotor becomes $\sigma = 0.0477$ and is representative of utility-scale wind turbines. Figure 3.20 shows the radial distribution of blade twist, with a total of $\Delta\beta_{Twist} = 9°$ twist between blade root and tip. Of interest is the "reverse" blade twist in the tip region, which is a result of the close-to linear $c/R$ distribution in the tip region, a behavior associated with the aerodynamic optimum (including tip loss).

Non-dimensional chord, blade twist, and blade tip pitch as well as the airfoil distribution from Table 3.1 are defined in the *XTurb* input file *PSU.inp*. The airfoil definition in *PSU.inp* is shown next, with the dimensionless RAIRF stations describing the radial breakpoint at which the associated airfoil entry in AIRFDATA begins.

---

**XTurb Example 3.6  PSU 1.5-MW Turbine (Airfoil Definition in PSU.inp)**

```
    NAIRF    = 10,

    RAIRF    = 0.0444,
                0.09,
                0.11,
                0.13,
                0.15,
                0.18,
                0.4,
                0.5,
                0.6,
                0.825,

    AIRFDATA = 'Cylinder.polar',
                'Cylinder1.polar',
                'Cylinder7.polar',
                'Cylinder9.polar',
                '00W240103.polar',
                '00W235003.polar',
                '97W30003.polar',
                '91W225003.polar',
                '93W21003.polar',
                '95W18003.polar',
```

---

The resulting speed and pitch-controlled dimensional power curve, that is, $P$ versus $V_0$, of the generic PSU 1.5-MW turbine is shown in Figure 3.21. The rated wind speed for this turbine is $V_{rated} = 11\frac{m}{s}$ and marks the beginning of Region III of the power curve.

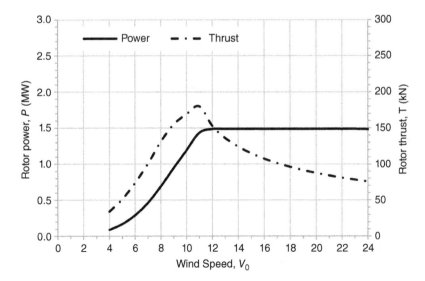

**Figure 3.21** Rotor power, $P$, and thrust, $T$, versus wind speed, $V_0$ (PSU 1.5-MW turbine).

Also shown in Figure 3.21 is the dimensional rotor thrust, $T$, versus wind speed, $V_0$. It can be seen that the maximum rotor thrust force, $T$, occurs at $V_0 = V_{rated}$, an important observation for loads design purposes and at first counterintuitive as one might expect rotor thrust to be proportional to $V_0^2$. While this is correct, see basic definition in Eq. (2.15), one has to consider the greatly reduced thrust coefficient, $C_T$, in Region III as a result of pitch control. Furthermore, it is worth mentioning that the same observation is true for the root-flap bending moment, $BE$, as documented in *XTurb_Output.dat* when run in PREDICTION mode.

Rotor tip speed ratio, $\lambda$, and blade tip pitch, $\beta_{Tip}$, are plotted in Figure 3.22 versus the wind speed, $V_0$. Here it is important to realize that in Region II, that is, $V_0 < V_{rated}$, the objective is typically to operate the rotor at zero blade tip pitch and at the tip speed ratio associated with the maximum power coefficient, $C_{P,max}$. This necessitates an increase in rotor speed, rpm, for increasing wind speed, $V_0$, until a constraint in tip speed, $V_{Tip}$, is reached. In practice, given a rotor diameter, $D$, a maximum rotor speed can be easily determined for a constraint in $V_{Tip}$, and the associated tip speed ratio becomes the design tip speed ratio. As stated previously in Section 3.7.4, the generic PSU 1.5-MW turbine is a rather conservative design with respect to $V_{Tip}$ and consequently, the maximum rotor speed occurs at $V_0 < V_{rated}$ as can be inferred from Figure 3.22. As for the blade tip pitch, $\beta_{Tip}$, in Figure 3.22, there is zero tip pitch throughout Region II and progressively higher blade pitch in Region III to control rotor power at maximum rotor speed where actual $\beta_{Tip}$ (or PITCH) values can be looked at in conjunction with Figure 3.17.

Shown next is part of the resulting *XTurb_Output.dat* and it can be seen that the power coefficient does indeed decrease for increasing Region III wind speed. Of further note is that all of $C_T$, $C_P$, and $C_B$ practically reduce by an order of magnitude between Region II and Region III, thus outweighing the $V_0^2$ effect between dimensionless coefficients and dimensional quantities.

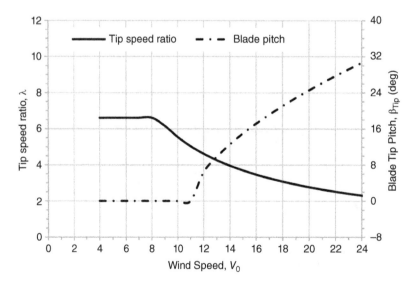

**Figure 3.22** Tip speed ratio, $\lambda$, and blade tip pitch, $\beta_{Tip}$, versus wind speed, $V_0$ (PSU 1.5-MW turbine).

# XTurb Example 3.7  PSU 1.5-MW Turbine – XTurb_Output.dat (PSU.inp)

```
3.7.6 - PSU 1.5-MW                ***** XTurb V1.9    -  OUTPUT *****
    Blade Number        BN =   3

PREDICTION                  +     BLADE ELEMENT MOMENTUM THEORY (BEMT)    +
    Blade Radius     BRADIUS   =   38.5 [m]
    Air Density      RHOAIR    =   1.225 [kg/m**3]
    Air Dyn. Visc.   MUAIR     =   0.000018 [kg/(m*s)]
    Number of Cases  NPRE      =   25

    Thrust          T  = 0.5*RHOAIR*VWIND**2.*(pi*BRADIUS**2.)*CT
    Power           P  = 0.5*RHOAIR*VWIND**3.*(pi*BRADIUS**2.)*CP
    Torque          TO = P / RPM
    Bending Moment  BE = 0.5*RHOAIR*VWIND**2.*(pi*BRADIUS**2.)*BRADIUS*CB

Number VWIND[m/s] RPM[1/min] TSR   PITCH[deg]    CT       CP       CB         T[N]          P[W]          TO[Nm]        BE[Nm]
   8    8.0000   13.1100   6.6070   0.0000     0.7161  -0.4705  -0.1556   130722.4816   -687144.9506   -500514.9312  -1093817.2931
  14   14.0000   13.7500   3.9597  12.5685     0.2220  -0.1897  -0.0431   124090.3092  -1484926.1961  -1031272.7461   -927191.0700
  20   20.0000   13.7500   2.7718  24.5408     0.0765  -0.0651  -0.0114    87330.2567  -1485574.1686  -1031722.7593   -501733.9134
```

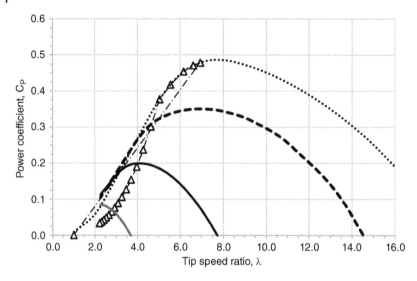

**Figure 3.23** Rotor power coefficient, $C_p$, versus tip speed ratio, $\lambda$ (PSU 1.5-MW turbine).

While rotor performance in terms of dimensional power, $P$, and thrust, $T$, is useful for purposes of power prediction and AEP computation, illustrating dimensionless coefficients of power, $C_P$, and thrust, $C_T$, is helpful from a designer's perspective. Figures 3.23 and 3.24 show the respective rotor performance coefficients versus the tip speed ratio, $\lambda$, with the blade tip pitch, $\beta_{Tip}$ (or PITCH), being a variable parameter. Here turbine start-up is indicated by a line connecting a symbol at $\lambda = 1$ to the respective symbol at $\lambda = \lambda_{max}$. With respect to the power coefficient, $C_P$, it can be seen in Figure 3.23 that the generic PSU 1.5-MW turbine may not exploit its full performance potential at zero blade pitch in Region II due to a rather conservative restriction on $V_{Tip}$. In general, the reader should note in Figures 3.23 and 3.24 that pitch control allows the turbine to not

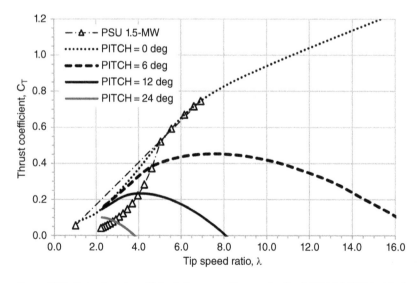

**Figure 3.24** Rotor thrust coefficient, $C_T$, versus tip speed ratio, $\lambda$ (PSU 1.5-MW turbine).

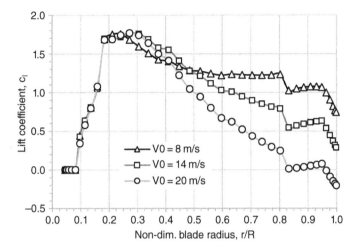

**Figure 3.25** Radial distribution of lift coefficient, $c_l$ (PSU 1.5-MW turbine).

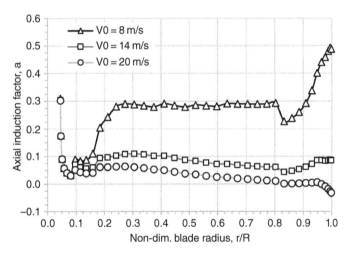

**Figure 3.26** Radial distribution of axial induction factor, $a$ (PSU 1.5-MW turbine).

operate on the "far left" sides of the $C_P$, $C_T$ versus $\lambda$ curves, thus avoiding blade stall and showing the true design benefit of pitch control.

This is further illustrated in Figure 3.25 where we can see that for $V_0 > V_{rated}$, pitch control as per Figure 3.22 effectively controls the section lift coefficient, $c_l$, in the outer blade half, that is, $r/R > 0.5$. Figure 3.25 also shows a rather abrupt breakpoint to the tip airfoil at $r/R = 0.82$, which could be improved by means of an appropriate transition airfoil section. Radial distributions of the axial induction factor, $a$, are shown in Figure 3.26. Of note is the quite uniform axial induction along the blade radius, except for the cylindrical root section, and we (hopefully) remember that uniform induction is typically associated with close-to ideal performance in momentum theory. In fact, an average $a$ value from Figure 3.26 provides a good first-order estimate for the thrust coefficient, $C_T$, using the simple momentum theory Eq. (2.15). In addition, $a \approx 0.3$ for most of the blade radius at $V_0 = 8 m/s$, which indicates that the rotor is operating here

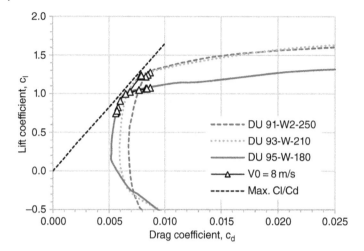

**Figure 3.27** Blade operating polar, $c_l$ versus $c_d$ (PSU 1.5-MW turbine, $0.5 \leq r/R \leq 1.0$).

close to its optimal tip speed ratio, $\lambda$. The behavior in the tip region at $V_0 = 8$ m s$^{-1}$ in Figure 3.26 is a result of higher induction due to tip-loss effects, a behavior that is not easily optimized "by hand" but that we will reiterate in Chapter 8 on optimization and inverse design.

Another interesting question that may arise with respect to assessing how close-to-optimal the generic PSU 1.5-MW turbine operates at $V_0 = 8$ m s$^{-1}$ ($\lambda = 6.6$) concerns the blade section lift-to-drag ratio, $c_l/c_d$. We found from simplified BEMT in Section 3.6 that optimal rotor performance can be achieved for the operating section lift coefficient, $c_l$, being close to $(c_l/c_d)_{max}$ of the local airfoil. Figure 3.27 shows $c_l$ versus $c_d$ airfoil polars for the three outer airfoils ($0.5 \leq r/R \leq 1.0$), along with the actual operating values for $c_l$ and $c_l/c_d$ taken from the corresponding *XTurb_Output_Method.dat*. Also shown in Figure 3.27 is a dashed line approximately tangent to the polars, thus indicating the maximum achievable lift-to-drag ratio, $(c_l/c_d)_{max}$, along the outer blade half. It can be seen that the blade operating points at $V_0 = 8$ m s$^{-1}$ are actually located in the vicinity of the $(c_l/c_d)_{max}$ line, thus supporting the fact that the generic PSU 1.5-MW turbine is a quite good baseline design, though still with a bit of room for further improvement in the tip region. Figure 3.28 shows an alternative way of looking at essentially the same, by plotting $c_l$ versus $c_l/c_d$ of the same data. It is again apparent that the turbine blade is a good baseline design. The nice aspect of Figure 3.28 is that one can actually quantify how much $\Delta(c_l/c_d)$ is not used in the tip region, and the integrated effect on the power coefficient, $C_P$, could be at least estimated using for example, Eq. (3.79).

In Section 3.3, we derived a relation between the optimum blade parameter, $\sigma' c_l$, and the blade flow angle, $\phi$, in the absence of tip-/root losses. Figure 3.29 plots the respective groupings of $\sigma' c_l$ and $2 \sin \phi \tan \phi$ on a double-logarithmic scale, with a straight dashed line of slope one indicating the (inviscid) optimum without tip-/root losses. It is interesting to see that most of the operating points at $V_0 = 8$ m s$^{-1}$ are very close to the optimum dashed line, with some discrepancies at blade root and tip, a result associated with the fact that the optimum dashed line is invalid in these regions.

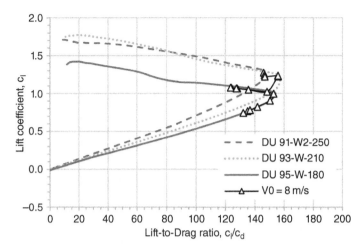

**Figure 3.28** Blade operating polar, $c_l$ versus $c_l/c_d$ (PSU 1.5-MW turbine, $0.5 \leq r/R \leq 1.0$).

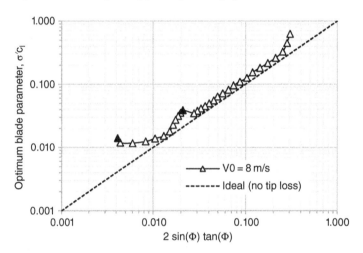

**Figure 3.29** Optimum blade parameter, $\sigma' c_l$, at $C_{P,max}$ (PSU 1.5-MW turbine).

The interested reader can actually verify independently that the blade operating points do not fall close to the dashed line for wind speeds, $V_0$, where the respective tip speed ratio, $\lambda$, is not close to the optimal one at that particular blade tip pitch, $\beta_{Tip}$.

## 3.8 Validity of BEMT

BEMT is and most likely will always be the primary design and analysis method used in the wind energy community. Given standard computational resources and processing power of desktop computers in use today, a full wind turbine aerodynamics analysis can be done in a fraction of a second with practically all types of BEM solution algorithms. The reason for its computational efficiency is rooted in its simple formulation as two-dimensional strip theory, with blade elements operating essentially independent of each other. Hence, BEMT is a truly two-dimensional modeling approach, though it allows for tip-/root effects to be incorporated by appropriate loss factors, some of which have been described in this chapter. While the strip theory (i.e. blade element) approach is the primary strength of BEMT, it is at the same time its major weakness with respect to its physical validity. The inherent assumption of zero radial flow in a rotating system does, by all means, work quite well for most of the blade radius, but cannot by itself predict any type of three-dimensional flow effects, unless further correction models are applied to the equations.

The first correction model is concerned with the formation and rollup process of both tip and root vortices as a consequence of a finite number of blades. After more than one hundred years, predicting blade tip loads is still a modeling challenge and an active research area. This is because vortex formation, separation, and rollup are complicated fluid-mechanic processes that depend on practically all possible parameters, including blade flow angle, tip shape/curvature, solidity, blade pitch, tip speed ratio, Reynolds number, thrust loading, tip airfoil, and so on. It is still challenging today to find a universal cure to the tip problem capable of predicting tip loads of new rotor blades to satisfactory accuracy.

A second physical effect not captured by classical BEMT is that of stall delay (or rotational augmentation) at inboard blade sections, see Chapter 5 dedicated in part to this subject. Inboard stall delay is derived from the "Himmelskamp effect" first seen on rotating propellers. This effect is due to the Coriolis force alleviating the adverse pressure gradient of upper-surface airfoil flows, thus delaying separation and stall. It is here where three-dimensional corrections are applied to airfoil tables used in BEM analyses as the theory itself is not capable of predicting those. Correction models are typically deduced from experimental data and more recently from full three-dimensional and rotating CFD analyses.

The third physical effect that BEMT is invalid for without further action is that of blade section separation and stall. Depending on the rotor operating condition of tip speed ratio and blade tip pitch, separation and stall can occur first either in the root or tip region. Typically, once flow separation occurs, a radial flow is initiated and the separated region propagates outboard along the blade radius, thus violating the strip-theory assumption. This effect is also known as centrifugal pumping. Some stall-delay models account for this effect at inboard blade sections, while outboard blade stall does in general not occur for modern speed- and pitch-controlled wind turbines. However, this is different for stall-controlled wind turbines, a prominent example being the NREL Phase VI rotor and the long history of BEM methods having difficulty in predicting blade loads and power at high wind speeds.

In general, though, BEMT does still perform quite well and is a fast and accurate design and analysis tool when used with the appropriate caution and suitable correction models.

### 3.8.1 Summary – BEMT

The following gives a brief summary of BEMT:

- Assumptions: Two-dimensional, steady, no radial flow, root-/tip losses
- Includes effect of airfoil lift-to-drag ratio: $c_l/c_d$
- Rotor Thrust: $T = T(V_0^2, D^2, C_T(\lambda, c_l/c_d))$

- Rotor Power: $P = P(V_0^3, D^2, C_P(\lambda, c_l/c_d))$
- Wind turbine power production driven by …
  - Wind resource, $V_0$
  - Rotor size, $D$
  - Blade/rotor design, $c_l$ close to $(c_l/c_d)_{max}$
  - Achievable $C_{P,\,max}$ function of $\sigma' c_l$ & $(c_l/c_d)_{max}$

## References

Anderson, J.D. (2001). *Fundamentals of Aerodynamics*, McGraw Hill Series in Aeronautical and Aerospace Engineering, 3e, 304. New York: McGraw Hill.

Basom, B. J. (2010) *Inviscid Wind-Turbine Analysis Using Distributed Vorticity Elements*. M.S. Thesis (The Pennsylvania State University).

Betz, A. (1919) *Schraubenpropeller mit geringstem Energieverlust*. Dissertation. Göttinger Nachrichten, Goettingen.

Bramesfeld, G. and Maughmer, M.D. (2008). Relaxed-wake vortex-lattice method using distributed vorticity elements. *AIAA Journal of Aircraft* 45 (2): 560–568.

Branlard, E. (2017). *Wind Turbine Aerodynamics and Vorticity-Based Methods - Fundamentals and Recent Applications*. London: Springer.

Branlard, E. and Gaunaa, M. (2014). Development of new tip-loss corrections based on vortex theory and vortex methods. The Science of Making Torque from Wind. *Journal of Physics: Conference Series* 555: 012012.

Branlard, E. and Gaunaa, M. (2016). Superposition of vortex cylinders for steady and unsteady simulation of rotors of finite tip-speed ratio. *Wind Energy* 19 (7): 1307–1323.

Branlard, E., Dixon, K., and Gaunaa, M. (2013). Vortex methods to answer the need for improved understanding and modelling of tip-loss factors. *IET Renewable Energy Generation* 7 (4): 311–320.

Brent, R.P. (1971). An algorithm with guaranteed convergence for finding a zero of a function. *The Computer Journal* 14 (4): 422–425.

Buhl, M. L. (2005) *A New Empirical Relationship between Thrust Coefficient and Induction Factor in the Turbulent Wake State*. Technical Report NREL/TP-500-36834, National Renewable Energy Laboratory.

Burton, T., Jenkins, N., Sharpe, D., and Bossanyi, E. (2011). *Wind Energy Handbook*, 2e, 66–68. Chichester: Wiley.

De Vries, O. (1979). *Fluid Dynamic Aspects of Wind Energy Conversion*. AGARD-AD No. 243. ISBN: 92-835-1326-6.

Eggleston, D.M. and Stoddard, F.S. (1987). *Wind Turbine Engineering Design*, 30–35. New York: Van Nostrand Reinhold Co, 58.

Glauert, H. (1926). *The Analysis of Experimental Results in the Windmill Brake and Vortex Ring States of an Airscrew*. Report 1026. Aeronautical Research Committee Reports and Memoranda. London: Her Majesty's Stationery Office.

Glauert, H. (1935). Airplane propellers. In: *Aerodynamic Theory*, vol. 4, Division L (ed. W.F. Durand), 169–360. Berlin: Julius Springer.

Hoerner, S.F. (1965). Pressure drag on rotating bodies. In: *Fluid Dynamic Drag* (ed. S.F. Hoerner), 3–14. Hoerner Fluid Dynamics (Published by the author).

Jonkman, J., Butterfield, S., Musial, W., and Scott, G. (2009) *Definition of a 5-MW Reference Turbine for Offshore System Development*. Technical Report NREL/TP-500-38060, National Renewable Energy Laboratory.

Kooijman, H. J. T., Lindenburg, C., Winkelaar, D., and van der Hooft, E. L. (2003) *DOWEC 6 MW Pre-Design: Aero-elastic modeling of the DOWEC 6 MW pre-design in PHATAS*. DOWEC Dutch Offshore Wind Energy Converter 1997–2003 Public Reports [CD-ROM], DOWEC 10046_009, ECN-CX–01-135, Petten, the Netherlands: Energy Research Center of the Netherlands.

Lock, C.N.H., Batemen, H., and Townsend, H.C.H. (1926). *An Extension of the Vortex Theory of Airscrews with Applications to Airscrews of Small Pitch, Including Experimental Results*. Report 1014. Aeronautical Research Committee Reports and Memoranda. London: Her Majesty's Stationery Office.

Maniaci, D. C. (2013) *Wind Turbine Design using a Free-Wake Vortex Method with Winglet Application*. Ph.D. Dissertation (The Pennsylvania State University).

Maniaci, S. and Schmitz, S. (2016). Extended Glauert tip correction to include vortex rollup effects. The Science of Making Torque from Wind. *Journal of Physics: Conference Series*, 753,: 022051. https://doi.org/10.1088/1742-6596/753/2/022051.

Manwell, J.F., McGowan, J.G., and Rogers, A.L. (2009). *Wind Energy Explained: Theory, Design, and Application*, 2e, 140. Chichester: Wiley.

McWilliam, M. and Crawford, C. (2011). The behavior of fixed-point iteration and Newton–Raphson methods in solving the blade element momentum equations. *Wind Engineering* 35 (1): 17–32.

Micallef, D. (2012) *3D Flows Near a HAWT Rotor: A Dissection of Blade and Wake Contributions*. Ph.D. Dissertation (Delft University of Technology).

Montgomerie, B. (1995) *De-camber: Explanation of an effect of lift reduction near the tip caused by the local flow around airplane wings or wind turbine tips*. In: Proceedings of Plenary Meeting of the group for Dynamic Stall and 3D Effects. A European Union Joule 2 Project, Cranfield Institute of Technology.

Moriarty, P. J. and Hansen, A. C. (2004) *AeroDyn Theory Manual*. Technical Report NREL/EL-500-36881, National Renewable Energy Laboratory.

Ning, S.A. (2014). A simple solution method for the blade element momentum equations with guaranteed convergence. *Wind Energy* 17 (9): 1327–1345.

Ramdin, S. F. (2017) *Prandtl Tip Loss Factor Assessed*. M.S. Thesis (Delft University of Technology).

Rohrbach, C. (1976) *Experimental and Analytical Research on the Aerodynamics of Wind Turbines*. Mid-Term Technical Report COO-2615-76-T-1, Energy Research and Development Administration.

Rohrbach, C. and Worobel, R. (1975) *Performance Characteristics of Aerodynamically Optimum Turbines for Wind Energy Generators*. Proceedings of the American Helicopter Society 31st Annual National Forum, Washington D.C., May 1975.

Schmitz, S. (2012). *XTurb-PSU: A Wind Turbine Design and Analysis Tool*. The Pennsylvania State University.

Schmitz, S. (2015) *PSU Generic 1.5-MW Turbine*. DOI: 10.13140/RG.2.2.22492.18567.

Schmitz, S. and Maniaci, D. (2016). Analytical method to determine a tip loss factor for highly-loaded wind turbine rotors. AIAA-2016-0752.

Schmitz, S. and Maniaci, D. (2017). Methodology to determine a tip-loss factor for highly loaded wind turbine. *AIAA Journal* 55 (2): 341–351. https://doi.org/10.2514/1.J055112.

Shen, W.Z., Mikkelsen, R., Sørensen, J.N., and Bak, C. (2005a). Tip loss corrections for wind turbine computations. *Wind Energy* 8 (4): 457–475.

Shen, W.Z., Sørensen, J.N., and Mikkelsen, R. (2005b). Tip loss correction for actuator/navier-stokes computations. *Journal of Solar Energy Engineering* 127: 209–213.

Sørensen, J.N. (2016). *General Momentum Theory for Horizontal Axis Wind Turbines*, Chapter 8. London: Springer.

Sørensen, J.N., Dag, K.O., and Ramos-Garcia, N. (2014). A new tip correction based on the decambering approach. The Science of Making Torque from Wind. *Journal of Physics: Conference Series*, 524,: 012097. https://doi.org/10.1088/1742-6596/524/1/012097.

Sørensen, J.N., Dag, K.O., and Ramos-Garcia, N. (2015). A refined tip correction based on decambering. *Wind Energy* 19 (5): 787–802.

Wilson, R.E. (1994). Aerodynamic behavior of wind turbines, Chapter 5. In: *Wind Turbine Technology* (ed. D.A. Spera), 231–232. New York, NY: The American Society of Mechanical Engineers.

Wilson, R.E. and Lissaman, P.B.S. (1974). *Applied Aerodynamics of Wind Power Machines*. Oregon State University.

Wilson, R.E., Lissaman, P.B.S., and Walker, S.N. (1976). *Aerodynamic Performance of Wind Turbines*. Energy Research and Development Administration, ERDA/NSF/04014-76/1.

Wood, D.H., Okulov, V.L., and Bhattacharjee, D. (2016). Direct calculation of wind turbine tip loss. *Renewable Energy* 95: 269–276.

## Further Reading

Burton, T., Jenkins, N., Sharpe, D., and Bossanyi, E. (2011). *Wind Energy Handbook*, 2e. Chichester: Wiley.

Hansen, M.O.L. (2008). *Aerodynamics of Wind Turbines*. London: Earthscan.

Leishman, J.G. (2000). *Principles of Helicopter Aerodynamics*. Cambridge: Cambridge University Press.

Manwell, J.F., McGowan, J.G., and Rogers, A.L. (2009). *Wind Energy Explained: Theory, Design, and Application*, 2e. Chichester: Wiley.

Sørensen, J.N. (2016). *General Momentum Theory for Horizontal Axis Wind Turbines*, Chapter 8. London: Springer.

# 4

# Wind Turbine Airfoils

*It's easy to explain how a rocket works, but explaining how a wing works takes a rocket scientist.*

– Philippe Spalart

## 4.1  Fundamentals of Airfoil Theory

Airfoil lift and drag constitute the blade section force triangle, and are hence the foundation for blade element momentum (BEM) theory as discussed in Chapter 3. The associated airfoil nomenclature has been introduced in Section 3.1.1, where it is important to realize that airfoil lift and drag coefficients, $c_l$ and $c_d$, are normalized by the dynamic pressure force at the respective blade element. This enables the use of essentially two-dimensional airfoil tables in BEM solution methods, and actual dimensional lift and drag forces are obtained by multiplying dimensionless force coefficients with a dynamic pressure force that is a function of the iteratively computed local axial-/angular induction.

In this chapter, we will learn about the basic airfoil aerodynamics and dependencies that determine airfoil lift and drag coefficients, $c_l$ and $c_d$. In particular, we will gain an understanding of the dependence of $c_l$ and $c_d$ on angle of attack, $\alpha$, the relative airfoil camber, $d/c$, and the chord-based Reynolds number, $Re$. Figure 4.1 provides a qualitative, though compelling, introduction to the $Re$ dependence of airfoil properties and the intersection of inviscid and viscous airfoil theory. The Reynolds number, $Re$, is often defined as the ratio of inertial to viscous forces, thus leading to the perception of the limit of "inviscid" flow in the case of high $Re$ and "viscous" flow for low $Re$, respectively. A better qualitative description with respect to Figure 4.1 is to understand the Reynolds number, $Re$, as a quantitative measure of the distance normal to the airfoil surface up to which viscous diffusion processes are dominant, see also McLean (2013).

Consequently, the action of viscosity is confined to a limited region around the airfoil, that is, the airfoil boundary layer, and its viscous wake downstream of the airfoil. Outside the viscous region (boundary layer + wake), the flow is essentially inviscid in the sense that viscous diffusion effects are negligible in their action on the flow when compared to the dominating pressure-gradient and convective acceleration terms. For high $Re$ flows such as occurring at blade sections of large utility-scale wind turbines, the viscous region is considerably thinner than the airfoil chord or even thickness (see

*Aerodynamics of Wind Turbines: A Physical Basis for Analysis and Design*, First Edition. Sven Schmitz.
© 2020 John Wiley & Sons Ltd. Published 2020 by John Wiley & Sons Ltd.
Companion website: www.wiley.com/go/schmitz/wind-turbines

Wind Turbines:

Small-Scale Testing (D ≈ 0.1 m)  Model-Scale Testing (D ≈ 1-10 m)  Utility-Scale (D ≈ 10-100 m)

Insects: $Re_c = O(10^3)$          Model Airplanes: $Re_c = O(10^5)$     Large Airplanes: $Re_c = O(10^7)$

**Figure 4.1** *Re*-dependence of viscous flow region (gray) around an airfoil.

Figure 4.1). This, as we shall see, has notable implications on airfoil characteristics such as the important lift-to-drag ratio, $c_l/c_d$, and the maximum lift coefficient, $c_{l,max}$, of a given airfoil. As indicated in Figure 4.1, airfoil drag is somewhat proportional to the size of the viscous wake, which has significant effects on the maximum attainable power coefficient, $C_{P,max}$, of a wind turbine rotor (see Section 3.7) and presents difficulties in model-scale blade designs for wind-tunnel testing, as will be discussed in more detail in Chapter 8. In this context, it is fascinating to realize the disparity in scales between an atmospheric boundary layer $\sim O(10^3$ m) and that of a high-*Re* airfoil boundary layer $\sim O(10^{-3}$ m) such as seen at radial blade sections along a large utility-scale wind turbine blade. Though qualitative in nature, Figure 4.1 also implies that there are more opportunities for airfoil design at high-*Re* where the entire airfoil contour can affect both extent and properties of the viscous flow regime, as opposed to at low-*Re* where the progression of the viscous flow regime is governed primarily by the fore part of the airfoil.

Figure 4.2 brings wind turbine aerodynamics in relation to other lifting devices as well as birds/insects by plotting the actual flight or relative wind speed over the respective airfoil cross sections versus the chord-based Reynolds number. Here it is important to realize once more the range of scales associated with wind turbine aerodynamics, that is, ranging from small model airplanes and birds to ultra-light aircraft such as sailplanes.

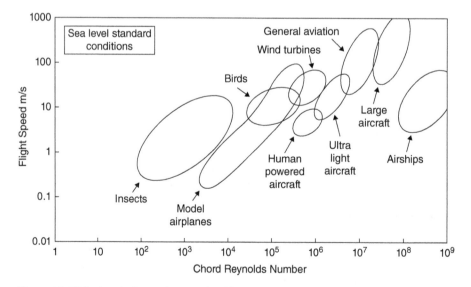

**Figure 4.2** Flight (or wind) speed versus chord-based Reynolds number, $Re_c$.

Fundamentally, airfoil lift and drag forces derive from integrating pressure, $p$, and wall shear stress (skin friction), $\tau_w$, over the closed airfoil contour, $C$. The total force vector, $\vec{F}$, is hence defined as:

$$\text{Airfoil Force Vector}: \vec{F} = \oint_C \{-(p-p_\infty)\vec{n} + \vec{\tau}_w\}ds = D\vec{i} + L\vec{k} \qquad (4.1)$$

In Eq. (4.1), $\vec{n}$ is the outer unit normal vector along $C$, $p_\infty$ is the freestream pressure, and $\vec{i}$ and $\vec{k}$ are unit vectors in the respective streamwise ($x$) and normal ($z$) directions. In the following, we separate the contributions of pressure and wall shear stress and find associated lift and drag coefficients, $c_l$ and $c_d$.

### 4.1.1 Inviscid Flow: Thin-Airfoil Theory

Let us begin by considering inviscid flow in the absence of viscous stresses, $\vec{\tau}$, in the flowfield. In this case, the flow is "frictionless" and slips along the airfoil contour, thus making the airfoil contour a streamline itself. Hence, inviscid flow is concerned with integrating the pressure term $(p - p_\infty)$ over the closed airfoil contour, $C$. We introduce a dimensionless pressure coefficient, $c_p$, according to $c_p = (p - p_\infty)/(\frac{1}{2}\rho_\infty U_\infty^2)$. Consequently, the lift coefficient, $c_l$, in inviscid theory becomes:

$$\text{Lift Coefficient}: c_l = -\oint_C c_p \vec{n} ds \qquad (4.2)$$

Here $s$ is the dimensionless arc length along $C$. Bernoulli's equation holds in steady, incompressible, inviscid, and irrotational flow:

$$\text{Bernoulli Equation}: p_\infty + \frac{1}{2}\rho_\infty U_\infty^2 = p + \frac{1}{2}\rho_\infty U^2 \qquad (4.3)$$

In Eq. (4.3), $U^2 = (U + u')^2 + w'^2 \approx \left(1 + 2\frac{u'}{U_\infty}\right)U_\infty^2$, assuming small disturbances $u'$, $w' \ll U_\infty$ such that higher-order terms can be neglected. Consequently, the pressure coefficient, $c_p$, can be approximated using the following relation:

$$\text{Pressure Coefficient (Small Disturbances)}: c_p \approx -2\frac{u'}{U_\infty} \qquad (4.4)$$

Hence the pressure coefficient becomes related to a perturbation velocity, $u'$, in the streamwise direction. This gives rise to examining the difference $u'_+ - u'_-$ just above/below the mean camber line. For this reason, we introduce the concept of the vortex sheet, which is central to thin-airfoil theory and shown in Figure 4.3. For thin airfoils, we assume that a variable jump in velocity, $< u' > = u'_+ - u'_-$ occurs across the mean camber line. A varying jump in velocity, $<u'>$, along the mean camber line is realized by a sheet of elemental potential vortices of strength (vorticity), $\gamma = <u'>$. Note that a single vortex sheet along the airfoil camber line is a special case of the more general concept of a vortex sheet wrapped around the entire airfoil contour (as is used in panel-based vortex methods).

The total circulation, $\Gamma$, "bound" to the airfoil can be determined by integrating the vortex-sheet strength, $\gamma$, along the airfoil chord-line coordinate, $x$, assuming small disturbances and "thin" airfoils on the order of 10% thickness in reference to the airfoil

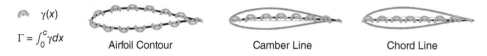

$\gamma(x)$

$\Gamma = \int_0^c \gamma dx$

Airfoil Contour         Camber Line          Chord Line

**Figure 4.3** Vortex-sheet representation of an airfoil.

chord, $c$:

$$\text{Airfoil Bound Circulation}: \Gamma = \int_0^c \gamma dx \tag{4.5}$$

Note that Eq. (4.5) is essentially a one-dimensional form of the more general Stokes' theorem, which relates a closed-contour circulation integral to the vorticity, $\vec{\omega}$, contained in the area, $A$, enclosed by contour, $C$:

$$\text{Stokes' Theorem}: \Gamma = -\oint_C \vec{U}.\vec{ds} = -\oint\!\!\!\oint_A \vec{\omega}.\vec{dA} \tag{4.6}$$

The task is now to find a suitable distribution of $\gamma(x)$ along the airfoil chord line that represents inviscid (or very high-$Re$) flow around a thin airfoil. Before solving for $\gamma(x)$, however, let us first relate the findings thus far to a classical theorem in aerodynamics.

#### 4.1.1.1 Kutta–Joukowski Lift Theorem

The vortex-sheet strength, $\gamma(x)$, describes a jump in velocity, $<u'>$, across the vortex sheet. Using Eq. (4.4), one can write a change in pressure coefficient, $\Delta c_p$, from upper to lower airfoil surface as $\Delta c_p = -2\gamma/U_\infty$. Hence the lift coefficient, $c_l$, in Eq. (4.2) becomes

$$\text{Lift Coefficient}: c_l = \frac{L'}{\frac{1}{2}\rho_\infty U_\infty^2 c} = -\int_0^1 \Delta c_p d\frac{x}{c} = 2\frac{\Gamma}{U_\infty c} \tag{4.7}$$

after using the relation for the bound circulation in Eq. (4.5). Next, the Kutta–Joukowski (KJ) lift theorem can be easily obtained by solving Eq. (4.7) for the lift force, $L'$, per unit width of an airfoil element:

$$\text{KJ Lift Theorem}: L' = \rho_\infty U_\infty \Gamma \tag{4.8}$$

Equation (4.8) is the classical KJ Lift Theorem for inviscid and irrotational flow. He we arrived at the KJ Lift Theorem from the vortex-sheet concept, while it is generally derived from potential-flow problems such as inviscid and irrotational flow around a rotating cylinder. Figure 4.4 illustrates the relationship between lift and circulation around a rotating cylinder and airfoil.

Nevertheless, the problem remains as to what is the actual bound circulation, $\Gamma$, (and vortex-sheet strength $\gamma(x)$), for a given airfoil shape as a function of angle of attack, $\alpha$. In the following, we will derive the fundamental equation of thin-airfoil theory and introduce its infinite-series solution.

#### 4.1.1.2 Symmetric-/Cambered Thin Airfoil

In general, thin-airfoil theory considers the flow being a superposition of a thickness and lifting problem. It can be shown using a distribution of potential source/sink terms that in the case of "thin" airfoils, the "thickness problem" actually results in zero lift and drag. The interested reader is referred to associated book chapters in, for example,

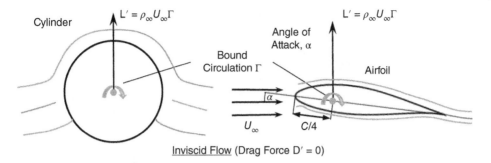

Figure 4.4 Inviscid flow – Kutta–Joukowski lift theorem and d'Alembert's paradox.

Anderson (2001), Katz and Plotkin (2001), or Chattot and Hafez (2015) for a more in-depth treatment on both thickness and lifting problems than the brief analysis presented next.

We define "thin airfoils" as those being on the order of 10% thick with respect to the airfoil chord, *c*, and definitely not notably more. As mentioned in the previous section, the free-slip condition on the airfoil surface makes the airfoil itself become a streamline. Hence, assuming "thin" airfoils, the flow is tangent to the mean camber line, $d(x)$, everywhere (at least to first order) as a result of the cumulative induced flow due to all vortex-sheet elements of strength, $\gamma(x)$. Mathematically, the vortex-sheet strength, $\gamma(x)$, is such that integrated induced velocities cancel the component of the freestream flow, $U_{\infty, n}$, normal to a local element along the mean camber line. In other words, there is no flow through the camber line because it is a streamline. This property is described in the fundamental equation as:

$$\text{Fundamental Equation 1}: \quad -\frac{1}{2\pi}\int_0^c \frac{\gamma(\xi)}{x-\xi}d\xi = U_{\infty}[d'(x) - \alpha] \qquad (4.9)$$

In Eq. (4.9), $d'(x)$ is the slope of the mean camber line, $d(x)$, and $\alpha$ is the airfoil angle of attack. Note that the integral on the left-hand-side is a principal-value integral excluding the singularity at $\xi = x$. A unique solution to the fundamental lifting equation of thin-airfoil theory is obtained when a applying an observed boundary condition at the trailing edge (TE):

$$\text{Kutta Condition}: \quad \gamma(c) = 0 \qquad (4.10)$$

In words, the Kutta condition in Eq. (4.10) states that "…the flow leaves the trailing edge smoothly," that is, the perturbation velocities, $u'_+$ and $u'_-$, are the same at the airfoil TE. The result of this simple, though brilliant, boundary condition is that it enforces the flow to stagnate at a finite-angle TE, and results in a smooth slip condition in the case of a cusp, thus resulting in a unique vortex-strength distribution, $\gamma(x)$, and hence bound circulation, $\Gamma$. In the following, a transformation is useful that reads:

$$x(\Theta) = \frac{c}{2}(1 - \cos\Theta), \ 0 \le \Theta \le \pi \qquad (4.11)$$

Using Eq. (4.11) in (4.9), we obtain the following:

$$\text{Fundamental Equation 2}: \quad -\frac{1}{2\pi}\int_0^\pi \frac{\gamma(\Theta)\sin\Theta}{-\cos\phi + \cos\Theta}d\Theta = U_{\infty}[d'(x) - \alpha] \qquad (4.12)$$

Using the "principal-value" integral of Glauert (see e.g. Appendix E in Abbott and von Doenhoff (1959))

$$\int_0^\pi \frac{\cos n\Theta}{-\cos\phi + \cos\Theta} d\Theta = \frac{\pi \sin n\phi}{\sin\phi} \tag{4.13}$$

and the relation $\sin n\Theta \sin\Theta = \frac{1}{2}[\cos(n-1)\Theta - \cos(n+1)\Theta]$, a general solution for the vortex-sheet strength, $\gamma$, can be constructed using an infinite Fourier series of the following form:

$$\text{Vortex-Sheet Strength}: \gamma[x(\Theta)] = 2U_\infty \left[ A_0 \frac{1+\cos\Theta}{\sin\Theta} + \sum_{n=1}^\infty A_n \sin n\Theta \right] \tag{4.14}$$

Substitution of Eq. (4.14) into (4.12) and subsequent integration provides the mechanism to determine the Fourier coefficients $A_0$ and the remaining $A_n$, noting that the solution has to be valid at all chordwise stations $x$, respectively, $\Theta$, for a given mean camber line, $d[x(\Theta)]$:

$$\text{Fourier Coefficients}: A_0 = \alpha - \frac{1}{\pi}\int_0^\pi d'[x(\Theta)]d\Theta$$

$$A_n = \frac{2}{\pi}\int_0^\pi d'[x(\Theta)]\cos n\Theta d\Theta \tag{4.15}$$

In general, Eq. (4.14) has a singularity at the leading edge ($\Theta = 0$) that does not exist for a special case of $A_0 = 0$, a condition that can be achieved at only one angle of attach, $\alpha_{adapt}$, (angle of adaptation). For a given mean camber line, $d(x)$, the simple relations $\cos\Theta = 1 - 2\frac{x}{c}$ and $\sin\Theta = 2\sqrt{\frac{x}{c}\left(1-\frac{x}{c}\right)}$ and their higher modes are useful to solving the Fourier coefficients in Eq. (4.15). The total airfoil bound circulation, $\Gamma$, can now be found by substituting Eq. (4.14) into (4.5) to find after integration:

$$\text{Airfoil Bound Circulation}: \Gamma = \pi U_\infty c \left( A_0 + \frac{A_1}{2} \right) \tag{4.16}$$

Hence the airfoil lift coefficient, $c_l$, is obtained after using Eq. (4.7) to find the following:

$$\text{Lift Coefficient (Thin Airfoil)}: c_l = 2\pi \left( A_0 + \frac{A_1}{2} \right) \tag{4.17}$$

As $A_0$ is the only coefficient that is a function of the airfoil angle of attack, $\alpha$, we can easily compute the lift-curve slope, $dc_l/d\alpha$, as:

$$\text{Lift-Curve Slope (Thin Airfoil)}: \frac{dc_l}{d\alpha} = 2\pi \tag{4.18}$$

Note that $\alpha$ is in units [rad] in the preceding relations. Consequently, a change in angle of attack of $\Delta\alpha = 1°$ [deg] results in $\Delta c_l = 0.11$ more lift on an inviscid lift curve of a thin airfoil. The corresponding aerodynamic moment about the airfoil leading edge is found as:

$$\text{Moment about Leading Edge}: M_0 = \rho_\infty U_\infty \int_0^c \gamma(x) x dx \tag{4.19}$$

Equation (4.19) can be non-dimensionalized by $1/2\rho_\infty U_\infty^2 c^2$ to obtain the respective moment coefficient, $c_{m,0}$, about the leading edge. Subsequent substitution of Eq. (4.14) and integration results in the following:

$$\text{Moment Coefficient (Thin Airfoil)} : c_{m,0} = -\frac{\pi}{2}\left(A_0 + A_1 - \frac{A_2}{2}\right) \tag{4.20}$$

It is interesting to see that the lift coefficient, $c_l$, depends only on $A_0$ and $A_1$, while the moment coefficient about the leading edge, $c_{m,0}$, is a function of $A_0$, $A_1$, and $A_2$, all of which can be determined from a given mean camber line, $d(x)$, through Eq. (4.15). The aerodynamic moment can be transferred to any other location along the airfoil chord line using the following formula:

$$\text{Transfer of Moment} : c_{m,D} = c_{m,0} + \frac{x_D}{c}c_l \tag{4.21}$$

Using the transfer of moment in Eq. (4.21), one can find the "center of pressure," $x_{c.p.}/c$, defined as the chordwise location about which the moment coefficient is zero:

$$\text{Center of Pressure} : \frac{x_{c.p.}}{c} = \frac{A_0 + A_1 - \frac{A_2}{2}}{4\left(A_0 + \frac{A_1}{2}\right)} = -\frac{c_{m,0}}{c_l} \tag{4.22}$$

For thin airfoils, the fact that only $A_0$ depends on the angle of attack, $\alpha$, lets us use the transfer of moment in Eq. (4.21) to find a chordwise location about which the moment coefficient is independent of $\alpha$. This is an important relationship not only for aircraft stability in general, but also for wing/blade torsion. The respective location is called the "aerodynamic center," $x_{a.c.}/c$, and is found by postulating $dc_{m,a.c.}/d\alpha = 0$. The aerodynamic center can then be found to be:

$$\text{Aerodynamic Center (Thin Airfoils)} : \frac{x_{a.c.}}{c} = \frac{1}{4} \tag{4.23}$$

It is important to realize that the location of the aerodynamic center, $x_{a.c.}/c$, is at the 1/4-chord for thin airfoils. The resulting moment coefficient about the aerodynamic center, $c_{m,a.c.}$, becomes:

$$\text{Moment Coefficient (Aerod.Center)} : c_{m,a.c.} = -\frac{\pi}{4}(A_1 - A_2) \tag{4.24}$$

Note that $c_{m,a.c.}$ does obviously not depend on $A_0$. In the following, let us work out some simple examples:

**Flat Plate/Thin Symmetric Airfoil,** $d(x) = 0 : A_0 = \alpha, A_1 \ldots A_n = 0$

$$c_l = 2\pi\alpha, \quad \frac{x_{c.p.}}{c} = \frac{1}{4}, c_{m,a.c.} = 0 \tag{4.25}$$

**Parabolic Thin Airfoil,** $d(x) = 4\frac{d}{c}x\left(1 - \frac{x}{c}\right) : A_0 = \alpha, A_1 = 4\frac{d}{c}, A_2 \ldots A_n = 0$

$$c_l = 2\pi\left(\alpha + 2\frac{d}{c}\right), \quad \frac{x_{c.p.}}{c} = \frac{1}{4}\frac{\alpha + 4\frac{d}{c}}{\alpha + 2\frac{d}{c}}, c_{m,a.c.} = -\pi\frac{d}{c} \tag{4.26}$$

As for a flat plate or thin symmetric airfoil, the lift-curve slope equals $2\pi$, and the center of pressure coincides with the aerodynamic center at the 1/4-chord. This changes for a thin cambered airfoil with a parabolic camber line where camber acts as an effective angle of attack to result in augmented lift, yet the same lift-curve slope. The location of the center of pressure, $x_{c.p.}/c$, is variable, and there is a constant nose-down moment about the aerodynamic center at the 1/4-chord point. Note that in airfoil tables such as used in BEM codes, the documented moment coefficient is typically that about the airfoil 1/4-chord point. The reader has to further realize that relative airfoil camber, $d/c$, is also subject to the assumption of small disturbances in thin-airfoil theory. Typical values for $d/c$ range between 1 and 6% and can (in real viscous flow) not be considered without the associated thickness distribution that together define adverse pressure-gradient effects and hence separation-/stall behavior at higher angles of attack, see Section 4.2.

### 4.1.1.3 Effect of Airfoil Thickness on Lift

The effect of airfoil thickness on lift has to be addressed in the context of thick airfoil sections at inboard blade stations of modern wind turbines. Here we follow a second-order thin-airfoil solution from Drela (2014). The airfoil thickness distribution, $t(x)$, is represented by a potential source sheet of strength, $\lambda(x)$, along the $x$ axis between leading and trailing edges. The source sheet models the displacement of the streamline due to airfoil thickness and its solution is written as:

$$\text{Source-Sheet Strength} : \lambda[x(\Theta)] = \frac{2U_\infty}{\sin \Theta} \sum_{n=1}^{\infty} n\, B_n \cos n\Theta \tag{4.27}$$

The airfoil thickness, $t(x)$, itself can be expanded in a Fourier sine series as:

$$\text{Airfoil Thickness} : t[x(\Theta)] = c \sum_{n=1}^{\infty} B_n \sin n\Theta \tag{4.28}$$

Following, the Fourier coefficients can be found from a given thickness distribution, $t(x)$, according to:

$$\text{Fourier Coefficients} : B_n = \frac{2}{\pi} \int_0^\pi \frac{t[x(\Theta)]}{c} \sin n\Theta\, d\Theta \tag{4.29}$$

It is useful to realize that the airfoil cross-sectional area, $A_t$, is governed by the first mode only as:

$$\text{Airfoil Area} : A_t = \int_0^c t(x)dx = \frac{c}{2} \int_0^\pi t[x(\Theta)] \sin \Theta\, d\Theta = \frac{\pi}{4} c^2 B_1 \tag{4.30}$$

This means in particular that in the special case of an elliptical thickness distribution with $B_2 \dots B_n = 0$, the source-sheet strength, $\lambda$, in Eq. (4.27) simplifies significantly as it becomes essentially an exclusive function of the airfoil cross-sectional area, $A_t$. This has been shown a good approximation even for non-elliptical thickness distributions, see Appendix D.4 of Drela (2014). Subsequent superposition of vortex-/source-sheet strengths in the fundamental equation, it can be shown that for symmetric airfoils, the effect of airfoil thickness on the lift-curve slope is found to be:

$$\text{Lift-Curve Slope (Airfoil)} : \frac{dc_l}{d\alpha} = \frac{2\pi}{1 - (4/\pi)\, A_t/c^2} \tag{4.31}$$

This means in particular that additional flow acceleration due to thickness over the upper airfoil surface at positive angle of attack results in some additional lift, roughly

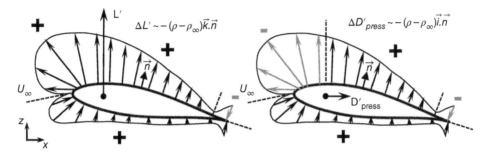

**Figure 4.5** Pressure integration around airfoil (inviscid).

proportional to the airfoil thickness. We have to note that Eq. (4.31) is strictly valid only for a symmetric airfoil with an elliptical thickness distribution and in inviscid flow; however, Eq. (4.31) has been shown to agree quite well with practically exact values computed by panel methods. In the end, however, the benefit of airfoil thickness on lift-curve slope is annihilated by boundary-layer displacement and the associated viscous decambering effect as discussed in Section 4.2.

### 4.1.1.4 d'Alembert's Paradox

A seemingly true paradox that baffled aerodynamicists for generations is due to d'Alembert who found that the drag force, $D$, (or streamwise resistance force) on a closed body immersed in inviscid and irrotational flow is always zero. This was indeed a paradox before the viscous flow equations were derived by Navier and Stokes more than one hundred years later, the reason being that it stood in contrast to observations.

Figure 4.5 illustrates that integrating the pressure distribution around an airfoil in the streamwise direction is a fine balance of positive and negative contributions, while contributions of suction-/pressure sides are (in general) additive for the lift force.

In a frictionless flow, the pressure distribution does indeed integrate to zero drag at all times. While it seems surprising at first, one can also make a simple argument that the drag force on a body immersed in inviscid and irrotational flow has to be zero as no work is being done in the streamwise direction of speed $U_\infty$.

### 4.1.2 Viscous Flow: Boundary-Layer Theory

This section gives an introduction to boundary-layer flows relevant to airfoil aerodynamics. The reader will learn about the primary boundary-layer parameters and behavior as well as how both determine viscous losses and profile drag. The boundary layer around an airfoil is either *laminar* and/or *turbulent*, with the former being characterized by smooth shear flow with parallel streamlines and the latter by turbulent mixing and associated statistical variations about mean flow speeds. Figure 4.6 illustrates the development of the boundary layer around an airfoil and in the wake.

Most streamlined bodies such as airfoils have *laminar* boundary layers developing from the leading-edge stagnation point. *Transition* to a *turbulent* boundary layer occurs when relative viscous stresses are not strong enough to damp flow perturbation frequencies that promote eddy length scales on the order of the local boundary-layer height (or thickness), $\delta$, see also Section 4.1.2.6. The chordwise location at which *laminar to turbulent transition* occurs can be to some extent controlled by appropriate airfoil design but

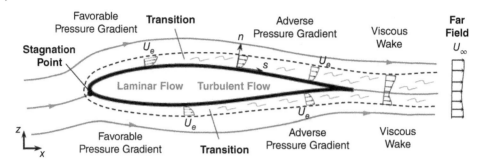

**Figure 4.6** General viscous flow around an airfoil.

occurs, in general, more aft on the lower than on the upper airfoil surface. As a general rule-of-thumb, flat-plate boundary layers (zero pressure gradient) transition from laminar to turbulent flow at a local Reynolds number of $Re_l \approx 5 \times 10^5$, see Schlichting (1979). Both upper- and lower-surface boundary layers merge at the airfoil TE into a wake that is practically turbulent in most cases (except for very low $Re$). The airfoil profile drag, $c_d$, is related to the momentum deficit in the far-downstream wake.

Let us proceed by considering the Navier–Stokes equations for two-dimensional, steady, incompressible, and viscous flow:

$$\text{Navier-Stokes equations : } \vec{\nabla}.\vec{u} = 0 \tag{4.32}$$

$$\vec{u}.\vec{\nabla}\vec{u} = -\frac{1}{\rho_\infty}\vec{\nabla}p + v\vec{\nabla}^2\vec{u} \tag{4.33}$$

Here, the velocity vector, $\vec{u} = (U, w')^T$, with $U = U_\infty + u'$ and $u'$, $w'$ being perturbation velocities. Note that $\vec{\nabla} = \partial/\partial x\, \vec{i} + \partial/\partial y\, \vec{j} + \partial/\partial z\, \vec{k}$ is the nabla operator. Furthermore, $v = (\mu + \mu_t)/\rho_\infty$ is the total kinematic viscosity composed of the sum of molecular and turbulent-eddy dynamic viscosities, $\mu$ and $\mu_t$, and normalized by the freestream density, $\rho_\infty$. Normalizing length scales by the airfoil chord, $c$, velocities by the freestream speed, $U_\infty$, and the pressure by a reference dynamic head, $\rho_\infty U_\infty^2$, the continuity equation is unaltered but the momentum Eq. (4.33) can be written as:

$$\text{Momentum Equation (Normal) : } \underline{\vec{u}.\vec{\nabla}\,\vec{u}} = -\underline{\vec{\nabla}p} + \frac{1}{Re_c}\underline{\vec{\nabla}^2\vec{u}} \tag{4.34}$$

In Eq. (4.34), $Re_c = U_\infty c/v$ is the airfoil chord-based Reynolds number as referenced in Figure 4.1. The normalized momentum equation is helpful in understanding the relative importance of the viscous diffusion term $\sim 1/Re_c$ that becomes of the order "one" compared to the convective acceleration and pressure-gradient terms if the curvature of the velocity-vector field normal to the airfoil surface is of order $Re_c$ itself. Given a no-slip condition with $\vec{u} = \vec{0}$ on the airfoil surface, this implies that the extent of the viscous region normal to the airfoil surface is inversely proportional to $Re_c$. In other words, the larger $Re_c$, the thinner the viscous region, that is, the boundary layer. Nevertheless, the no-slip condition ensures that a boundary layer exists at all times in viscous flow, even in the limit of very high Reynolds number.

### 4.1.2.1 Boundary-Layer Displacement Effect

The boundary-layer thickness, $\delta$, describes the extent of the viscous flow region normal to the airfoil surface. Typically, the exact criterion defines $\delta$ being the normal distance from the airfoil surface to where the shear-layer velocity becomes 99% that of the outer (external to boundary layer) inviscid-flow velocity, $U_e$. An order-of-magnitude estimate for $\delta$ can be obtained by normalizing the vertical coordinate, $z$, by the boundary-layer thickness, $\delta$, in Eq. (4.33) and postulating that, in the boundary layer, convective acceleration and pressure terms are of the same order as the viscous diffusion term, that is, $U^2/x \sim \nu U/\delta^2$. Hence the boundary-layer thickness, $\delta$, along a solid surface can be simply estimated for a laminar boundary layer to be:

$$\text{Boundary-Layer Thickness (Laminar)} : \delta(x) \sim \frac{x}{\sqrt{Re_x}} \tag{4.35}$$

For turbulent flow, on the other hand, the boundary layer is typically thicker due to the action of the turbulent-eddy viscosity and associated mixing of momentum from the outer flow down into the boundary layer. As evidenced by Eq. (4.35), high Reynolds number boundary layers are thin compared to the streamwise reference length scale. This allows making some useful thin shear-layer approximations:

$$\text{Thin Shear-Layer Approximations} : w' \approx \frac{\delta}{c} \cdot U$$

$$\frac{\partial U}{\partial s} \approx \frac{\partial U}{\partial n}$$

$$\frac{\partial p}{\partial n} \approx 0 \tag{4.36}$$

In Eq. (4.36), we introduced local element coordinates, with $s$ being the coordinate tangential to a surface element and $n$ normal to a local surface element, respectively. The thin shear-layer approximations in Eq. (4.36) are a direct consequence of the normalized Navier–Stokes equations (4.32)–(4.33), with the primary observations that the vertical (normal) perturbation velocity, $w'$, is proportional to $\delta$ and that the pressure is nearly constant through the boundary layer, with the latter enabling viscous-inviscid solution methodologies as discussed further next. The boundary-layer equations, as a reduced form of the Navier–Stokes equations subject to the thin shear-layer approximations, then become:

$$\text{Boundary-Layer Equations} : \frac{\partial U}{\partial s} + \frac{\partial w'}{\partial n} = 0$$

$$\rho_\infty U \frac{\partial U}{\partial s} + \rho_\infty w' \frac{\partial U}{\partial n} = \rho_\infty U_e \frac{dU_e}{ds} + \frac{\partial \tau}{\partial n}$$

$$\tau = (\mu + \mu_t) \frac{\partial U}{\partial n} \tag{4.37}$$

In Eq. (4.37), the pressure-gradient term was replaced by convective acceleration of the boundary-layer edge velocity, $U_e$, that is, $\partial p/\partial s \approx \rho_\infty U_e dU_e/ds$. The boundary-layer equations are notably easier to solve than the full Navier–Stokes equations and some self-similar solutions to the boundary-layer equations do exist, see Schlichting (1979). For example, the boundary-layer thicknesses along laminar/turbulent flat plates under zero pressure gradient become:

$$\text{Laminar Flat Plate } (\partial p/\partial x = 0) : \delta(x) \approx \frac{5.0x}{\sqrt{Re_x}} \tag{4.38}$$

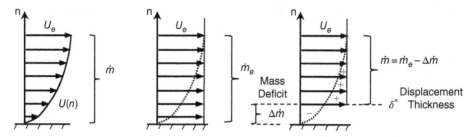

**Figure 4.7** Displacement body model in viscous flow.

$$\text{Turbulent Flat Plate } (\partial p/\partial x = 0) : \delta(x) \approx \frac{0.37x}{Re_x^{1/5}} \text{ (empirical)} \tag{4.39}$$

Though Eqs. (4.38) and (4.39) are strictly valid only for a zero pressure gradient, they can serve as estimates for the physical boundary-layer thickness as a function of chord-based Reynolds number, $Re_c$. As a rule-of-thumb, a freely developing boundary layer stays laminar until $Re_x \cong 5 \times 10^5$ along an airfoil ($0 \le x \le c$) and for notably higher $Re_x$ for carefully designed airfoils.

Due to the action of the viscous no-slip condition with $\vec{U} = \vec{0}$ on the airfoil surface, the viscous boundary layer transports less mass flow in the streamwise direction when compared to an inviscid (frictionless) free-slip condition, see Section 4.1.1. For quantitative comparisons between inviscid and viscous flow involving control-volume analysis, it is therefore convenient to "displacing" the original airfoil surface by a distance, $\delta^*$, normal to the surface such that the mass deficit is accounted for. An illustration of the displacement thickness, $\delta^*$, is given in Figure 4.7.

A similar "displacement" argument can be made for a deficit in transported flow momentum due to the action of the no-slip condition. Definitions for displacement thickness, $\delta^*$, and momentum thickness, $\Theta$, respectively, are as follows:

$$\text{Displacement Thickness} : \delta^* = \int_0^\delta \left(1 - \frac{U}{U_e}\right) dy \tag{4.40}$$

$$\text{Momentum Thickness} : \Theta = \int_0^\delta \frac{U}{U_e}\left(1 - \frac{U}{U_e}\right) dy \tag{4.41}$$

A direct consequence of boundary-layer displacement is that in a coupled viscous-inviscid flow field, the resultant pressure becomes that around the "displaced" body rather than the original body (or airfoil). Hence, under a very good approximation that the pressure is constant across the boundary-layer height, d'Alembert's paradox in inviscid flow no longer holds and a so-called pressure drag (or form drag) results from boundary-layer displacement. Integrating the boundary-layer equations from the surface to the edge of the boundary layer and using some algebra results in the momentum integral relation:

$$\text{Momentum Integral Relation} : \frac{d\Theta}{ds} + (2 + H)\frac{\Theta}{U_e}\frac{dU_e}{ds} = \frac{c_f}{2} \tag{4.42}$$

In Eq. (4.42), $H$ is the shape factor and $c_f$ is the skin friction coefficient defines as:

$$\text{Shape Factor} : H = \frac{\delta^*}{\Theta} \tag{4.43}$$

$$\text{Skin Friction Coefficient}: c_f = \frac{\tau_w}{\frac{1}{2}\rho_\infty U_e^2} = \frac{\mu(\partial U/\partial n)_w}{\frac{1}{2}\rho_\infty U_e^2} \tag{4.44}$$

The momentum integral relation in Eq. (4.42) is the basis for viscous-inviscid inter-action codes, such as XFOIL (Drela 1989). Let us remember that the boundary-layer equations are much easier to solve than the Navier–Stokes equations. As the pressure is effectively constant across the boundary layer, following the thin shear-layer approximation in Eq. (4.36), the pressure around the displaced airfoil can be computed from inviscid theory, that is, either potential-flow theory or the Euler Equations (inviscid Navier–Stokes with free-slip boundary condition). As a next step, the boundary-layer equations can be solved with a given velocity, $U_e$, at the edge of the boundary layer, and it can be checked as to whether or not the displacement thickness, $\delta^*$, integrated from the current boundary-layer solution is equal to the one assumed previously in the outer inviscid flow solution. This is the basic principle for all viscous-inviscid solution methodologies used in airfoil/wing aerodynamics. Typical values for the shape factor are $H = 2.6$ at the edge of a laminar boundary layer, while $H \approx 1.4$ at the edge of a turbulent boundary layer. The total profile drag is the sum of pressure (form) drag and skin friction drag, see also Eq. (4.1) and Section 4.1.2.5.

### 4.1.2.2 Viscous Lift Theorem

In addition to its effect on airfoil drag, the viscous boundary layer also has some notable effects on the generation of lift, which are discussed in this section. Let us consider a simply connected domain, $\Omega$, around an airfoil as depicted in Figure 4.8, with boundary $\partial\Omega = (\Sigma I) + (\Sigma O) + (\Sigma B) + (\Sigma W)$. Here $(\Sigma I)$ denotes the inviscid-flow part of $\partial\Omega$, $(\Sigma O)$ the viscous wake plane, $(\Sigma B)$ an equal-and-opposite branch cut, and $(\Sigma W)$ the airfoil surface.

Following control-volume analysis as, for example, in Wu et al. (2006), Marongiu and Tognaccini (2010), or Schmitz (2014), and using Stokes' theorem and the definition

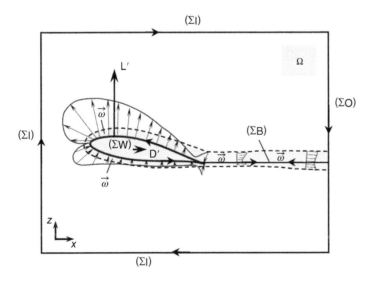

**Figure 4.8** Control-volume analysis of viscous airfoil flow.

of the vorticity vector defined as $\vec{\omega} = \vec{\nabla} \times \vec{U}$, the lift force, $L'$, can be found as a surface/volume integral of the vortex force as:

$$\text{Vortex Lift} : L' = -\rho_\infty \vec{k} \cdot \oiint_\Omega (\vec{\omega} \times \vec{U}) \cdot d\vec{\Omega} = \rho_\infty \oiint_\Omega U\omega_y d\Omega \qquad (4.45)$$

In Eq. (4.45), the Lamb vector $\vec{\omega} \times \vec{U}$ is essentially an elemental Kutta–Joukowski lift theorem, though not applied along a vortex-sheet analogy as introduced in inviscid theory in Section 4.1.1 but to all fluid particles in the thin viscous boundary layer where the flow is rotational, that is, $\vec{\omega} \neq \vec{0}$. Considering the governing component of the vorticity vector, $\omega_y = \frac{\partial u'}{\partial z} - \frac{\partial w'}{\partial x}$, it becomes clear that the boundary layer itself is a thin layer (or sheet) of vorticity responsible for the generation of lift, though distributed off the airfoil surface, with the side effect of an associated drag force. Note that $\omega_y$ is governed by the viscous shear gradient, $\partial u'/\partial z$, being maximum at the airfoil surface and tending to zero at the edge of the boundary layer. Using $U = U_\infty + u'$ and Stokes' theorem, a viscous lift theorem can be formulated as:

$$\text{Viscous Lift Theorem} : L' = \rho_\infty U_\infty (\Gamma + \Delta\Gamma)$$

$$\Delta\Gamma = \frac{1}{U_\infty} \oiint_\Omega u'\omega_y d\Omega \qquad (4.46)$$

In Eq. (4.46), $\Delta\Gamma$ is a viscous circulation correction resulting from a perturbation Lamb vector. Following the work of Schmitz (2014), using mass conservation with $\vec{\nabla}.\vec{U} = 0$, the integrand $u'\omega_y$ can be written in conservation form as:

$$u'\omega_y = -\frac{\partial}{\partial x}(u'w') + \frac{1}{2}\frac{\partial}{\partial z}(u'^2 - w'^2) \qquad (4.47)$$

Subsequent application of Greens' theorem yields:

$$\Delta\Gamma = -\frac{1}{U_\infty} \left[ \int_{(\Sigma I)} \frac{1}{2}(u'^2 - w'^2)dx + u'w'dz + \int_{(\Sigma O)} u'w'dz \right] \qquad (4.48)$$

For $(\Sigma I)$ being far away from the airfoil in inviscid flow, $u'$ and $w'$ behave as $\sim 1/R$ according to potential-flow theory where in this case, we refer to $R$ being the distance from the airfoil quarter-chord, for example. As the path of integration along $(\Sigma I)$ is $\sim R$, the integral of second-order perturbation terms along $(\Sigma I)$ thus tends to zero with $1/R$ or faster. This, however, is not true for laminar and/or turbulent wakes, see Schlichting (1979). Hence the viscous circulation correction becomes an integral along the viscous wake plane, $(\Sigma O)$, as:

$$\text{Viscous Circulation Correction} : \Delta\Gamma = -\frac{1}{U_\infty} \int_{(\Sigma O)} u'w'dz \qquad (4.49)$$

Next, using the thin shear-layer approximation with $w' \sim U(\delta/c)$ and assuming that the streamwise velocity in the viscous part along $(\Sigma O)$ behaves as $U \sim U_e(z/\delta)^a$ with $a > 1$, one can estimate the viscous circulation correction, $\Delta\Gamma$, at the airfoil TE as:

$$\text{Viscous Circulation Correction (TE)} : \Delta\Gamma \sim -\frac{U_e^2 c}{U_\infty}\left(\frac{\Delta\delta_{TE}}{c}\right)^2 \qquad (4.50)$$

In Eq. (4.50), we denote by $\Delta\delta_{TE} = \delta_u - \delta_l$ the difference between upper and lower boundary-layer thickness, $\delta_u$ and $\delta_l$, at the airfoil TE. Subsequent application of Eq. (4.7) results in an order-of-magnitude relation for a corresponding change in airfoil lift coefficient, $\Delta c_l$, due to viscous boundary-layer displacement:

$$\text{Viscous } c_l \text{ correction} : \Delta c_l \sim -2 \left( \frac{U_e}{U_\infty} \right)^2 \left( \frac{\Delta \delta_{TE}}{c} \right)^2 \tag{4.51}$$

Equation (4.51) reveals that lift reduction due to boundary-layer displacement is essentially a function of the extent of the viscous region and hence chord-based Reynolds number, $Re_c$, but also of whether or not the boundary layer is attached or separated. The viscous circulation and $c_l$ corrections can also be interpreted as a quantification of the viscous decambering effect, briefly described in the next section. For attached flow and $Re_c = O(10^6)$, the viscous circulation and lift corrections are on the order of 1% of the total circulation, see Schmitz (2014).

### 4.1.2.3 Viscous Decambering Effect

A consequence of the viscous circulation correction is that the lift-curve slope in viscous flow deviates from its inviscid thin-airfoil value of $2\pi$. For a lift-generating airfoil in viscous flow, the boundary-layer thickness at the TE of the upper surface is usually larger than its counterpart on the lower surface, that is, $\delta_u > \delta_l$. One way to interpret this is to consider the upper surface as more "displaced" due to higher velocity, $U_e$, at the edge of the boundary layer, and hence the effective TE being moved a bit upward. Consequently, this can be either seen as a reduction in angle of attack or a "viscous decambering" effect, with Eq. (4.51) giving some order-of-magnitude idea of the resulting effect. Figure 4.9 shows a qualitative viscous lift curve compared to an inviscid lift curve of an airfoil with the same camber.

The progressive deviation from the inviscid lift curve can be seen in Figure 4.9. The viscous decambering effect increases notably once flow separation and stall occur over the upper airfoil surface, leading to a significant increase in boundary-layer thickness at the TE of the upper airfoil surface. Also shown in Figure 4.9 are dashed lines going through sample points with the inviscid thin-airfoil $2\pi$ slope, thus indicating the effective loss of camber (or increased zero-lift angle) associated with viscous effects.

### 4.1.2.4 Flow Separation and Stall

In viscous flow, the lift curve shown in Figure 4.9 is subject to a viscous decambering effect that increases with the angle of attack, $\alpha$, as the surface boundary layer(s) thicken.

**Figure 4.9** Lift curve in inviscid and viscous flow.

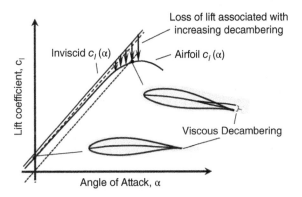

For wind turbine airfoils, the flow typically starts to separate from the upper-surface TE, and the separation point moves toward the leading edge with increasing angles of attack, $\alpha$. In general, flow separation starts to occur prior to reaching the maximum lift coefficient, $c_{l,\,max}$. Stall is associated with a rapid forward movement of the separation point and a notable loss in lift post $c_{l,\,max}$. Wind turbine airfoils are usually designed to have a rather soft stall behavior (see Section 4.2), which comes in many cases at the expense of a lower $c_{l,\,max}$ compared to airfoils experiencing sharp leading-edge stall. In the post-stall regime, the lift coefficient increases again with angle of attack, practically following the behavior of a stalled flat plate at high incidence.

In the following, let us dissect the physics of flow separation. A separated flow region is characterized by a separation (or recirculation) bubble, with reverse flow (opposite to freestream flow) close to the airfoil surface. In essence, a separation bubble is an enclosed flow region "separated" from the outer flow and practically a large displacement of the original body/airfoil when viewed from the outer inviscid flow. In general, a boundary layer around an airfoil can separate due to the prolonged action of an adverse pressure gradient. An illustration is given in Figure 4.10.

*Favorable Pressure Gradient, $\partial p/\partial s < 0$.* This corresponds to an acceleration of the outer (external to the boundary layer) flow; that is, $dU_e/ds > 0$. As the pressure is approximately constant across the boundary-layer depth, the same accelerating force applies to all fluid particles inside the boundary layer, though affecting the lower-momentum fluid elements close to the surface relatively more than those at the edge of the boundary layer. A favorable pressure gradient occurs over the fore part of an airfoil and adds momentum and energy to the boundary layer.

*Adverse Pressure Gradient, $\partial p/\partial s > 0$.* An adverse pressure gradient has to always follow a favorable pressure gradient as the flow has to be recompressed to a rear stagnation

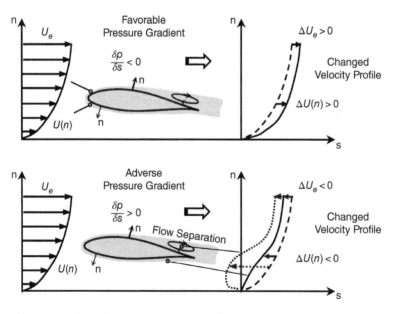

**Figure 4.10** Effect of pressure gradient on airfoil boundary-layer dynamics.

point. Hence an adverse pressure gradient corresponds to a deceleration of the outer inviscid flow, that is, $dU_e/ds < 0$. Again, the same pressure gradient acting on fluid particles across the boundary layer has a larger effect closer to the surface than at the boundary-layer edge. If a sufficiently strong adverse pressure gradient acts over large enough chordwise distance, boundary-layer flow close to the surface may have lost all its streamwise momentum, thus ultimately resulting in *flow reversal* and hence separation. From a displacement perspective, reversed flow leads to a large mass deficit in the boundary layer and thus naturally to a notable increase in the displacement thickness with all its effects on lift and drag.

As the angle of attack, $\alpha$, increases, so do magnitude of favorable and adverse pressure gradient over the upper airfoil surface, with the region of favorable pressure gradient moving progressively closer to the leading edge. The consequence is a prolonged region of adverse pressure gradient that will eventually cause flow separation for all lifting bodies. It is here where airfoil design methodologies do have an impact on separation/stall behavior by affecting the extent of the pressure-recovery region as well as the magnitude of the adverse pressure gradient by, for example, moving the chordwise location of maximum thickness toward the TE. If properly designed, such an airfoil may exhibit softer separation/stall behavior but may, at the same time, have a reduced maximum lift coefficient, $c_{l,max}$.

### 4.1.2.5 Understanding Profile Drag: Pressure and Skin Friction

In reference to Eq. (4.1), profile drag results from integrating pressure, $p$, and wall (surface) shear stress, $\tau_w$, over the closed airfoil contour. Note that in inviscid (frictionless, $\tau_w = 0$) flow, the profile drag remains zero due to d'Alembert's paradox, see Section 4.1.1.4. In viscous flow, however, the original body (airfoil) is displaced as a result of a mass deficit in the boundary layer, and the resulting pressure distribution along the closed airfoil contour is conceptually that around a "displaced" airfoil. Consequently, the fine balance of positive/negative elemental pressure forces around the airfoil, resulting in d'Alembert's paradox in inviscid flow, no longer holds as the pressure does not change across the boundary layer, that is, from the outer "displaced" pressure to the actual airfoil surface. The result is what is referred to as form (pressure) drag and, as it is related to the displacement thickness, increases with angle of attack. In general, form (pressure) drag is $\sim c_l^2$ as a good approximation in attached flow. This is shown quantitatively in Figure 4.11 for a NACA 4412 airfoil at $Re = 1 \times 10^6$ computed with *XFOIL* (Drela 1989).

On the other hand, skin-friction drag due to wall (surface) shear stress, $\tau_w$, remains fairly constant with $c_l$, respectively, $\alpha$, and is primarily inversely proportional to some power of the Reynolds number, $Re$. As the flow separates and stalls, it is interesting to note that skin-friction drag might actually decrease a bit due to the associated reverse flow and hence reverse wall (surface) shear stress, $\tau_w$. Though somewhat appearing as a secondary effect, skin friction and viscous-stress diffusion in the boundary are the physical cause of generating vorticity and hence airfoil lift (Stokes' theorem). Note that the vorticity-vector component $\omega_y$ is governed by the viscous shear gradient, $\partial u'/\partial z$, being maximum at the surface and tending to zero at the edge of the boundary layer. Thus, the boundary layer can also be interpreted as a (thin) distributed vortex sheet off the airfoil surface. Hence viscous stresses are the actual physical mechanism to generate

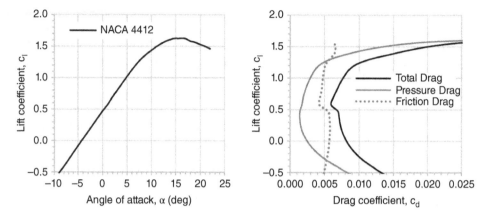

**Figure 4.11** Integrating surface pressure and viscous stress (NACA 4412, $Re = 1 \times 10^6$, computed with *Xfoil*).

vorticity, vortex sheets, and a resulting total circulation, $\Gamma$, "bound" to the airfoil. In the viscous wake downstream of the TE, no additional vorticity is generated but the merged vertical (rotational) shear layers from upper-/lower airfoil surfaces are then subject to viscous diffusion and dissipation.

### 4.1.2.6 Laminar-Turbulent Transition

Transition from laminar to turbulent boundary layers on airfoils is, at least in most cases, caused by time-varying external freestream disturbances or by surface vibrations themselves. A stable laminar shear layer is able to dissipate flow disturbances up to a certain development length (or $Re_{X_{tr}}$), beyond which flow disturbances start amplifying and providing the momentum and energy to generate turbulent-eddy length scales on the order of the boundary-layer thickness, $\delta$. This is in general described as the *receptivity problem* (Saric et al. 2002). In the following, let us introduce the three broadly defined types of transition mechanisms relevant to wind turbine aerodynamics:

*Forced Transition.* Any geometric feature on the surface with a height on the order of the boundary-layer thickness, $\delta$, acts as a strong receptor such that external flow disturbances can enter the boundary layer. This can be caused, for example, by surface contamination (insects, dirt, rivet lines) or an intentionally placed trip strip, all of which trigger laminar-turbulent transition a short chordwise distance downstream.

*Natural Transition.* This type of transition occurs in general on smooth surfaces in quiet (low-turbulence) flow. Here external disturbances are very weak, and any initial disturbances can only be amplified by natural *flow instabilities* whose amplitude subsequently increases exponentially downstream and becomes chaotic at the actual transition point. One type of exponentially growing instabilities are two-dimensional *Tollmien-Schlichting (TS) waves*, see also Schlichting (1979). They are essentially sinusoidal oscillations of pressure and velocity perturbations inside the boundary layer. These are initially (meaning close to the leading edge) very weak oscillations, but can grow exponentially with chordwise distance until they are strong enough to trigger laminar-turbulent transition. The so-called "N-factor" quantifies the instability growth in an "envelope $e^N$" transition prediction method by an accumulating integral over the upstream surface, see some classical works of Smith and Gamberoni

(1956) and van Ingen (1956). In particular, $N$ is the ratio of the initial disturbance to the final one causing transition. An $e^N$ method relies on linear stability theory and the Orr-Sommerfeld equation. Transition occurs at the chordwise location where the "N-factor" reaches a critical value determined by the ambient noise and disturbance level. A typical value is $N = 9$ such that the final disturbance is $e^N = 8103$ times the initial one. The shape factor, $H$, in Eq. (4.43) plays an important role in empirical relations for the integral that determines the amplification factor, see for example, Drela and Giles (1987). As an approximate rule-of-thumb and typical $H$ values, the streamwise Reynolds number at which laminar-turbulent transition occurs is around $Re_{x_{tr}} \approx 5 \times 10^5$ on a flat plate and can increase notably in the presence of a favorable pressure gradient on a well-designed airfoil. This has important implications on scaled wind turbine aerodynamics. The interested reader is also referred to Reed and Saric (1996) for a comprehensive review on the linear stability theory of boundary layers.

*Bypass Transition.* This type of transition occurs only in very noisy external flows with high inflow turbulence that is sufficiently strong by itself to enter the boundary layer. Here the unstable TS waves are simply "bypassed" such that turbulent spots and streaks are formed directly in the boundary layer. A well-known example in aeronautics is that of bypass transition in turbomachinery. For wind turbines operating in the atmospheric surface layer, bypass transition can occur if sufficient spectral energy is in eddy frequencies that the blade/airfoil boundary layer responds to, particularly in an unstable atmosphere or for wind turbines operating in the wakes of upstream turbines.

For utility-scale wind turbines, laminar-turbulent transition will occur due to one of the mechanisms described previously. It is therefore important to have at least a basic understanding of the effects of transition on airfoil characteristics. The chordwise transition location has a considerable effect on drag, and also on maximum lift and stall behavior (though not for all airfoils). Note that a turbulent boundary layer is in general associated with a higher skin-friction coefficient, $c_f$, compared to its laminar counterpart, simply because of a "fuller" velocity profile close to the surface. A simple example is shown in Figure 4.12 for a NACA 4412 airfoil polar at $Re = 1 \times 10^6$ computed with *XFOIL*.

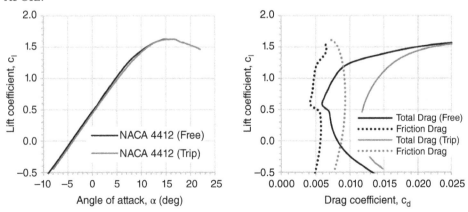

**Figure 4.12** Effect of free versus forced transition on airfoil drag coefficient, $c_d$, and lift-to-drag ratio, $c_l/c_d$ (NACA 4412, $Re = 1 \times 10^6$, computed with *XFOIL*).

The effect of forced versus free transition is seen as a notable reduction in the maximum lift-to-drag ratio, $c_l/c_d$, which can be estimated by drawing a tangent to the "total drag" curves in Figure 4.12. This in turn can have some significant impact on the maximum attainable power coefficient, $C_{P,max}$, as shown in Figure 3.14, and hence on the annual energy production (AEP) of a given wind turbine design.

## 4.2 Design Characteristics of Wind Turbine Airfoils

One primary advantage of wind produced energy is its scalability, being able to power single homes with a small wind turbine to entire neighborhoods with one utility-scale wind turbine. The rotor diameter and turbine hub height range from less than 10 m to more than 100 m. For an optimal tip speed ratio, $\lambda$, around 6–8, a small wind turbine spins at several hundred rpm, while a utiliy-scale wind turbine rotor operates around 10 rpm. In the following, we will derive some simple relations that give us an idea of the radial variation of Reynolds number, $Re$, along wind turbine blades of different radius.

### 4.2.1 Radial Variation of the Reynolds Number

The local chord-based Reynolds number, $Re_c$, at a radial blade element was first defined in Eq. (3.2) as:

$$\text{Reynolds number}: Re_c = \frac{\rho\, V_{rel}\, c}{\mu}$$

In the following analysis, we assume close to sea-level conditions with an air density of $\rho = 1.225$ kg m$^{-3}$ and a dynamic viscosity of $\mu = 1.8 \times 10^{-5}$ kg ms$^{-1}$. Note that air density is actually a function of elevation, see Eq. (1.25), and that the dynamic viscosity, $\mu$, is primarily a function of temperature, $T$, and therefore also of elevation according to an assumed adiabatic lapse rate, see Eq. (1.20). The relative velocity magnitude, $V_{rel}$, at a given blade element is determined by Eq. (3.6) as:

$$\text{Relative Velocity}: V_{rel}(r) = V_0\sqrt{(1-a(r))^2 + (1+a'(r))^2\lambda_r^2}$$

As representative first-order examples, let us consider two blade lengths:

- Small Scale (S): $R_S = 5$ m
- Utility Scale (U): $R_U = 50$ m

We further assume that the root radius, $r_{Root}$, is approximately 15% of the blade radius (or length), $R$. Some additional typical operating conditions and blade parameters are:

**Operating Conditions**

- Wind Speed: $V_0 = 7$ m s$^{-1}$
- Tip Speed Ratio: $\lambda = 6$

**Blade Parameters**

- Root Radius: $r_{Root}/R = 0.15$
- Root Chord: $c_{Root}/R = 0.15$

- Tip Radius: $r_{Tip}/R = 1.00$
- Tip Chord: $c_{Tip}/R = 0.05$

Note that values listed previously are used for the sole purpose of $Re$ estimates along the blade radius, with no particular blade planform in mind. As far as axial-/angular induction factors, $a$ and $a'$, at the blade root and tip are concerned, we assume ideal flow and use Figure 2.10 to find:

**Induction Factors (Ideal, $\lambda = 6$)**

- Blade Root: $a' = 0.2$, $a = 0.31$
- Blade Tip: $a' = 0.01$, $a = 0.33$

In general, we denote a dimensionless radial station by $r/R$ and blade chord by $c/R$. The corresponding blade section Reynolds number then becomes:

$$\text{Blade Section } Re_c : Re_c = \frac{1}{2}\frac{c}{R}\sqrt{(1 - a(r))^2 + (1 + a'(r))^2\lambda_r^2} \cdot Re_D \tag{4.52}$$

In Eq. (4.52), $\lambda_r = \lambda\frac{r}{R}$ is the local speed ratio and $Re_D = (2\,\rho\,V_0\,R)/\mu \approx 4.75 \times 10^5\,D$ is a Reynolds number based on the rotor diameter. Note that the actual maximum Reynolds number along the blade radius occurs around the $r/R = 0.75$ station, assuming a linear taper distribution between blade root and tip. Hence Eq. (4.52) can be used to compute some general $Re$ estimates along the blade radius as:

$$\text{Blade Root } Re : Re_{Root} \approx 0.10Re_D$$
$$\text{Blade Tip } Re : Re_{Tip} \approx 0.15Re_D$$
$$\text{Max. Blade } Re : Re_{75\%R} \approx 0.20Re_D \tag{4.53}$$

For an assumed small-scale (S) and utility-scale (U) wind turbine, we thus obtain:

$$\text{Small-Scale (S)} : Re_{Root} \approx 4.8 \times 10^5, \ \ Re_{Tip} \approx 7.1 \times 10^5 \tag{4.54}$$

$$\text{Utility-Scale (U)} : Re_{Root} \approx 4.8 \times 10^6, \ \ Re_{Tip} \approx 7.1 \times 10^6 \tag{4.55}$$

Note again that Eqs. (4.54) and (4.55) should be understood as "rough" estimates to some extent; however, they provide an accurate order-of-magnitude analysis that has to be seen in conjunction with associated viscous displacement effects, profile drag, and maximum attainable lift-to-drag ratio, $c_l/c_d$. It is important to realize that small-scale wind turbines operate at section Reynolds numbers around a typical laminar-turbulent transition Reynolds number, $Re_{Xtr}$. Hence a notable run of laminar flow can be expected, which is reduced (though still present and important for $c_l/c_d$ considerations) along utility-scale wind turbine blades.

## 4.2.2  Force/Torque and Velocity Triangle Along the Blade Radius

Prior to specifying general design criteria for wind turbine airfoils, it is useful to consider again the force/torque and velocity triangles local to a blade element, see Figure 3.3, though with some differentiation between the blade root, mid-blade stations, and blade tip. Figure 4.13 shows the respective triangles, assuming a constant axial induction factor, $a$, along the blade radius.

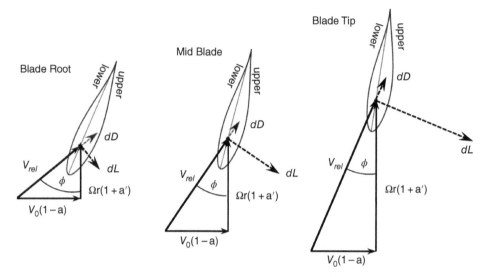

**Figure 4.13** Force/torque and velocity triangles along blade radius.

Here we bear in mind that simplified BEM theory (see Section 3.6) suggests that the power coefficient, $C_p$, can be maximized for a high lift-to-drag ratio, $c_l/c_d$, really though for minimizing $c_d/c_l \cot \phi$. Assuming constant axial induction along the blade radius, it can be seen from Figure 4.13 that local force/torque and velocity triangles change their respective orientation primarily due to the local speed ratio, $\lambda_r$, and associated angular induction, $a'$. At the blade tip and at mid-blade stations, projecting section lift and drag forces into the rotor plane (i.e. in the direction of torque) does indeed suggest that high lift-to-drag ratios, $c_l/c_d$ are desirable as the profile-drag component has to be overcome to have a torque surplus to generate power. This situation is not as clear, though, at the blade root where the section drag force has a comparatively smaller torque-reducing component compared to mid-blade and tip sections. This is actually helpful with respect to structural design considerations and "thick" inboard airfoils that naturally have higher profile drag. We further note that inboard blade sections experience a substantially lower relative velocity head, $V_{rel}^2$. Hence section lift forces, in particular, are scaled down as $\sim(\lambda_r/\lambda)^2$; however, part of this deficit can be overcome by operating inboard blade section at higher design lift coefficients, $c_l$, compared to outer blade sections, also because the resulting additional increase in profile drag does not seem to affect power generation too adversely.

### 4.2.3 Airfoil Design Criteria for Wind Turbine Blades

From simple arguments on sectional force/torque and velocity triangles, we can therefore conclude with some preliminary design criteria for wind turbine airfoils:

**Basic Airfoil Design Criteria (Variable-Speed/Pitch Rotors)**

- *Blade Root.* High $c_{l,max}$
- *Mid-Blade.* High $c_{l,max}$, High $c_l/c_d$
- *Blade Tip.* High $c_l/c_d$

In this regard, an interesting observation can be made for the radial $c_l$ distribution along the Penn State University (PSU) 1.5-MW turbine blade in Region II/III, see Figure 3.25 in Section 3.7.6, in conjunction with Figure 4.13. As the blade "pitches in" to control rotor power in Region III, the lift coefficient, $c_l$, decreases notably for $r/R > 0.5$; however, the lift coefficient stays high for the inboard blade half. Note that in Region III, the rotor operates at constant rpm. Consequently, the increased blade pitch inboard (reducing section $\alpha$ and $c_l$) is actually compensated by a lower $\lambda_r$ (increasing section $\alpha$ and $c_l$). This behavior at inboard blade sections of pitch-controlled wind turbines supports the design of thick airfoils with high $c_{l,\,max}$ and benign separation/stall behavior.

Though being counterintuitive at first, a reduction in blade weight can be achieved using thick(er) airfoils compared to general aviation applications, despite the higher profile drag associated with an increased thickness ratio. The reason for this lies in the optimum blade parameter (see Section 3.3), $\sigma'\,c_l$, which has a desired distribution along the blade radius. Therefore, operating at relatively high design lift coefficients, $c_l$, often ranging from 1.0...1.25, results in lower blade section solidity, $\sigma'$, and hence blade chord and weight. The reduced chord, however, may present structural issues to carry the loads. This can be overcome by choosing airfoils with a higher thickness ratio, $t/c$, an efficient way of increasing the area moment of inertia about the chord line as $\sim(t/c)^3$. The fact of design lift coefficients being relatively high can actually (nearly) outweigh the increased profile drag such that no, or only a small, penalty in lift-to-drag ratio, $c_l/c_d$, is obtained. A further aspect of high design lift coefficients and associated reduced blade chord (and overall rotor solidity, $\sigma$) are storm loads on parked rotors where structural loads benefit directly from a reduced blade area. As far as stall behavior is concerned, a sharp leading-edge stall has to be avoided at all costs due to its adverse effects on stall behavior and associated blade fatigue loads. In some cases, thick airfoils with the chordwise location of maximum thickness being aft of $0.30c$ typically have a comparatively benign turbulent trailing-edge stall behavior; that is, $c_l$ remains fairly close to its design point while $c_d$ is of course increased.

Today's utility-scale wind turbine blades have a relative thickness of about 18% at the tip, 25% at mid-blade stations, and up to 40% at the blade root where the respective airfoil blends into a cylindrical root-hub section. In general, *inboard* $(r/R < 0.50)$ blade sections require a high maximum lift coefficient, $c_{l,\,max}$, to generate a notable contribution to rotor torque within the constraints of local values for $V_{rel}^2$. Again, considering an optimum blade parameter, $\sigma'\,c_l$, designing closer to $c_{l,\,max}$ results in smaller absolute blade chord and hence reduced storm loads in parked conditions and easier transportation. For *outboard* $(r/R \geq 0.50)$ blade sections, the design lift coefficient at which the airfoil has its maximum lift-to-drag ratio, $(c_l/c_d)_{max}$, should be no more than 0.4 less than the airfoil's $c_{l,\,max}$ to avoid both excessive loads/torque in gusts and significant blade stall itself. Moreover, the effective angle of attack can be reduced by several degrees for high surface roughness associated with strong airfoil surface contamination, thus also reducing the airfoil stall angle. Hence staying $0.25 - 0.4$ in $c_l$ below (but no more) $c_{l,\,max}$ is desirable. Note that *inboard* thick airfoils can be designed to have a higher $c_{l,\,max}$ than *outboard* airfoils as well as a fairly benign stall behavior. In summary, a list of desired airfoil characteristics can be summarized as follows:

**Desired Airfoil Characteristics**

- High Lift-to-Drag Ratio, $(c_l/c_d)_{max}$
- High Design Lift Coefficient to reduce blade weight
- Thick(er) Airfoils to carry structural loads
- Low collection efficiency for reduced surface contamination
- Benign stall characteristics for reduced fatigue loads

Lastly, the effect of Reynolds number, $Re$, has to be mentioned. Given sectional Reynolds numbers along utility-scale wind turbine blades from Eq. (4.55), the wind energy community is challenged with testing new airfoil designs at the correct high $Re$ range that the blades experience in field operation. Indeed, most airfoil wind-tunnel tests are conducted at chord-based Reynolds numbers $Re_c \leq 3 \times 10^6$. Consequently, tested wind-tunnel performance may not scale as expected to full-scale conditions. The opposite can also be true, that is, one has to be very careful in using airfoils designed for $Re = O(10^6)$ on small-scale test turbines with $Re = O(10^4 - 10^5)$ as the resulting scaled wind turbine performance may be full of surprises if chosen airfoils perform poorly at significantly lower $Re$ than they were designed for. We will reiterate some aspects of the aforementioned in Chapter 8.

## 4.3 Development of Wind Turbine Airfoils

In the following, a brief design history of wind turbine airfoils is presented. Note that the design of new airfoil families has to be seen in conjunction with the design evolution of modern wind turbines, particularly the increase in size, see Figure 1.7. The material presented next is drawn from a number of sources, most notably Spera (2009), Burton et al. (2011), and Manwell et al. (2009), and augmented by other sources and catalogs of wind turbine airfoils as noted.

### 4.3.1 A Brief Historical Review of Wind Turbine Airfoils

During the early days of wind turbine development up to the mid-1900s, wind turbine designers were drawing from basic airfoil lift principles developed in the aeronautics community (see Section 1.1.3), which was still a quantum leap at that time in terms of achievable rotor power coefficients. The following brief review begins in the 1970s and attempts to show a shift in airfoil design paradigms largely due to the increasing scale of wind turbine rotors.

*1970s and Early 1980s.* The trend at that time was to design wind turbine blades operating at the minimum drag coefficient everywhere along the blade. The primary turbine/airfoil requirements can be summarized as follows:
- High $c_{l,max}$, Low $c_m$, Low $c_{d,min}$
- Emphasis on point design for blade twist and chord
- Choice from available general aviation airfoils (NACA 4- and 5-digit series)

While point designs perform very well at one specific condition, the resulting $C_P$ versus $\lambda$ curve becomes quite steep around the design point, and hence particularly stall-controlled turbine performance is notably degraded when operating off-design.

Furthermore, the turbulent atmospheric surface-layer inflow to the rotor disk was largely not understood in its adverse effects on blade fatigue loads, airfoil stall behavior, airfoil surface contamination, and so on.

*Early 1980s.* The wind energy community became increasingly aware of the adverse effects associated with airfoil surface contamination caused primarily by dirt, insects, and small amounts of rime ice. The effect of the associated leading-edge roughness is premature laminar-turbulent boundary-layer transition, thus increasing profile drag and reducing the lift-to-drag ratio, $c_l/c_d$, and power coefficient, $C_P$, respectively. For example, the NACA 230XX series airfoils experience a strong dependence of the maximum lift coefficient, $c_{l,max}$, on airfoil surface contamination/fouling. As blade cleaning is both very expensive in the process itself and associated with a notable downtime and hence loss in AEP, one has to simply wait for a strong rain shower to clean the blades that can take up to several months depending on the wind site. The experiences gained in field operation thus resulted in an additional turbine/airfoil requirement:

- Airfoils with reduced sensitivity to leading-edge roughness

Here the NACA 63-XXX series have demonstrated good overall performance, with reasonable changes associated with airfoil surface contamination/fouling. Another airfoil family is the LS(1)-04XX (McGhee et al. 1979) series that was designed to better tolerate changes in surface roughness; however, airfoils in this family suffered from large nose-down pitching moment coefficients, $c_m$, with adverse effects on elastic blade twist along thin blades. At the time, pitch-controlled wind turbines had not been that popular, largely due to the added complexity, weight, and cost. Instead, many wind turbine rotors with diameters on the order of $20 - 30$ m in the early 1980s (see Figure 1.7) were designed as stall-controlled rotors where rotor power in Region III is controlled passively by a soft airfoil/blade stall. Stall-controlled airfoils, for example the S809 airfoil used on the National Renewable Energy Laboratory (NREL) Phase VI rotor (Hand et al. 2001), do have a separation behavior that does not result in classical leading-edge stall, thus keeping the lift coefficient close to $c_{l,max}$ up to high angles of attack. While the idea of stall-controlled turbines (and airfoils) is appealing, the concept does not scale well to larger turbines with >30 m diameter due to higher fatigue loading compared to pitch-controlled machines, thus adversely affecting turbine lifetime and overall cost of energy.

*Late 1980s and 1990s.* During this time, a paradigm shift occurred that considered pitch-controlled in addition to stall-controlled turbines. This, at least to some extent, allowed again more of a point-design methodology in Region II with the highest wind probability. Hence the overall goal for airfoil performance became:

- High-performance (lift-to-drag ratio, $c_l/c_d$) that are fairly insensitive to changes in surface roughness

Airfoil design for wind turbines experienced a *renaissance* as efficient computational inverse airfoil design codes/methods were applied, a quantum leap compared to, for example, early NACA series airfoils with prescribed thickness and camber distributions. In inverse airfoil design, desired airfoil criteria can be specified, and an airfoil is designed through iterative shape and boundary-layer optimization. The most prominent airfoil analysis and design codes are listed next:

- Eppler Code – Inverse Design and Analysis of Low-Speed Airfoils (Eppler and Somers 1980; Eppler 1993)

- XFOIL – Viscous/Inviscid Airfoil Analysis and Design (Drela 1989)
- RFOIL – Version of XFOIL including Rotational Effects (van Rooij 1996)
- PROFOIL – Multi-Point Inverse Design (Selig and Tangler 1995)

Some examples of airfoil families designed in the 1980s and 1990s with emphasis on low sensitivity to leading-edge roughness and a benign stall behavior are the S-series of airfoils designed by D. Somers of *Airfoils Incorporated* (Port Matilda, PA) between 1995 and 2005 (Somers 1997a–c, 2004a,b, 2005a–h) and the Solar Energy Research Institute (SERI), see Tangler and Somers (1985, 1986) and Tangler (1987). These airfoils were intended to operate on stall-controlled machines, though with differentiation on airfoils designed for root, mid-blade, and tip regions. Parallel developments in Europe at the University of Stuttgart in Germany (Althaus and Wortmann 1981; Althaus 1984), Delft University of Technology (DUT) in the Netherlands, and the Risø National Laboratories in Denmark, focused on airfoil design for use on pitch-controlled wind turbines, resulting in airfoil series with $(c_l/c_d)_{max}$ occurring only a few degrees angle-of-attack below $c_{l, max}$. The work of Timmer and van Rooij (2003), among others, is a comprehensive reference on the DUT airfoil family, with relatively sharp trailing-edge airfoils for the tip region progressively transitioning into finite-thickness (blunt) TEs mid-blade and at the blade root for structural support. One main characteristic of the DUT airfoils, for example when compared to earlier NACA 5-digit-series airfoils, is an increased leading-edge thickness and less upper-surface thickness, see the references to the work of Timmer and his colleagues between 2003 and 2010 (Timmer and van Rooij 2003; Van Rooij and Timmer 2003; Timmer 2009, 2010), both of which affect the adverse pressure gradient along the airfoil in a way to promote reduced sensitivity to surface roughness, soft stall, and to increase the deep-stall angle, respectively.

*Late 1990s and 2000s.* Again with reference to the evolution of wind turbine size in Figure 1.7, increased turbine diameter and tower height required advances in both materials and aerodynamics. At this stage, closer attention was given to blade root airfoils with the following requirements:

Blade Root Airfoils

- Thicker airfoils (up to 40 % $t/c$) to carry higher loads/moments for structural strength
- High $c_{l, max}$ to produce sufficient torque at low wind speeds
- Low sensitivity to leading-edge roughness

This is when a transition occurred from moderately blunt trailing-edge airfoils (TE thickness <5 % $c$) to thick blunt (or flatback) airfoils (TE thickness $5 - 20 \% c$). The benefits of these types of airfoils are their increased cross-sectional area (and associated area moment of inertia) to handle high structural loads at reduced maximum stresses and easier manufacturing and handling/transportation compared to thick airfoils with sharp TEs. These airfoils have a higher $c_{l, max} \approx 1.5 - 2.0$ compared to their sharp trailing-edge counterparts but a base profile drag that can be five times as high, that is, $c_d \approx 0.1$, due to increased form (pressure) drag as a result of the blunt TE and associated separated shedding behavior. Note that the higher profile drag results in a reduction of $(c_l/c_d)_{max}$ down to only $25 - 30$; however, it turns out that operation at a design $c_l \approx 1.4$ outweighs the drag penalty due to the nature of the sectional velocity triangle at the blade root, see Figure 4.13, and these airfoils also reduce material stresses (Griffith and Richards 2014).

Note that the original concept of flatback airfoils dates back to efforts in Göttingen, for example, the Go-490 airfoil (Hoerner 1965). In the early 2000s, the development of flatback airfoils was reinvigorated by Sandia and UC Davis (Standish and van Dam 2003; van Dam et al. 2005, 2008; Baker et al. 2006, 2008) in the United States and at the Energy Research Centre of the Netherlands (ECN) (Grasso and Ceyhan 2015).

### 4.3.2 Catalog of Wind Turbine Airfoils

In the following, a comprehensive list of airfoils is presented that are either known to have been used or are currently being used on wind turbine blades. Emphasis is given to airfoils that were actually tested at Reynolds numbers of $Re = 3 \times 10^6$ and higher and those that were specifically designed for wind turbines with smaller chord-based Reynolds numbers. A very good airfoil database for all kinds of applications is hosted by Professor M. Selig at the University of Illinois at Urbana-Champaign (UIUC). In this context, it is always important to be aware of the actual "intended" use of specific airfoils, and the author wants to emphasize to the readers that off-$Re$ airfoil performance characteristics can result in unforeseen surprises, both with respect to scaling up to larger and scaling down to smaller rotors.

For benchmarking purposes of new airfoil designs, Table 4.1 lists a few NACA (Abbott and von Doenhoff 1959), FX/AH series (Althaus and Wortmann 1981; Althaus 1984), and S-series (Somers 1997c) airfoils, most of which have been used on wind turbines or at least share the primary desired performance characteristics of wind turbine airfoils and have measured data available at $Re = 3 \times 10^6$, see also the chapter on wind turbine airfoil characteristics by W. A. Timmer in Brønsted and Nijssen (2013). Note that the FX 77 airfoils in Table 4.1 do have an unsatisfactory rough behavior, a shortcoming that has been addressed to some extent with modern thick airfoils listed in the remaining tables in this section. Other reference airfoils are those of the FFA-W3 series (Bjørck 1990) although no measured data are available for Reynolds numbers larger than $1.6 \times 10^6$.

The S-series airfoils listed in Table 4.2 were designed by Airfoils Incorporated and the NREL, formerly the SERI. Starting from 1984 until the early 2000s, a total of nine

**Table 4.1** Benchmark airfoils for utility-scale wind turbines with available measured data at $Re = 3 \times 10^6$.

| Airfoil | Max. $t/c$ % | $c_{l,max}$ | Design $c_l$ | Max. $c_l/c_d$ |
|---|---|---|---|---|
| NACA 4418 | 18 | 1.43 | 1.16 | 110 |
| NACA $63_3$-618 | 18 | 1.39 | 1.10 | 138 |
| NACA $64_3$-618 | 18 | 1.37 | 1.09 | 156 |
| FX S03-182 | 18.2 | 1.48 | 1.04 | 157 |
| NACA $63_4$-421 | 21 | 1.33 | 1.04 | 127 |
| NACA $64_4$-421 | 21 | 1.34 | 1.04 | 142 |
| S814 | 24.1 | 1.41 | 1.11 | 116 |
| AH 93-W-257 | 25.7 | 1.41 | 1.20 | 121 |
| FX 77-W-270 S | 27.0 | 1.94 | 1.54 | 118 |
| FX 77-W-343 | 34.3 | 2.01 | 1.72 | 96 |

**Table 4.2** S-series airfoils.

| Airfoil | Type | Design $Re \times 10^{-6}$ | t/c % | Design D [m] | Design r/R | Design specifications | $c_{l,max}$ | $c_{d,min}$ | $c_{m,0}$ |
|---------|------|------|-------|------|------|------|------|------|------|
| S822 | Pitch | 0.60 | 16 | 2–10 | 0.90 | Thick, low tip $c_{l,max}$ | 1.00 | 0.010 | −0.07 |
| S823 | Pitch | 0.40 | 21 | 2–10 | 0.40 | Thick, low tip $c_{l,max}$ | 1.20 | 0.018 | −0.15 |
| S834 | Pitch | 0.40 | 15 | 2–10 | 0.95 | Thick, low noise, $c_{l,max}$ | 1.00 | Low | −0.15 |
| S833 | Pitch | 0.40 | 18 | 2–10 | 0.75 | Thick, low noise, $c_{l,max}$ | 1.10 | Low | −0.15 |
| S835 | Pitch | 0.25 | 21 | 2–10 | 0.40 | Thick, low noise, $c_{l,max}$ | 1.20 | Low | −0.15 |
| S803 | Pitch | 2.60 | 11.5 | 10–20 | 0.95 | Thin, high tip $c_{l,max}$ | 1.50 | 0.006 | −0.15 |
| S801 | Pitch | 2.00 | 13.5 | 10–20 | 0.75 | Thin, high tip $c_{l,max}$ | 1.50 | 0.007 | −0.15 |
| S804 | Pitch | 0.8 | 18.0 | 10–20 | 0.30 | Thick, high tip $c_{l,max}$ | 1.50 | 0.012 | −0.15 |
| S806A | Stall | 1.3 | 11.5 | 10–20 | 0.95 | Thin, low tip $c_{l,max}$ | 1.10 | 0.004 | −0.05 |
| S805A | Stall | 1.0 | 13.5 | 10–20 | 0.75 | Thin, low tip $c_{l,max}$ | 1.20 | 0.005 | −0.05 |
| S807 | Stall | 0.8 | 18.0 | 10–20 | 0.30 | Thick, low tip $c_{l,max}$ | 1.40 | 0.010 | −0.10 |
| S808 | Stall | 0.4 | 21.0 | 10–20 | 0.20 | Thick, low tip $c_{l,max}$ | 1.20 | 0.012 | −0.12 |
| S820 | Stall | 1.3 | 16.0 | 10–20 | 0.95 | Thick, low tip $c_{l,max}$ | 1.10 | 0.007 | −0.07 |
| S819 | Stall | 1.0 | 21.0 | 10–20 | 0.75 | Thick, low tip $c_{l,max}$ | 1.20 | 0.008 | −0.07 |
| S821 | Stall | 0.8 | 24.0 | 10–20 | 0.40 | Thick, low tip $c_{l,max}$ | 1.40 | 0.014 | −0.15 |
| S810 | Stall | 2.0 | 18.0 | 20–30 | 0.95 | Thick, low tip $c_{l,max}$ | 0.90 | 0.006 | −0.05 |
| S809 | Stall | 2.0 | 21.0 | 20–30 | 0.75 | Thick, low tip $c_{l,max}$ | 1.00 | 0.007 | −0.05 |
| S814 | Stall/Pitch | 1.4 | 24.0 | 20–30 | 0.40 | Thick, low tip $c_{l,max}$ | 1.30 | 0.012 | −0.15 |
| S815 | Stall/Pitch | 0.3 | 26.0 | 20–30 | 0.30 | Thick, low tip $c_{l,max}$ | 1.10 | 0.014 | −0.15 |
| S813 | Stall | 2.0 | 16.0 | 20–30 | 0.95 | Thick, low tip $c_{l,max}$ | 1.10 | 0.007 | −0.07 |
| S812 | Stall | 2.0 | 21.0 | 20–30 | 0.75 | Thick, low tip $c_{l,max}$ | 1.20 | 0.008 | −0.07 |
| S826 | Pitch | 1.5 | 14.0 | 20–40 | 0.95 | Thin, high tip $c_{l,max}$ | 1.60 | 0.006 | −0.14 |
| S825 | Pitch | 2.0 | 17.0 | 20–40 | 0.75 | Thick, high tip $c_{l,max}$ | 1.60 | 0.008 | −0.14 |
| S817 | Stall | 3.0 | 16.0 | 30–50 | 0.95 | Thick, low tip $c_{l,max}$ | 1.10 | 0.007 | −0.07 |
| S816 | Stall | 4.0 | 21.0 | 30–50 | 0.75 | Thick, low tip $c_{l,max}$ | 1.20 | 0.008 | −0.07 |
| S818 | Stall | 2.5 | 24.0 | 30–50 | 0.40 | Thick, low tip $c_{l,max}$ | 1.30 | 0.012 | −0.15 |
| S828 | Stall | 3.0 | 16.0 | 40–50 | 0.95 | Thick, low tip $c_{l,max}$ | 0.90 | 0.007 | −0.07 |
| S827 | Stall | 4.0 | 21.0 | 40–50 | 0.75 | Thick, low tip $c_{l,max}$ | 1.00 | 0.008 | −0.15 |
| S832 | Pitch | 2.5 | 15.0 | 40–50 | 1.00 | Thick, high tip $c_{l,max}$ | 1.4 | Low | −0.15 |
| S831 | Pitch | 3.5 | 18.0 | 40–50 | 0.90 | Thick, high tip $c_{l,max}$ | 1.50 | Low | −0.15 |
| S830 | Pitch | 4.0 | 21.0 | 40–50 | 0.75 | Thick, high tip $c_{l,max}$ | 1.60 | Low | −0.15 |

airfoil families were designed (Somers and Tangler 1995; Somers 1997–2005) for different sized rotors, using the design method developed by Eppler (1990, 1993). Note that airfoils listed in Table 4.2 are designed for both pitch- and stall-regulated machines. Some of the most prominent S-series airfoils are shown in Figure 4.14.

The Risø National Laboratories in Denmark designed a total of three airfoil families for use on horizontal-axis wind turbines starting in the 1990s. A comprehensive summary

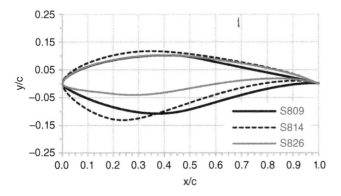

**Figure 4.14** Examples of S-series airfoils (S809, S814, S826).

**Table 4.3** Risø-A airfoil family.

| Airfoil | $Re \times 10^{-6}$ | Max. t/c % | x/c at $t_{max}$ | TE y/c × $10^2$ | $\alpha_0$ [deg] | $c_{l,max}$ | Design $\alpha$ [deg] | Design $c_l$ | Max. $c_l/c_d$ |
|---------|------|-----|-------|------|------|------|------|------|------|
| Risø-A1-15 | 3.00 | 15 | 0.325 | 0.25 | −4.0 | 1.50 | 6.0 | 1.13 | 168 |
| Risø-A1-18 | 3.00 | 18 | 0.336 | 0.25 | −3.6 | 1.53 | 6.0 | 1.15 | 167 |
| Risø-A1-21 | 3.00 | 21 | 0.298 | 0.50 | −3.3 | 1.45 | 7.0 | 1.15 | 161 |
| Risø-A1-24 | 2.75 | 24 | 0.302 | 1.00 | −3.4 | 1.48 | 7.0 | 1.19 | 157 |
| Risø-A1-27 | 2.75 | 27 | 0.303 | 1.00 | −3.2 | 1.44 | 7.0 | 1.15 | − |
| Risø-A1-30 | 2.50 | 30 | 0.300 | 1.00 | −2.7 | 1.35 | 7.0 | 1.05 | − |
| Risø-A1-33 | 2.50 | 30 | 0.304 | 1.00 | −1.6 | 1.20 | 7.0 | 0.93 | − |

**Table 4.4** Risø-P airfoil family.

| Airfoil | $Re \times 10^{-6}$ | Max. t/c % | x/c at $t_{max}$ | TE y/c × $10^2$ | $\alpha_0$ [deg] | $c_{l,max}$ | Design $\alpha$ [deg] | Design $c_l$ | Max. $c_l/c_d$ |
|---------|------|-----|-------|------|------|------|------|------|------|
| Risø-P-15 | 3.00 | 15 | 0.328 | 0.25 | −3.5 | 1.49 | 6.0 | 1.12 | 173 |
| Risø-P-18 | 3.00 | 18 | 0.328 | 0.25 | −3.7 | 1.50 | 6.0 | 1.15 | 170 |
| Risø-P-21 | 3.00 | 21 | 0.323 | 0.50 | −3.5 | 1.48 | 6.0 | 1.14 | 159 |
| Risø-P-24 | 2.75 | 24 | 0.320 | 1.00 | −3.7 | 1.48 | 6.0 | 1.17 | 156 |

can be found in Bertagnolio et al. (2001) and Fuglsang and Bak (2004). The design tools used at Risø were again the Eppler code in conjunction with higher-fidelity computational fluid dynamics (CFD) analyses using the Ellipsys-2D-CFD code developed at the Technical University of Denmark, see for example, Sørensen (1995). The Risø-A family, see Table 4.3, was designed in the 1990s for stall-regulated wind turbines, while the Risø-P series in Table 4.4 of only four airfoils, on the other hand, was designed later for variable pitch/speed wind turbines, see Bak et al. (2004, 2008). The later Risø-B series of airfoils focused on thicker airfoils and higher design lift coefficient, see Table 4.5.

**Table 4.5** Risø-B airfoil family.

| Airfoil | $Re \times 10^{-6}$ | Max. $t/c$ % | $x/c$ at $t_{max}$ | TE $y/c \times 10^2$ | $\alpha_0$ [deg] | $c_{l,max}$ | Design $\alpha$ [deg] | Design $c_l$ | Max. $c_l/c_d$ |
|---|---|---|---|---|---|---|---|---|---|
| Risø-B1–15 | 6.00 | 15 | 0.278 | 0.60 | −4.1 | 1.92 | 6.0 | 1.21 | 157 |
| Risø-B1–18 | 6.00 | 18 | 0.279 | 0.40 | −4.0 | 1.87 | 6.0 | 1.19 | 166 |
| Risø-B1–21 | 6.00 | 21 | 0.278 | 0.50 | −3.6 | 1.83 | 6.0 | 1.16 | 139 |
| Risø-B1–24 | 6.00 | 24 | 0.270 | 0.70 | −3.1 | 1.76 | 6.0 | 1.15 | 120 |
| Risø-B1–30 | 6.00 | 30 | 0.270 | 1.00 | −2.1 | 1.61 | 5.0 | 0.90 | – |
| Risø-B1–36 | 6.00 | 36 | 0.270 | 1.20 | −1.3 | 1.15 | 5.0 | 0.90 | – |

**Table 4.6** DU airfoil family.

| Airfoil | $Re \times 10^{-6}$ | Max. $t/c$ % | $x/c$ at $t_{max}$ | TE $y/c \times 10^2$ | $\alpha_0$ [deg] | $c_{l,max}$ | Design $\alpha$ [deg] | Design $c_l$ | Max. $c_l/c_d$ |
|---|---|---|---|---|---|---|---|---|---|
| DU 96-W-180 | 3.00 | 18 | 0.300 | 0.18 | −2.7 | 1.26 | 6.59 | 1.07 | 145 |
| DU 00-W-212 | 3.00 | 21.2 | 0.300 | 0.23 | −2.7 | 1.29 | 6.50 | 1.06 | 132 |
| DU 91-W2250 | 3.00 | 25 | 0.300 | 0.54 | −3.2 | 1.37 | 6.68 | 1.24 | 137 |
| DU 97-W-300 | 3.00 | 30 | 0.300 | 0.48 | −2.2 | 1.56 | 9.30 | 1.39 | 98 |
| DU 00-W-350 | 3.00 | 35 | 0.300 | 1.00 | −2.0 | 1.39 | 7.00 | 1.13 | 81 |
| DU 00-W-401 | 3.00 | 40.1 | 0.300 | 1.00 | −3.0 | 1.04 | 5.00 | 0.82 | 54 |

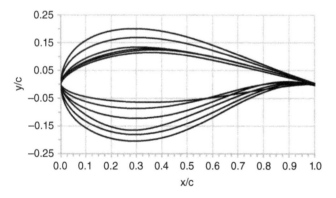

**Figure 4.15** DU series airfoils.

The DUT in the Netherlands has designed an independent airfoil family designated as the DU airfoils (Timmer and van Rooij 2003; Timmer 2009–2010). Table 4.6 lists the prominent airfoils and their performance characteristics. Modification enhancements to the DU airfoils are ongoing, with some notable new developments by the ECN, specifically the G2 series (Grasso 2012). The DU airfoils are illustrated in Figure 4.15. The primary design method used at Delft has been the RFOIL code (van Rooij 1996).

Airfoil design and validation continues, for example Shen et al. (2014) and Zahle et al. (2014), although many new designs are proprietary and, therefore, not publicly available.

## References

Abbott, I.H. and von Doenhoff, A.E. (1959). *Theory of Wing Sections*. New York: Dover Publications.

Althaus, D. (1984). *Niedriggeschwindigkeitsprofile – Airfoils and Experimental Results from the Laminar Wind Tunnel of Institute for Aerodynamics and Gas Dynamics of the Stuttgart University*. Germany: Stuttgart University.

Althaus, D. and Wortmann, F.X. (1981). *Stuttgarter Profilkatalog I*. Braunschweig/Wiesbaden: Friedrich Vieweg & Sohn.

Anderson, J.D. (2001). *Fundamentals of Aerodynamics*, McGraw Hill Series in Aeronautical and Aerospace Engineering, 3e. New York: McGraw Hill, Inc.

Bak, C., Fuglsang, P., Gaunaa, M., and Antoniou, I. (2004). Design and verification of the Risø-P airfoil family for wind turbines. In: *Proceedings of Torque 2004: The Science of Making Torque from Wind*. The Netherlands: Delft.

Bak, C., Andersen, P. B., Madsen, H. A., Gaunaa, M., Fuglsang, P., and Bove, S. (2008) Design and verification of airfoils resistant to surface contamination and turbulence intensity. *26th AIAA Applied Aerodynamics Conference (AIAA-2008-7050)*, Honolulu HI.

Baker, J.P., Mayda, E.A., and van Dam, C.P. (2006). Experimental analysis of thick blunt trailing-edge wind turbine airfoils. *Journal of Solar Energy Engineering* 128: 422–431.

Baker, J.P., van Dam, C.P., and Gilbert, B.L. (2008). *Flatback Airfoil Wind Tunnel Experiment*. Sandia National Laboratories SAND2008-2008.

Bertagnolio, F., Sørensen, N., Johansen, J., and Fuglsang, P. (2001). *Wind Turbine Airfoil Catalogue*. Roskilde (Denmark): Risø National Laboratory, Risø-R-1280(EN).

Bjørck, A. (1990). *Coordinates and Calculations for the FFA-Wl-xx, FFA-W2-xx and FFA-W3-xx Series of Airfoils for Horizontal Axis Wind Turbines*. The Aeronautical Research Institute of Sweden, FFA TN 1990-15.

Brønsted, P. and Nijssen, R.P.L. (2013). *Advances in Wind Turbine Blade Design and Materials*. Oxford: Elsevier Science.

Burton, T., Jenkins, N., Sharpe, D., and Bossanyi, E. (2011). *Wind Energy Handbook*. Chichester: Wiley.

Chattot, J.J. and Hafez, M. (2015). *Theoretical and Applied Aerodynamics and Related Numerical Methods*. Dordrecht: Springer.

Drela, M. (1989). X-foil: an analysis and design system for low Reynolds number airfoils. In: *Low Reynolds Number Aerodynamics*, vol. 54, Springer-Verlag Lecture Notes in Engineering. Godalming: Springer-Verlag.

Drela, M. (2014). *Flight Vehicle Aerodynamics*. Cambridge MA: The MIT Press.

Drela, M. and Giles, M.B. (1987). Viscous-inviscid analysis of transonic and low Reynolds number airfoils. *AIAA Journal* 25 (10): 1347–1355.

Eppler, R. (1990). *Airfoil Design and Data*. Berlin: Springer-Verlag.

Eppler, R. (1993) *Airfoil Program System. User's Guide*. Eppler, R. Airfoil Program System, User's Guide, Institut A fuer Mechanik, Universitaet Stuttgart (1988).

Eppler, R. and Somers, D.M. (1980). *A Computer Program for the Design and Analysis of Low-Speed Airfoils*. National Aeronautics and Space Administration, NASA TM-80210.

Fuglsang, P. and Bak, C. (2004). Development of the Risø wind turbine airfoils. *Wind Energy* 7: 145–162.

Grasso, F. (2012) Design of a family of new advanced airfoils for low wind class turbines. *Proceedings of Torque 2012: The Science of Making Torque from Wind*, Oldenburg, Germany.

Grasso, F. and Ceyhan, O. (2015) Non-conventional flat back thick airfoils for very large offshore wind turbines. *33rd Wind Energy Symposium, AIAA SciTech Forum (AIAA 2015-0494)*, Kissimmee, FL, USA.

Griffith, T.D. and Richards, P.W. (2014). *The SNL100–03 Blade: Design Studies with Flatback Airfoils for the Sandia 100-Meter Blade*. Sandia National Laboratories, SAND2014-18129.

Hand, M.M., Simms, D.A., Fingersh, L.J. et al. (2001). *Unsteady Aerodynamics Experiment Phase VI: Wind Tunnel Test Configurations and Available Data Campaigns*. National Renewable Energy Laboratory, NREL/TP-500-29955.

Hoerner, S.F. (1965). *Fluid-Dynamic Drag. Hoerner Fluid Dynamics*. New Jersey: Self-published by S. F. Hoerner.

Katz, J. and Plotkin, A. (2001). *Low-Speed Aerodynamics*, 2e. Cambridge: Cambridge University Press.

Manwell, J.F., McGowan, J.G., and Rogers, A.L. (2009). *Wind Energy Explained: Theory, Design, and Application*, 2e. Chichester: Wiley.

Marongiu, C. and Tognaccini, R. (2010). Far-field analysis of the aerodynamic force by Lamb vector integrals. *AIAA Journal* 48 (11): 2543–2555.

McGhee, R.J., Beasley, W.D., and Whitcomb, R.T. (1979). *NASA Low- and Medium-Speed Airfoil Development*. National Aeronautics and Space Administration, NASA-TM-78709.

McLean, D. (2013). *Understanding Aerodynamics: Arguing from the Real Physics*. Chichester: Wiley.

Reed, H.L. and Saric, W.S. (1996). Linear stability theory applied to boundary layers. *Annual Review of Fluid Mechanics* 28: 389–428.

Saric, W.S., Reed, H.L., and Kerschen, E.J. (2002). Boundary layer receptivity. *Annual Review of Fluid Mechanics* 34: 291–319.

Schlichting, H. (1979). *Boundary-Layer Theory*, 7e. New York: McGraw Hill.

Schmitz, S. (2014). Finite domain viscous correction to the Kutta–Joukowski theorem in incompressible flow. *AIAA Journal* 52 (9): 2079–2083.

Selig, M.S. and Tangler, J.L. (1995). Development and application of a multipoint inverse design method for horizontal-axis wind turbines. *Wind Engineering* 19 (2): 91–105.

Shen, W.Z., Zhu, W.J., Fischer, A. et al. (2014). Validation of the CQU-DTU-LN1 series of airfoils. *Journal of Physics: Conference Series* 555: 012093. https://doi.org/10.1088/1742-6596/555/1/012093.

Smith, A. M. O. and Gamberoni, N. (1956) Transition, pressure gradient, and stability theory. Report ES 26388, Douglas Aircraft Co.

Somers, D.M. (1997a). *Design and Experimental Results for the S805 Airfoil*. National Renewable Energy Laboratory, NREL/SR-440-6917.

Somers, D.M. (1997b). *Design and Experimental Results for the S809 Airfoil*. National Renewable Energy Laboratory, NREL/SR-440-6918.

Somers, D.M. (1997c). *Design and Experimental Results for the S814 Airfoil*. National Renewable Energy Laboratory, NREL/SR-440-6919.

Somers, D.M. (2004a). *The S814 and S815 Airfoils*. National Renewable Energy Laboratory, NREL/SR-500-36292.

Somers, D.M. (2004b). *The S816, S817, and S818 Airfoils*. National Renewable Energy Laboratory, NREL/SR-500-36333.

Somers, D.M. (2005a). *Effects of Airfoil Thickness and Maximum Lift Coefficient on Roughness Sensitivity*. National Renewable Energy Laboratory, NREL/SR-500-36336.

Somers, D.M. (2005b). *The S829 Airfoil*. National Renewable Energy Laboratory, NREL/SR-500-36337.

Somers, D.M. (2005c). *The S819, S820, and S821 Airfoils*. National Renewable Energy Laboratory, NREL/SR-500-36334.

Somers, D.M. (2005d). *The S830, S831, and S832 Airfoils*. National Renewable Energy Laboratory, NREL/SR-500-36339.

Somers, D.M. (2005e). *The S822 and S823 Airfoils*. National Renewable Energy Laboratory, NREL/SR-500-36344.

Somers, D.M. (2005f). *The S827 and S828 Airfoils*. National Renewable Energy Laboratory, NREL/SR-500-36343.

Somers, D.M. (2005g). *The S825 and S826 Airfoils*. National Renewable Energy Laboratory, NREL/SR-500-36344.

Somers, D.M. (2005h). *Design and Experimental Results for the S827 Airfoil*. National Renewable Energy Laboratory, NREL/SR-500-36345.

Somers, D. M. and Tangler, J. L. (1995) Wind-Tunnel Test of the S814 Thick-Root Airfoil. Fourteenth ASME-ETCE Wind Energy Symposium, Houston TX. Published as: National Renewable Energy Laboratory, NREL/TP-442-7388.

Sørensen, N.N. (1995). *General Purpose Flow Solver Applied to Flow over Hills*. Denmark: Risø National Laboratories, Risø-R-827(EN).

Spera, D.A. (ed.) (2009). *Wind Turbine Technology – Fundamental Concepts of Wind Turbine Engineering*, 2e. New York: ASME Press.

Standish, K.J. and van Dam, C.P. (2003). Aerodynamic analysis of blunt trailing edge airfoils. *Journal of Solar Energy Engineering* 125: 479–487.

Tangler, J.L. (1987). *Status of Special-Purpose Airfoil Families*. National Renewable Energy Laboratory, SERI/TP-217-3264.

Tangler, J. L. and Somers, D. M. (1985) Advanced Airfoils for HAWTS. Proceedings of the Wind Power '85 Conference, Washington, D.C., USA. SERI/CP-217-2902, pp. 45–51.

Tangler, J. L. and Somers, D. M. (1986) A Low Reynolds Number Airfoil Family for Horizontal Axis Wind Turbines. Proceedings of the International Conference on Aerodynamics at Low Reynolds Numbers, London, UK.

Timmer, W. A. (2009) An overview of NACA 6-digit airfoil series characteristics with reference to airfoils for large wind turbine blades. 47th AIAA Aerospace Sciences Meeting (AIAA-2009-0268), Orlando, FL, USA.

Timmer, W.A. (2010). Aerodynamic characteristics of wind turbine blade airfoils at high angles of attack. In: *Proceedings of Torque 2010: The Science of Making Torque from Wind*, 71–97. Heraklion (Greece): European Wind Energy Association (EWEA).

Timmer, W.A. and van Rooij, R.P.J. (2003). Summary of Delft University wind turbine dedicated airfoils. *Journal of Solar Energy Engineering* 125: 488–496.

Van Dam, C. P., Mayda, E., Chao, D., Jackson, K., Zutek, M., and Berry, D. (2005) Innovative Structural and Aerodynamic Design Approaches for Large Wind Turbine Blades. 43rd AIAA Aerospace Sciences Meeting and Exhibit (AIAA-2005-0973).

Van Dam, C.P., Kahn, D.L., and Berg, D.E. (2008). *Trailing Edge Modifications for Flatback Airfoils*. Sandia National Laboratories, SAND2008-1781.

Van Ingen, J. L. (1956) A suggested semi-empirical method for the calculation of the boundary layer transition region. Report VTH-74, Delft University of Technology, Department of Aerospace Engineering, The Netherlands.

Van Rooij, R. P. J. O. M.. (1996) Modification of the boundary layer in XFOIL for improved airfoil stall prediction. Report IW-96087R, Delft University of Technology, The Netherlands.

Van Rooij, R. P. J. O. M. and Timmer, W. A. (2003) Roughness sensitivity considerations for thick rotor blade airfoils. 41st Aerospace Sciences Meeting (AIAA-2003-0350), Reno, NV, USA.

Wu, J.Z., Lu, X.Y., and Zhuang, L.X. (2006). *Vorticity and Vortex Dynamics*, 627–633. New York: Springer.

Zahle, F., Bak, C., Sørensen, N.N. et al. (2014). Design of the LRP series airfoils using 2D CFD. *Journal of Physics: Conference Series* 524: 012020. https://doi.org/10.1088/1742-6596/524/1/012020.

## Further Reading

Abbott, I.H. and von Doenhoff, A.E. (1959). *Theory of Wing Sections*. New York: Dover Publications.

Anderson, J.D. (2001). Fundamentals of aerodynamics. In: *Mc-Graw Hill Series in Aeronautical and Aerospace Engineering*, 3e. New York: McGraw Hill, Inc.

Bertagnolio, F., Sørensen, N., Johansen, J., and Fuglsang, P. (2001). *Wind Turbine Airfoil Catalogue*. Roskilde (Denmark): Risø National Laboratory, Risø-R-1280(EN).

Brønsted, P. and Nijssen, R.P.L. (2013). *Advances in Wind Turbine Blade Design and Materials*. Oxford: Elsevier Science.

Burton, T., Jenkins, N., Sharpe, D., and Bossanyi, E. (2011). *Wind Energy Handbook*. Chichester: Wiley.

Chattot, J.J. and Hafez, M. (2015). *Theoretical and Applied Aerodynamics and Related Numerical Methods*. Dordrecht: Springer.

Drela, M. (1989). X-foil: an analysis and design system for low Reynolds number airfoils. In: *Low Reynolds Number Aerodynamics*, vol. 54, Springer-Verlag Lecture Notes in Engineering. Godalming: Springer-Verlag.

Schlichting, H. (1979). *Boundary-Layer Theory*, 7e. New York: McGraw Hill.

Somers, D.M. and Maughmer, M.D. (2002). *Theoretical Aerodynamic Analyses of Six Airfoils for Use on Small Wind Turbines*. National Renewable Energy Laboratory, NREL/SR-500-33295.

Spera, D.A. (ed.) (2009). *Wind Turbine Technology – Fundamental Concepts of Wind Turbine Engineering*, 2e. New York: ASME Press.

# 5

# Unsteady Aerodynamics and 3-D Correction Models for Airfoil Characteristics

*The Nebulous Art of Using Wind-Tunnel Airfoil Data for Predicting Rotor Performance [NREL/CP-500-31243]*

– James L. Tangler

## 5.1  Unsteady Aerodynamics on Wind Turbine Blades

In the previous chapter, we covered steady aerodynamics in which the effect of the wake vortex sheet is neglected, simply because the wake vorticity has zero strength for a constant (steady) bound circulation. If, however, the flow and/or airfoil motion are time dependent, the bound circulation changes with time as a result of a time-varying angle of attack. Here, the result is an elemental "starting" vortex being shed and advected into the wake at each time step according to Kelvin's circulation theorem. In this context, we talk about "unsteady aerodynamics" to broadly characterize time-dependent aerodynamic responses of an airfoil that do not follow steady (or static) airfoil data as introduced in Chapter 4; and we refer to a "quasi-steady" (time-dependent) behavior as one where the operating point of an airfoil follows very closely its steady characteristics and where, consequently, the wake effect is small.

A wind turbine operating in the atmospheric boundary layer (ABL) is subject to a number of time-varying processes that either derive from a time-dependent wind inflow to the rotor-disk area or from blade motion itself, see for example, Leishman (2002). An illustration is given in Figure 5.1. In particular, the rotor diameter of utility-scale wind turbines is large enough for the turbine to experience variations in axial wind speed, $V_0$, over time, $t$, and with height, $z$, due to a sheared velocity profile in the lower ABL, with shear coefficients depending on the ABL stability state and surface roughness (see Section 1.2.2). In general terms, the mean wind speed and direction change with time, $t$, and over the rotor-disk coordinates, $r$ (radial) and $\psi$ (azimuthal), resulting in locally yawed inflow to the rotor blades, $V_0(t, r, \psi)$; furthermore, the ABL stability state determines temporal and spatial scales of large coherent turbulent structures, affecting the local blade-element inflow at different strength and frequency. As for the rotor blades themselves, they may be subject to "flapping" out of the rotor-disk plane (or "plunging" from an airfoil perspective), experience a "torsional response" (or "pitching" from an airfoil perspective), or even have an "edgewise" (or "lead-lag" from an airfoil

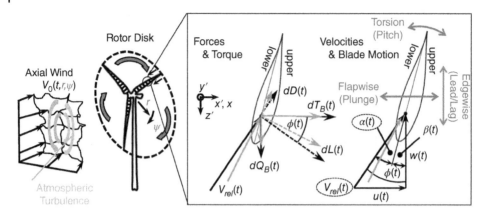

**Figure 5.1** Force/torque and velocity triangles in time-dependent flow.

perspective) motion within the rotor-disk plane. All of these are forced by changes in the local inflow to the blade element and/or rotor speed itself. In the end, though, it is reasonable to argue that the primary combined effect reveals itself in local blade aerodynamics as a time-varying angle of attack, $\alpha(t)$, and local relative velocity, $V_{rel}(t)$, from the point-of-view of local rotor inflow and, in a highly simplified manner, without considering apparent mass and inertia terms. The question then arises whether (or not) individual and/or combined effects result in an "unsteady" blade load response along turbine blades and what operating conditions and radial blade stations are most susceptible to unsteady aerodynamics. These are important considerations with respect to possible load fluctuations that affect blade fatigue.

In the following, we will approach this seemingly complex problem from the standpoint of good engineering judgment, that is, dissecting the problem into its parts and estimating from classical theory whether or not unsteady aerodynamics may have to be taken into account. Let us begin by looking at classical unsteady aerodynamics theories for two-dimensional (2-D) incompressible flow, which had been developed by the 1940s. In particular, classical works by Wagner (1925), Küssner (1935), Theodorsen (1935), and von Kármán and Sears (1938) are landmarks in unsteady aerodynamic theory. In the following, a very brief introduction to Theodorsen's theory is given.

### 5.1.1 Fundamentals of Unsteady Aerodynamics – Theodorsen's Theory

The single most important dimensionless parameter describing the "degree of unsteadiness" is the reduced frequency, $k$. In general, the resultant force acting on an airfoil of chord, $c$, subject to a uniform inflow, $U_\infty$, is a function of the Reynolds number, Mach number, and when oscillating at an angular frequency (or circular frequency), $\omega$, also becomes a function of the "reduced frequency" defined as:

$$\text{Reduced Frequency (Airfoil)} : k = \frac{\omega c}{2U_\infty} \tag{5.1}$$

Note that in Eq. (5.1), the angular frequency $\omega = 2\pi f$ where $f$ [Hz] is the excitation frequency. It is trivial to note that $k = 0$ in the steady case. One can also think of the reduced frequency, $k$, as a time scale normalized by a reference time of $c/(2U_\infty)$ it takes for the freestream, $U_\infty$, to convect over half of the airfoil chord; here low $\omega$ are associated with large time scales compared to the reference time scale, approaching the steady case of $k \to 0$. In general, the following classification is used to quantify unsteady aerodynamics:

- Steady: $k = 0$
- Quasi-Steady: $0 \leq k \leq 0.05$
- Unsteady: $0.05 < k \leq 0.2$ (Notable hysteresis response)
- Highly Unsteady: $k > 0.2$ (Dominant hysteresis response)

A resultant "hysteresis" behavior of a lift (or blade load) response is described further in Section 5.1.1.2 and is a consequence of the airfoil bound circulation, $\Gamma$, and hence lift not responding instantaneously to angle-of-attack (AoA) changes with time, but lagging behind its static values. As for wind turbine blades, Eq. (5.1) states that for a given angular frequency, $\omega$, the reduced frequency, $k$, increases with chord, $c$, and lower reference freestream speed, $U_\infty$. This in turn indicates that unsteady aerodynamics is likely to be more important at inboard compared to outboard wind turbine blade sections where the local blade chord is higher and the local relative velocity is lower.

### 5.1.1.1 Flow Model – Unsteady Thin-Airfoil Theory

The mathematical treatment of unsteady thin-airfoil theory is very involved, and the interested reader is referred to the original literature and Leishman (2006) for a more in-depth treatment of the subject matter. Here, we assume a very special case of pure AoA variations in the freestream velocity that can be caused by time-varying atmospheric inflow and turbulence, turbine yaw, and blade-tower interaction. In this respect, apparent mass contributions from plunging/pitching airfoils (or turbine blade bend/twist coupling) are not considered for the sole purpose of simplicity. For this special case, Theodorsen's Theory is rooted in a simple flow model shown in Figure 5.2 where both the airfoil and shed wake are represented as planar vortex sheets, with the latter starting at the trailing edge and extending to downstream infinity. The airfoil bound circulation, $\Gamma$, is now a function of time according to:

$$\text{Airfoil Bound Circulation} : \Gamma(t) = \int_0^c \gamma(x,t)\,dx \tag{5.2}$$

In Eq. (5.2), $\gamma(x, t)$ denotes the vorticity distribution bound to the airfoil (see also Section 4.1.1). At the trailing edge, the Kutta condition holds as:

$$\text{Kutta Condition} : \gamma(c,t) = 0 \tag{5.3}$$

It is important in this context to be aware of the assumptions of 2-D, inviscid, and incompressible flow subject to small perturbations in Theodorsen's theory. The basic driver of unsteady aerodynamics is that the wake vorticity, $\gamma_w(x, t)$, induces a restoring

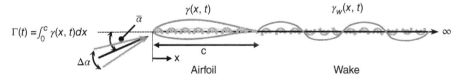

**Figure 5.2** Unsteady thin-airfoil theory – mathematical model (freestream angle-of-attack variation).

effect on the airfoil vorticity, $\gamma(x, t)$, such that the (steady) fundamental equation (4.9) (see Section 4.1.1.2) of unsteady thin-airfoil theory now becomes:

$$\text{Fundam. Eqn.}: \; -\frac{1}{2\pi}\int_0^c \frac{\gamma(\xi, t)}{x - \xi}d\xi - \frac{1}{2\pi}\int_c^\infty \frac{\gamma_w(\xi, t)}{x - \xi}d\xi = U_\infty[d'(x) - \alpha] \quad (5.4)$$

Following Kelvin's theorem of circulation conservation, the wake vorticity shed at the trailing edge, $\gamma_w(c, t)$, is proportional to the time-rate-of-change of the airfoil bound circulation, $d\Gamma/dt$, as:

$$\text{Wake Vorticity}: \; U_\infty\gamma_w(c, t) = -\frac{d\Gamma(t)}{dt} \quad (5.5)$$

Here Eq. (5.5) also assumes that the shed wake vorticity, $\gamma_w(x, t)$, advects at a constant freestream speed, $U_\infty$. It is easy to understand from Eq. (5.5) that the wake vorticity, $\gamma_w$, has to be zero in the steady case when the airfoil bound circulation, $\Gamma$, is not a function of time.

The effect of the wake vorticity, $\gamma_w(x, t)$, in Eq. (5.4) is that it continuously changes the downwash onto the airfoil vortex sheet, the airfoil vorticity, $\gamma(x, t)$, the total airfoil bound circulation, $\Gamma(t)$, and again the shed wake vorticity, $\gamma_w(x, t)$. Solution of the fundamental equation (5.4) is not trivial in the general case, and we focus on the simplest possible case in the next section.

### 5.1.1.2 Special Case: Freestream Angle-of-Attack Oscillation

Let us focus on a pure AoA oscillation, $\alpha(t)$, of the following form:

$$\text{Freestream AoA Oscillation}: \; \alpha(t) = \bar{\alpha} + \Delta\alpha \sin(\omega t) \quad (5.6)$$

In Eq. (5.6), $\bar{\alpha}$ is the mean angle of attack and $\Delta\alpha$ is the amplitude of the AoA oscillation. The circulatory part of the lift-coefficient response is written as:

$$\text{Lift Response (Circulatory)}: \; c_l(t) = 2\pi|C(k)|[\bar{\alpha} + \Delta\alpha \sin(\omega t + \varphi) - \alpha_0] \quad (5.7)$$

Note that all angles in Eq. (5.7) are in [rad]. Also, $|C(k)| < 1$ is the amplitude of the complex Theodorsen function, resulting in a response of the lift coefficient, $c_l(t)$, with a slope less than the inviscid thin-airfoil result of $2\pi$; the angle, $\varphi$, describes a phase lag between the $\alpha(t)$ forcing and the $c_l(t)$ response, meaning that the time-dependent lift coefficient, $c_l(t)$, is lagging behind what it would be in the static (non-oscillating) case. Furthermore, $\alpha_0$ is the airfoil zero-lift angle of attack. The Theodorsen function is shown in Figure 5.3 from which amplitude and phase can be determined according to:

$$\text{Theodorsen Function (Amplitude)}: \; |C(k)| = \sqrt{F(k)^2 + G(k)^2} \quad (5.8)$$

$$\text{Theodorsen Function (Phase)}: \; \varphi(k) = \tan^{-1}\left(\frac{G(k)}{F(k)}\right) \quad (5.9)$$

Here, $F(k)$ and $G(k)$ are real and imaginary parts of the complex Theodorsen function. Also shown in Figure 5.3 are sample $c_l(t)$ responses for different values of the reduced frequency, $k$.

For $k = 0$, the steady lift behavior is exactly recovered, with $c_l(t)$ following the (steady) inviscid linear lift curve of slope $2\pi$ (see Chapter 4). As $k$ is increased to the unsteady regime ($k > 0.05$), however, "hysteresis" loops develop, signifying a phase lag between $\alpha(t)$ and $c_l(t)$. Note that the hysteresis loops are circumvented counter-clockwise, that is,

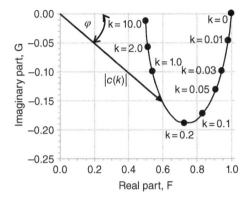
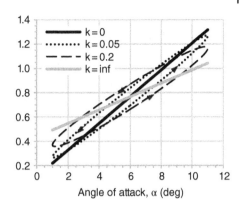

**Figure 5.3** Theodorsen function and effect on circulatory part of the unsteady lift response (example: $\alpha(t) = 6° + 5° \sin(\omega t)$, $\alpha_0 = -1°$).

the $c_l(t)$ response is always lagging behind the value that would be expected from a steady response based on the actual $\alpha(t)$. Following the Theodorsen function in Figure 5.3, it is interesting to note that the amplitude, $|C(k)|$ continues to decrease with increasing $k$ and is equal to 1/2 for the limiting case of $k \to \infty$ in which case the resultant lift-curve slope is equal to $\pi$. The behavior is different though for the phase angle, $\varphi$, which increases up to $k = 0.2$ (i.e. increasing hysteresis behavior) but then decreases to zero for $k \to \infty$.

## 5.1.2 Dynamic Stall Models

At higher angles of attack when time-dependent flow separation occurs during part of an AoA oscillation, a phenomenon known as "dynamic stall" may occur. In this case, both amplitude and phase of the airfoil lift response cannot be described by Theodorsen's theory, which is limited to unsteady attached flow conditions and small perturbations in general. In fact, the "hysteresis" behavior changes from counter-clockwise to a clockwise direction as the airfoil stalls, lift breaks down, and only recovers later during the oscillatory cycle. The subject of "dynamic stall" has been studied in great detail in the rotorcraft community, see Chapter 8 of Leishman (2006) for a comprehensive treatment on the subject. For rotorcraft, the combined effects of both leading- and trailing-edge stall are considered in conjunction with compressibility effects, having resulted in model parameters that describe particular events during a dynamic stall cycle. The most prominent dynamic stall model in the rotorcraft community is the "Leishman-Beddoes Model" (Leishman and Beddoes 1989) that also considers airfoil moment and drag coefficients, with the former being very important with respect to mechanical pitch-link loads on the retreating blade side of a helicopter in forward flight. In this context, it is important to understand that dynamic stall parameters are very sensitive to the actual airfoil shape, and one has to be very cautious in using rotorcraft airfoil parameters on wind turbine blades. This is also because wind turbine airfoils are typically governed by trailing-edge separation and stall, and the effect of dynamic stall on the lift coefficient is considered the most dominant. The most comprehensive database for dynamic stall including the (stall-controlled) S809 wind turbine airfoil was generated at the University of Glasgow (Galbraith et al. 1992; Sheng et al. 2006a,b), and the reader is encouraged to explore this truly unique database and resultant dynamic stall model for the stall-controlled S809 airfoil (Sheng et al. 2006c, 2008). In the wind energy community, the dynamic stall model of Øye (1991) has been widely adopted due to its generality and ease of implementation. Though it is not a physical model that captures details of particular airfoil-dependent events during a dynamic stall cycle, it is quite practical to at least capture general characteristics of a dynamic lift response. According to Øye (1991), the lift-coefficient response to an oscillatory AoA variation can be described as:

$$\text{Dynamic Stall (Øye)}: c_l(\alpha, t) = f_s(t)\, c_{l,inv}(\alpha) + (1 - f_s(t))\, c_{l,fs}(\alpha) \tag{5.10}$$

Here $c_{l,inv}(\alpha)$ is an equivalent inviscid airfoil lift coefficient, and $c_{l,fs}(\alpha)$ is the lift coefficient in fully separated flow, for example for a cambered flat plate with a sharp leading edge. The time-dependent function $f_s(t)$ is defined as a "separation function" and practically determines the temporal weighting of the airfoil lift coefficient between inviscid ($c_{l,inv}$) and fully separated ($c_{l,fs}$) flow. Note that the static lift coefficient, $c_l(\alpha)$, is recovered for $f_s(t) = f_s^{st}$ where $f_s^{st}$ is the static (or equilibrium) value that can be determined from a given $c_l(\alpha)$ by solving Eq. (5.10) such that

$$f_s^{st} = \frac{c_l - c_{l,fs}}{c_{l,inv} - c_{l,fs}} \tag{5.11}$$

For a given (static) airfoil $c_l(\alpha)$, distributions for $c_{l,inv}(\alpha)$ and $c_{l,fs}(\alpha)$ are given by:

$$c_{l,inv}(\alpha) = c_{l,\alpha}(\alpha - \alpha_0) \tag{5.12}$$

$$c_{l,fs}(\alpha) = \begin{cases} 0.5 \cdot c_{l,\alpha}(\alpha - \alpha_0), & \alpha < \alpha_1 \\ c_l, & \alpha \geq \alpha_1 \end{cases} \tag{5.13}$$

In the previous equations, $c_{l,\alpha}$ is the lift-curve slope in the linear (attached) flow regime of the airfoil, and $\alpha_0$ is the airfoil zero-lift angle, that is, $c_l(\alpha_0) = 0$. Note that Eqs. (5.12) and (5.13) are practically valid only for $\alpha > \alpha_0$, and $c_{l,fs}$ is initially approximated by a linear flat-plate type approximation with one-half of the (linear) lift-curve slope, $c_{l,\alpha}$. The angle $\alpha_1$ is defined by the intersection of the linear regime in Eq. (5.13) with the actual (stalled) $c_l$ curve. An example illustration for a wind turbine airfoil is shown in Figure 5.4. A more refined version of Eq. (5.13) can be for example, found in Branlard (2017).

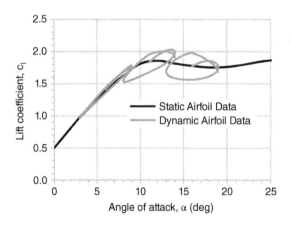

Figure 5.4 Illustration of Øye dynamic stall model – FFA-W3-360 Airfoil ($\alpha(t) = \bar{\alpha} + 3°\sin(\omega t), \omega = 0.94\ \text{s}^{-1}, \tau = 1\ \text{s}$).

Static Airfoil Data
Dynamic Airfoil Data

As for the separation function, $f_s(t)$, it is assumed that $f_s(t) \to f_s^{st}$ in a time response to a change in angle of attack, $\alpha(t)$. In other words, $f_s$ is always going to seek the static $f_s^{st}$ value at a given instantaneous $\alpha(t)$. This leads to the following simple ordinary differential equation for $f_s$:

$$\text{Differential Eq. for } f_s : \quad \frac{df_s}{dt} = \frac{f_s^{st} - f_s}{\tau} \tag{5.14}$$

In Eq. (5.14), $\tau = A \cdot c / V_{rel} (\approx A/\Omega \cdot c/r)$ is a characteristic time for the local airfoil flow, with $c$ being the airfoil chord length and $V_{rel}$ being the relative velocity magnitude experienced by a turbine blade section (or $U_\infty$ for airfoil flow). Note that for $A = 1$, the characteristic time, $\tau$, is equal to the time it takes the local relative velocity, $V_{rel}$, to traverse over the local airfoil chord, $c$. In general, $A$ ranges between 3 and 4 (Hansen et al. 2004). Eq. (5.14) can be easily integrated to obtain the following relation:

$$f_s(t + \Delta t) = f_s^{st} + (f_s(t) - f_s^{st}) e^{-\frac{\Delta t}{\tau}} \tag{5.15}$$

The example in Figure 5.4 reveals that the Øye dynamic stall model only becomes active in the separated flow regime when $f_s(t) \neq f_s^{st}$, and it is important to remember that it actually does not model physical dynamic events, but nonetheless is suitable in wind turbine performance analyses to capture the general behavior of a dynamic stall cycle for the lift coefficient. One can estimate an equivalent reduced frequency, $k$, based on the time constant, $\tau$, by setting $\omega = 2\pi/\tau$ to find that $k = \pi/A$ is of order $O(1)$, which is indeed in the (highly) unsteady regime.

### 5.1.3 Relevance of Atmospheric Boundary Layer on Unsteady Aerodynamics

The question arises as to how important unsteady aerodynamics is for wind turbine blades subject to time-varying inflow conditions in the ABL. In this respect, it is unfortunate that the term "unsteady aerodynamics" is often used incorrectly as a general term of time-dependent blade loads; however, as outlined in previous sections, "unsteady aerodynamics" truly means a non-static response to airfoil AoA changes. For this reason, let us develop a few simple relations to estimate reduced frequencies, $k$, (and hence unsteady aerodynamics) occurring on wind turbine blades. For a wind turbine blade section, a radial distribution of reduced frequency, $k(r)$, can be defined as:

$$\text{Reduced Frequency (Turbine)}: \quad k(r) = \frac{n\Omega \cdot c(r)}{2 V_{rel}(r)} \tag{5.16}$$

In Eq. (5.16), $n\Omega$ is a rational number multiple of the rotor angular velocity, $\Omega$. Hence a reference reduced frequency is based on the "one-per-revolution" frequency, that is, $n = 1$. Further assuming as a first-order approximation that the local relative velocity magnitude $V_{rel} \sim \Omega r$, Eq. (5.16) can be written as:

$$\text{Reduced Frequency 2 (Turbine)}: \quad k(r) \cong \frac{n}{2} \cdot \frac{c}{r} \tag{5.17}$$

In reference to classical unsteady airfoil aerodynamics in Section 5.1.1.2 and unsteady hysteresis effects in the lift response becoming notable for $k \geq 0.05$, unsteady aerodynamics may occur at radial blade stations when the following condition is satisfied:

$$\text{Unsteady Aerodynamics (Turbine)}: \quad \frac{c}{r} \geq 0.10 \frac{1}{n} \quad \text{or} \quad \frac{c}{R} \geq 0.10 \frac{1}{n} \frac{r}{R} \tag{5.18}$$

Equation (5.18) particularly means that in a given blade chord distribution, $c(r)$, is more likely to experience unsteady aerodynamics at frequencies higher than the rotor speed ($n > 1$) and at inboard blade sections where blade chord is higher and $V_{rel}$ is lower compared to outer blade sections. In the following, local time-varying blade loads are categorized into those that occur at the rotor frequency and those due to atmospheric turbulence.

### 5.1.3.1 Effect of Yawed Inflow, Mean Shear, and Tower Interaction

Let us focus first on time-varying blade loads due to the rotor angular speed, $\Omega$, (or $n = 1$). These include yawed inflow, sheared inflow, and blade-tower interaction. As for possible unsteady aerodynamic effects associated with the rotor angular speed, $\Omega$, the interested reader can verify that for typical low-solidity blade designs such as the Penn State University (PSU) 1.5-MW turbine (Schmitz 2015) or the National Renewable Energy Laboratory (NREL) 5-MW turbine (Jonkman et al. 2009), Eq. (5.18) is practically satisfied for $r/R < 0.50$. In other words, unsteady effects may occur from the blade root up to half of the blade radius, including possible "dynamic stall" responses as lift coefficients, $c_l$, are typically high(er) at inboard stations, see for example, Figure 3.25.

***Turbine Yaw*** In stationary yawed flow, the incoming axial wind speed, $V_0$, is oriented by a constant angle, $\gamma$, (about the tower axis) to the rotor disk (turbine plane), see Figure 5.5. In Blade-Element Momentum (BEM) theory, this means in particular that the axial wind speed (normal to the rotor disk) becomes $V_0 \cos \gamma$; furthermore, the azimuthal inflow to a blade element now also depends on the rotor/blade azimuthal (or circumferential) angle, $\psi$, because a rotating blade element is influenced by the corresponding $V_0 \sin \gamma$ component of the incoming wind speed, depending on the rotor/blade azimuthal angle, $\psi$. The following perturbation velocity vector $(V_0', v', w')^T$ can be defined using a suitable transformation matrix according to definitions in Figure 5.5:

$$\text{Transformation (Yaw)} : \begin{pmatrix} V_0' \\ v' \\ w' \end{pmatrix} = \begin{pmatrix} \cos \gamma & \sin \gamma & 0 \\ -\sin \gamma \sin \psi & \cos \gamma \sin \psi & \cos \psi \\ \sin \gamma \cos \psi & -\cos \gamma \cos \psi & \sin \psi \end{pmatrix} \begin{pmatrix} V_0 \\ 0 \\ 0 \end{pmatrix} \quad (5.19)$$

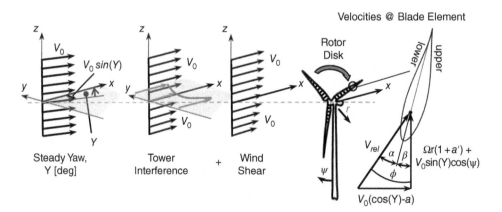

**Figure 5.5** Effect of yaw, tower interference, and wind shear on rotor inflow.

In reference to Figure 5.5, it becomes clear that turbine yaw results in "one-per-revolution" changes in the local angle of attack, $\alpha$, at a given blade element that are governed by changes in a blade element's azimuthal velocity component, $\Omega r(1+a') + V_0 \sin\gamma \cos\psi$. A simple estimate for AoA changes can be realized by assuming that the angular induction factor $a' \ll 1$ and by normalizing the azimuthal velocity component by $V_0$ such that the expression $V_0(\lambda_r + \sin\gamma \cos\psi)$ allows us to actually estimate changes in the local blade flow angle, $\phi$ (and hence angle of attack, $\alpha$, for constant local blade pitch, $\beta$). This can be done using $\cos\psi = \mp 1$ and essentially looking at the influence that $\sin\gamma$ has on $\lambda_r$ over one rotor revolution. Hence larger AoA changes are expected to occur for higher turbine yaw angle, $\gamma$, in general and at inboard blade stations with lower local speed ratio, $\lambda_r$, in particular. An example simple "napkin" estimate for expected "one-per-revolution" AoA changes, $\Delta\alpha$, is documented in Table 5.1 for a utility-scale wind turbine. It becomes clear that $\Delta\alpha$ increases with turbine yaw angle, $\gamma$, and is higher at inboard stations in general and lower tip speed ratio, $\lambda$, in particular. This is important to know as a turbine yaw angle of $\gamma > 10°$ may actually lead to dynamic stall events at inboard blade stations for Region III wind speeds.

*Tower Interference*  As far as tower interference (or blade-tower interaction) is concerned, each turbine blade passes in front of the tower (i.e. "upwind" configuration) once every rotor revolution. Here the tower acts as a "blockage" to the incoming wind speed, $V_0$, which is, at least to first order, the same for all radial blade sections; however, tower interference effects on local "one-per-revolution" AoA changes, $\Delta\alpha$, are expected to have most impact at inboard blade sections due to lower local $\lambda_r$. Turbine blades typically experience tower "blockage" over an azimuthal range of 30°, with the largest wind deficit when the blade passes exactly upstream (or downstream) of the tower. In that respect, tower interference is not really a harmonic one-per-revolution oscillation but more of a ramp or step-change response. Several tower interference models have been proposed in the literature. In this context, some examples are the works of Bak et al. (2001) and Chattot (2006). An even simpler model for a quick "napkin estimate" is that of superposition of a uniform flow and a potential doublet in classical incompressible potential-flow theory, see for example, Wilcox (2010). At a distance of one tower (cylinder) radius upstream of the tower (cylinder) surface, the freestream velocity along the stagnation streamline has reduced to 75% of its undisturbed (freestream)

**Table 5.1** "Napkin" estimate of periodic angle-of-attack variations (PSU 1.5-MW Turbine). Bold text emphasizes the occurrence of the unsteady effect.

| PSU 1.5-MW Turbine | $V_0 = 8\frac{m}{s}$ ($\lambda = 6.6$) | | $V_0 = 20\frac{m}{s}$ ($\lambda = 2.8$) | |
|---|---|---|---|---|
| | $r/R = 0.3$ | $r/R = 0.9$ | $r/R = 0.3$ | $r/R = 0.9$ |
| Reduced frequency | $\mathbf{k \approx 0.114}$ | $k \approx 0.017$ | $\mathbf{k \approx 0.114}$ | $k \approx 0.017$ |
| Mean angle of attack | $\bar{\alpha} = 11°$ | $\bar{\alpha} = 8°$ | $\bar{\alpha} = 14°$ | $\bar{\alpha} = -1.5°$ |
| Yawed flow ($\gamma = 10°$) | $\Delta\alpha \approx 1.5°$ | $\Delta\alpha \approx 0.25°$ | $\mathbf{\Delta\alpha \approx 5°}$ | $\Delta\alpha \approx 0.5°$ |
| Yawed flow ($\gamma = 30°$) | $\Delta\alpha \approx 5°$ | $\Delta\alpha \approx 0.75°$ | $\mathbf{\Delta\alpha \approx 10°}$ | $\Delta\alpha \approx 3.5°$ |
| Wind shear ($\alpha = 1/7$) | $\Delta\alpha < 1°$ | $\Delta\alpha \approx 1°$ | $\Delta\alpha \approx 1°$ | $\Delta\alpha \approx 2°$ |
| Tower interference | $\Delta\alpha \approx 5°$ | $\Delta\alpha \approx 2°$ | $\mathbf{\Delta\alpha \approx 10°}$ | $\Delta\alpha \approx 5°$ |

value. Some associated AoA variations, $\Delta\alpha$, are listed in Table 5.1, and it can be seen that inboard blade sections at a high Region III wind speed ($V_0 = 20$ m s$^{-1}$) experience highest values for $\Delta\alpha$. As inboard blade sections are already more prone to unsteady aerodynamics because of local values for the reduced frequency, $k(r)$, and operate at higher mean angle of attack, $\bar{\alpha}$, one has to be aware of the possibility for dynamic stall responses under these conditions.

*Wind Shear*  As far as mean wind shear is concerned, rotor blades sweep through a variable mean incoming wind speed as a function of height, $V_0(z)$, see also Figure 5.5. Note here that outer blade elements experience a larger variation over the course of one rotor revolution than do inboard blade elements. These steady wind speed variations with height can be estimated following, for example, the log-law and power-law extrapolations introduced in Section 1.2.2.1. Assuming an approximated turbine hub height of $z_{hub} = 80$ m and rotor radius of $R = 40$ m, it can be easily shown that the blade tip experiences variations in the incoming wind speed on the order of $\mp 0.1 \cdot V_0(z_{hub})$. The associated "one-per-revolution" AoA variations, $\Delta\alpha$, are listed in Table 5.1, and it can be seen that these are quite small. In fact, it is unlikely that either notable unsteady aerodynamic effects or dynamic stall events occur due to typical mean wind shear. One has to be aware, however, that "instantaneous" wind shear can look very different due to turbulent eddy scales in the ABL (see Section 5.1.3.2).

The reader has to be aware that the preceding simple estimates of unsteady aerodynamic effects occurring on wind turbine blades can only be accurate to first order. Nevertheless, they provide a basic understanding of the nature of unsteady aerodynamics and under what conditions it can be important for large utility-scale wind turbine blades. In reality, however, both wind speed and direction can vary notably in the ABL and hence over the rotor-disk area. The following section provides a brief introduction.

### 5.1.3.2 Effect of Atmospheric Turbulence

In reference to Section 1.2.2.2, the energy-dominant turbulent eddies in the lower ABL are typically of order 100 m in the vertical and 500 m in the horizontal direction, and are hence of the order of the rotor diameter of a utility-scale wind turbine. As such, it takes about 10 rotor revolutions for such a dominant eddying structure to pass through the rotor-disk area of a multi-MW wind turbine, assuming a mean convection wind speed of 10 m s$^{-1}$. Consequently, the associated low-frequency variations in wind speed are expected to result in quasi-steady responses of blade loads and rotor power as respective reduced frequencies, $k$, are definitely below 0.05 for $n < 1$ in Eq. (5.17). The culprit here, however, is that these energy-dominant eddies do not fill the rotor disk uniformly; specifically, the instantaneous inflow to the rotor disk may be in part within a low-/high-speed eddying structure, while the remaining part of the rotor disk is not, thus having wind-speed variations as large as 50% of the mean wind speed over parts of the rotor disk. This in turn can have notable effects on local ramp-like blade load responses that are not easily quantifiable by classical harmonics (Aguasvivas et al. 2015). A notable work in this context looked at a combination of unique datasets collected from a General Electric (GE) 1.5 MW wind turbine during the daytime and data from a Large-Eddy Simulation (LES) of an equivalent atmospheric boundary-layer state, with the rotor being modeled by an actuator-line method (see Chapter 7). This work by Nandi et al. (2017) analyzed time-varying characteristics of the local velocity field and computed correlation coefficients between met-tower anemometers, leading-/trailing-edge velocity probes at several radial blade sections, and electrical rotor power of the instrumented GE 1.5-MW turbine. A total of three different time scales were observed to be affecting rotor loads and electrical power:

- Energy-Dominant Eddies: $\approx 25s - 50s$
- Once-per-Revolution (1P): $\approx 3s - 6s$
- Sub-1P Scale (Internal Eddy Structure): $< 1s$

Computed and measured power spectra revealed that the "one-per-revolution" (1P) response is dominant, which is in fact not surprising as this corresponds to the time it takes each rotor blade to sweep once over inflow variations over the rotor-disk area. However, an additional interesting result was an observed broadening of spectra surrounding the 1P peak, which indicated local AoA variations due to horizontal (i.e. wind shear) velocity fluctuations. These can be attributed to eddy advection and both the intersection between larger low-/high-speed eddying structures and, furthermore, internal eddy structure itself. The effect can be both ramp-like and harmonic-like fluctuations and responses. In reference to the previous section, note that higher than 1P oscillations may actually result in "unsteady" responses along the entire blade length based on an equivalent reduced frequency, $k$. The question is then whether such higher than 1P frequency forcing is strong enough to lead to unsteady aerodynamic blade responses.

An introductory study to answer this question was conducted by Jha and Schmitz (2016) who focused on modeling a turbine-turbine interaction problem of two NREL 5-MW turbines (Jonkman et al. 2009) at a Region II mean wind speed and within a LES framework of both a neutral and moderately-convective ABL. In order to quantify the potential for unsteady aerodynamic effects due to ramp-like events with higher than 1P frequency, radial AoA distributions were sampled in real time along the rotating turbines blades modeled as actuator lines (see Chapter 7). The local rate-of-change, $\dot{\alpha}(r,t)$, was used to define some type of an instantaneous reduced frequency as:

$$\text{Reduced Frequency 3 (Turbine)}: k(r,t) = 2\pi|\dot{\alpha}(r,t)|\frac{c(r)}{2V_{rel}(r)} \tag{5.20}$$

Note that the definition in Eq. (5.20) does neither quantify the duration nor the strength of any harmonic or ramp-like event at a local blade section; however, it allows to quantify what "percentage of time" a given blade element operates under potentially unsteady conditions. It was found that for the upstream turbine, inboard blade sections may operate 30% of the time at (instantaneous) reduced frequencies in the "unsteady regime" (i.e. $k > 0.05$), while outboard blade sections are practically responding "quasi-steady" to inflow variations. It is interesting that this changes for the downstream turbine, which experiences higher turbulence intensity in the inflow and associated more energetic turbulent structures with length scales smaller than the rotor diameter. Indeed, for the downstream turbine, inboard blade sections operate as much as 50% of the time at $k > 0.05$, with outboard blade sections close to 10% of the time. Initial applications of the Øye dynamic stall model showed that some of the potential unsteady events do not have enough energy and strength to result in dominant unsteady responses from an aerodynamics perspective. Nevertheless, the question warrants further investigation (a hint to the interested reader), particularly during startup and high Region III wind speeds. Ultimately, the significance of unsteady aerodynamics on wind turbine blades has to be assessed in its impact on potentially increased blade fatigue due to fluctuating blade loads/moments and associated variations in blade torque and rotor power that may result in less-than-optimal operation.

## 5.2 Rotational Augmentation and Stall Delay

One of the many challenging aspects of BEM-type methods is the accurate prediction of three-dimensional (3-D) effects associated with a rotating blade reference frame. The difficulty of the problem-at-hand is further exacerbated by the need of correctly scaling 3-D effects from model-scale to utility-scale wind turbine blades. Rotational effects are known to "augment" blade loads at inboard and mid-span stations, hence the term "rotational augmentation," and do have to be taken into account for accurate power and load predictions; see Schreck and Robinson (2007), Breton (2008), and Breton et al. (2008). In general, rotational effects result in increased lift and a delay in static airfoil stall at inboard blade sections, caused by the so-called Himmelskamp effect.

### 5.2.1 Himmelskamp Effect

The first record of augmentation of aerodynamic loads on a rotating blade was an experiment on propeller blades conducted by Himmelskamp (1947) who hypothesized that the aerodynamic load augmentation was due to the combined action of centrifugal and Coriolis forces. Himmelskamp investigated the performance of a propeller blade equipped with a Göttingen 625 airfoil. Having knowledge of the (non-rotating) 2-D wind-tunnel airfoil data, pressure-tap measurements on the actual rotating blade indicated that the airfoil was exceeding the static maximum lift coefficient, $c_{l, max} = 1.4$, at the respective chord Reynolds number. Furthermore, he found that the measured $c_l$ versus $\alpha$ curves at different radial stations were approaching the 2-D wind-tunnel data at progressively more outboard stations. Figure 5.6 depicts Himmelskamp's original data.

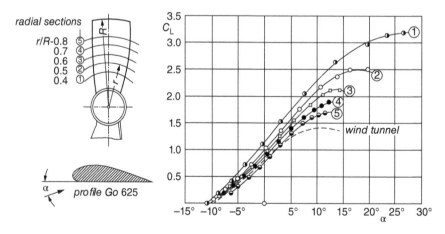

**Figure 5.6** Himmelskamp's measurements of rotational augmentation (or stall delay) on rotating propeller blades (Himmelskamp 1947).

Actual theoretical analyses followed almost two decades later by Banks and Gadd (1963) who found that the effect of blade rotation was to delay the laminar separation point or to prevent separation altogether and was caused by an alleviation of the adverse pressure gradient acting on the boundary layer. Similar theoretical work was conducted in the rotorcraft field by McCroskey and Yaggy (1968) and McCroskey (1971). Later work by Du and Selig (1998, 2000) confirmed that rotation does delay the separation point also on a wind turbine blade, and Madsen and Rasmussen (1988) had suggested earlier that stall delay on wind turbine blades is indeed due to a combined action of the Coriolis effect and centrifugal pumping.

### 5.2.2 Coriolis Effect and Centrifugal Pumping

Today, it is well accepted that the two primary contributors to the "Himmelskamp Effect" are (i) Coriolis force on the boundary layer and (ii) Centrifugal Pumping, both of which are described in more detail in the following sections.

#### 5.2.2.1 Coriolis Effect
The Coriolis effect is associated with a planar force acting perpendicular to the motion of a fluid parcel within a rotating reference frame. The effect is attributed to the work of Gaspard-Gustave Coriolis (1835) who discovered the effect in the context of large geophysical flows (e.g. the nature of hurricane trajectories is to a large extent governed by the action of the Coriolis force). The Coriolis force can be expressed as:

$$\text{Coriolis Force Vector}: \vec{F}_{Co} = -2m\vec{\Omega} \times \vec{v}_{rel} \tag{5.21}$$

Here, $m$ is the mass of a fluid parcel, $\vec{\Omega}$ is the angular velocity of the rotating reference frame, and $\vec{v}_{rel}$ is the velocity vector of the fluid parcel relative to the rotating system. On a rotating wind turbine blade, the velocity vector, $\vec{v}_{rel}$, relative to the rotating reference frame, is primarily in the chordwise direction; however, it has a radial-flow component both at the blade root and the blade tip, see Figure 5.7.

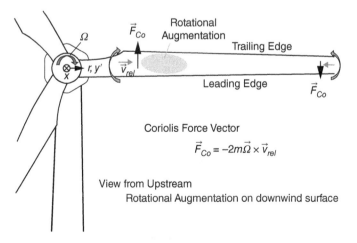

**Figure 5.7** Direction of Coriolis forces acting at blade root and tip.

Note that Figure 5.7 is viewed from upstream, with rotational augmentation occurring on the downstream (airfoil upper) surface. Hence the radial flow associated with the formation of the root vortex, that is, from the pressure (upstream) surface to the suction (downstream) surface, is directed outboard, while the opposite is true for the tip vortex. Application of Eq. (5.21) to the flow situation in Figure 5.7 reveals that the Coriolis force at the blade root has a component toward the trailing edge and at the blade tip toward the leading edge, respectively. The result is that the Coriolis force is accelerating the boundary-layer flow at inboard stations, while it is decelerating the former close to the blade tip. In reference to typical blade operating conditions, see Section 3.7.6, inboard blade sections operate at significantly higher lift coefficient, $c_l$, than their outboard counterparts, and one can therefore expect that the action of the Coriolis force has a more prominent effect at blade sections that already operate close to their respective performance limit with respect to $c_{l,\,max}$ and beginning flow separation. We therefore continue by focusing our attention on airfoil sections close to the blade root and perform a *Gedankenexperiment* as to the effect of the Coriolis force on the associated boundary-layer dynamics, see Figure 5.8.

As discussed in Section 4.1.2.4, the physical mechanism of flow separation and stall is a prolonged action of an adverse pressure gradient on the upper-surface boundary layer at higher angles of attack. Figure 5.8 (left) shows an example distribution of the upper-surface pressure coefficient, $-c_p$, for an airfoil flow with a notable amount of flow separation. A Coriolis force acting toward the trailing edge of the airfoil section is practically equivalent to adding some favorable pressure gradient to fluid parcels in the boundary layer, thus effectively mitigating the adverse pressure gradient causing flow separation at that angle of attack, see Figure 5.8 (right). Hence accelerating the boundary-layer flow due to the action of the Coriolis force will both increase airfoil lift (due to flow acceleration) and "delay" separation and stall to higher angles of attack, $\alpha$, by mitigating the adverse pressure gradient that causes the boundary-layer flow to ultimately reverse and thus separate.

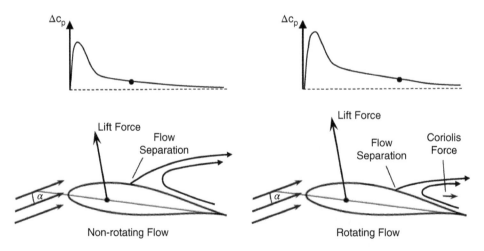

**Figure 5.8** Effect of Coriolis force on boundary-layer lift dynamics at inboard blade stations.

### 5.2.2.2 Centrifugal Pumping

The Coriolis effect alone, as outlined in the previous section, cannot fully explain rotational augmentation and stall delay effects, the reason being that the radial flow associated with the formation of the blade root vortex does not extend up to blade sections where rotational augmentation and stall delay are still observed, see for example, Himmelskamp's original data. Hence there must be an additional physical effect that supports the observations. For attached flow in a rotating reference frame, the centrifugal force acting on boundary-layer fluid parcels does not, at least in general, have a measurable effect on blade section flows (and fortunately so, as this would otherwise be the demise of BEM-type methods). Nevertheless, the centrifugal force does affect existing separation bubbles as fluid parcels that are trapped in a separated region do not possess the chordwise momentum and energy to not be affected in their pathlines by the centrifugal force. On the contrary, the centrifugal force initiates an effect known as "centrifugal pumping," generating an outboard pressure gradient that effectively elongates separated regions further outboard on the blade, see Figure 5.9.

The radial pressure gradient associated with centrifugal pumping is largest between the suction peak of the respective airfoil section(s) and the separation point. The associated radial flow prolongs the Coriolis effect, which by itself enhances the lift generation and acts as a limiter to a growing separated region. Figure 5.9 illustrates a radially growing, though decreasing in chordwise extent, separated region. We will learn about some general scaling relations and models for rotational augmentation in Section 5.1.3. Here it is worth capturing some of the literature discussing rotational augmentation on wind turbine blades. The reader has to be aware that all flow separation is practically 3-D, though wind-tunnel data are typically considered being practically 2-D high-AoA airfoil data of separated and stalled flow.

*Literature on the Coriolis effect and centrifugal pumping*: In general, there is consensus in the wind energy community that rotational augmentation and stall delay are due to the Coriolis effect and centrifugal pumping. In this context, some researchers hypothesize that the Coriolis effect dominates centrifugal pumping on wind turbine blades, see

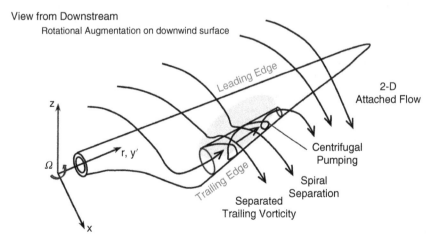

**Figure 5.9** Effect of centrifugal pumping on a rotating wind turbine blade.

Lindenburg (2004) and Carcangiu et al. (2007). However, the exact balance between both effects is predominantly a function of the rotor blade geometry, the blade operating conditions, and the general separation/stall behavior of the airfoils themselves. For example, the (stall-controlled) NREL Phase VI rotor shows rotational augmentation effects and, in addition, shear-layer impingement of the separated flow region on the upper blade surface as well as a standing vortex trailing from a radially extended separated flow region, see works by Schreck and Robinson (2002, 2007), Tangler and Kocurek (2005), Schmitz and Chattot (2006), Sant et al. (2006), Schreck et al. (2007, 2010, 2013), and Gonzalez and Munduate (2008). Nevertheless, rotational augmentation reveals itself in general by an increased lift force in the separated flow region at inboard blade stations. The effect of blade rotation on the drag force in the separated region, however, is still not fully understood and subject to continued debate. While some researchers argue that the drag force decreases in the separated flow region (due to a smaller chordwise separation bubble), see Du and Selig (1998) and Corten (2001), the majority of researchers suggest an increase in the drag force caused by rotational augmentation on a rotating wind turbine blade, see for example, Sørensen (2000), Sørensen et al. (2002), Chaviaropoulos and Hansen (2000), Tangler (2004), Bak and Fuglsang (2004), Gerber et al. (2005), and Breton et al. (2008). The effect of the general separation behavior of a given blade geometry (planform) was investigated by several researchers through the use of parked blade data, see for example, Madsen and Christensen (1990), Ronsten (1992), Eggers et al. (2003), Johansen et al. (2002), Schmitz and Chattot (2006), Schreck and Robinson (2007), and Gonzalez and Munduate (2008). Here, most of the work was applied to analyzing data from the NREL Phase VI rotor, and it was indeed found that the general separation behavior of a given blade planform remains consistent between parked and rotating conditions.

### 5.2.3 Stall Delay Models

The phenomenon of rotational augmentation and associated stall delay of airfoil performance has seen continued interest in the wind energy community. Since its discovery by Himmelskamp in the late 1940s, a variety of stall delay models have been developed.
What are stall delay models used for and why are they important?

- Incorporate 3-D rotational effects into BEM-type performance codes
- Knowledge of increased inboard blade loads due to rotational augmentation is important to estimating blade fatigue loads
- An efficient and accurate stall-delay model is useful in the design phase to avoid expensive fully resolved computational fluid dynamics (CFD) simulations
- Important for small-scale stall-controlled wind turbines with fixed blade pitch to control Region III power
- Also important for pitch-controlled wind turbines where inboard stations see, in general, higher angles of attack

In the following, the most prominent stall-delay models in use today are introduced, emphasizing the primary physical concept addressed by the respective model. Most contemporary stall-delay models follow a general approach with

$$\text{General Approach}: c_{l,3D} = c_{l,2D} + f_l \cdot \Delta c_l \qquad (5.22)$$

$$c_{d,3D} = c_{d,2D} + f_d \cdot \Delta c_d \tag{5.23}$$

where $c_{l,3D}$ and $c_{d,3D}$ are the 3-D corrected airfoil lift and drag coefficients, while $f_l$ and $f_d$ are functions of the actual stall-delay model. In Eq. (5.22), $\Delta c_l$ is the difference between the viscous airfoil $c_{l,2D}$ and that from inviscid thin-airfoil theory in Section 4.1.1.2; in Eq. (5.23), $\Delta c_d$ is the difference between the viscous airfoil $c_{d,2D}$ and the drag coefficient at zero angle of attack, that is, $c_{d,0} = c_d(\alpha = 0°)$.

### 5.2.3.1 Snel et al.

An order-of-magnitude analysis of the boundary-layer equations proposed by Snel (1991) and Snel et al. (1992, 1993a,b) was conducted and validated against experimental data obtained on the Aeronautical Research Institute of Sweden (FFA) 5WPX turbine. A simple model for $f_l$ was found to be

$$f_l = 3\left(\frac{c}{r}\right)^2 \tag{5.24}$$

where $c/r$ is the local (i.e. normalized by $r$) chord ratio. The original work of Snel (1991) and the identification of the local chord ratio, $c/r$, as a primary parameter was a pivotal contribution in the area of stall-delay modeling that served as the foundation for all subsequent modeling approaches to rotational augmentation.

### 5.2.3.2 Corrigan and Schillings

The empirical stall-delay model by Corrigan and Schillings (1994) is rooted in the earlier theoretical analysis by Banks and Gadd (1963), though with adjustments based on experimental helicopter data. In essence, the model is formulated as a delay in the static stall angle, that is, $\Delta \alpha$, as:

$$\Delta \alpha = (\alpha_{cl.max} - \alpha_0)\left(\left(\frac{K}{0.136}\frac{c}{r}\right)^n - 1\right) \tag{5.25}$$

In Eq. (5.25), $\alpha_{cl.max}$ is the angle of attack at $c_{l,max}$, $\alpha_0$ is the airfoil zero-lift angle, $K$ is an assumed linear adverse velocity gradient, $c/r$ is the local chord ratio, and the exponent $n$ is typically set to 1. The 2-D lift curve is then shifted according to

$$c_{l,3D}(\alpha + \Delta\alpha) = c_{l,2D}(\alpha) + \frac{\partial c_{l,pot}}{\partial \alpha} \cdot \Delta\alpha \tag{5.26}$$

where $\partial c_{l,pot}/\partial \alpha$ is the airfoil lift slope in the potential flow region. Note that the drag coefficient remains unaltered but is subject to the same shift $\Delta \alpha$, thus resulting in an effective lower $c_d$ at a given $c_l$. Tangler and Selig (1997) later evaluated the Corrigan and Schillings stall-delay model for horizontal-axis wind turbines.

### 5.2.3.3 Du and Selig

Du and Selig (1998, 2000) extended the original work by Snel et al. by performing an analysis of the 3-D integral boundary-layer equations. A modified tip speed ratio was defined according to

$$\Lambda = \Omega R / \sqrt{V_0^2 + (\Omega r)^2} \tag{5.27}$$

which is practically a tip speed ratio based on the relative velocity, $V_{rel}$, for a very lightly loaded rotor (i.e. zero induction). The following relations for $f_l$ and $f_d$ were suggested:

$$f_l = \frac{1}{2\pi} \left[ \frac{1.6\,(c/r)}{0.1267} \cdot \frac{a - (c/r)^{\frac{dR}{\Lambda r}}}{b + (c/r)^{\frac{dR}{\Lambda r}}} - 1 \right]$$

(5.28)

$$f_d = \frac{1}{2\pi} \left[ \frac{1.6\,(c/r)}{0.1267} \cdot \frac{a - (c/r)^{\frac{dR}{2\Lambda r}}}{b + (c/r)^{\frac{dR}{2\Lambda r}}} - 1 \right]$$

(5.29)

In Eqs. (5.28) and (5.29), the additional parameters are suggested to $a = b = d = 1$. The inclusion of the modified tip speed ratio, $\Lambda$, as an additional free parameter resulted in improved comparisons against experimental data on the FFA 5WPX wind turbine and the NREL Combined Experiment Rotor (CER).

### 5.2.3.4 Chaviaropoulos and Hansen

The model proposed by Chaviaropoulos and Hansen (2000) is based on the quasi-3D incompressible Navier–Stokes equations with focus on the radial momentum equation. They hypothesized that rotational augmentation is primarily a function of the local chord ratio, $c/r$, and the blade twist angle, $\beta$, and that the same correction is applied to both lift and drag coefficients according to:

$$f_l, f_d = a \left( \frac{c}{r} \right)^h \cos^n \beta$$

(5.30)

The free parameters in Eq. (5.30) are $a = 2.2$, $h = 1$, and $n = 4$ determined by comparisons against experimental data on a stall-regulated Bonus 300 Combi wind turbine.

### 5.2.3.5 Dumitrescu et al.

The model of Dumitrescu et al. (2007, 2013) is based on the existence of an inboard standing vortex in the blade region affected by rotational augmentation. An augmented local angle of attack, $\alpha_1$, is proposed as

$$\alpha_1 = \tan^{-1} \left( \frac{2}{3} \frac{V_0}{\Omega c_1} \right) - \beta_1$$

(5.31)

where $c_1$ and $\beta_1$ are actual local chord and twist angle. The corresponding $\Delta c_{l1}$ is still the difference between the potential-flow lift coefficient and that of the viscous airfoil polar, though evaluated at $\alpha_1$.

$$\Delta c_{l1} = 2\pi(\alpha_1 - \alpha_0) - c_{l,2D}(\alpha_1)$$

(5.32)

The actual 3D corrected lift coefficient, $c_{l,3D}$, is then computed via

$$c_{l,3D} = c_{l,2D} + \Delta c_{l1} \left[ 1 - \exp \left( -\frac{\gamma}{r/c - 1} \right) \right]$$

(5.33)

assuming a viscous decay of the standing vortex in the radial direction with $\gamma = 1.25$.

### 5.2.3.6 Eggers et al.

Eggers et al. (2003) proposed a model to correct normal and tangential force coefficients, $c_n$ and $c_t$, rather than lift and drag coefficients, $c_l$ and $c_d$. This model also relies on the (local) tip speed ratio, $\lambda_r$, as well as axial and angular induction factors, $a$ and $a'$. It is generally presented in the following form:

$$c_{n,3D} = c_{n,2D} + \Delta c_{n,3D} \tag{5.34}$$

$$c_{t,3D} = c_{t,2D} + 0.12 \times \Delta c_{n,3D} \tag{5.35}$$

$$\Delta c_{n,3D} = \frac{1}{2} \frac{(r_0^2 - r^2)\lambda^2}{(1-a)^2 + (1+a')^2\lambda_r^2} \tag{5.36}$$

In Eq. (5.36), the parameter, $r_0$, is the outermost radial blade location where deep stall or rotational augmentation occur. Note that rotational effects tend to zero either for $r \to r_0$ or in the limit of a non-rotating blade with $\lambda \to 0$.

### 5.2.3.7 Lindenburg

Lindenburg (2003, 2004) based his model on an analysis of separated flow at the trailing edge and respective action of centrifugal forces and resultant centrifugal pumping. Model parameters include the local chord ratio, $c/r$, and a modified tip speed ratio, $\Omega r/V_{rel}$. The centrifugal pumping model is formulated in terms of the normal force coefficient. The shift (or delay) of the separation point toward the trailing edge is expressed as a shift in the angle of attack through the following relation:

$$\alpha_{3D} = \alpha_{2D} + \frac{0.25}{2\pi}1.6\left(\frac{c}{r}\right)\left(\frac{\Omega r}{V_{rel}}\right)^2 \tag{5.37}$$

The rotating angle of attack, $\alpha_{3D}$, is then subsequently used to obtain corrections for the lift and drag coefficients according to:

$$c_{l,3D} = c_{l,2D} + 1.6\left(\frac{c}{r}\right)\left(\frac{\Omega r}{V_{rel}}\right)^2 [(1-f)^2 \cos(\alpha_{3D}) + 0.25 \cos(\alpha_{3D} - \alpha_0)] \tag{5.38}$$

$$c_{d,3D} = c_{d,2D} + 1.6 \sin(\alpha_{3D})(1-f)^2 \left(\frac{c}{r}\right)\left(\frac{\Omega r}{V_{rel}}\right)^2 \tag{5.39}$$

In Eqs. (5.38) and (5.39), the parameter $f$ is the fraction of the blade chord over which the flow is separated.

### 5.2.3.8 Dowler and Schmitz

The original work by Dowler (2013) and Dowler and Schmitz (2015) approaches the challenging problem of rotational augmentation and stall delay from classical fluid mechanics by identifying physical quantities that are relevant to the phenomenon. They are listed in Table 5.2, where $\Delta\beta$ is the total built-in blade twist from root to tip, $d\beta/dr$ is the local twist slope, $Vrel$ is the relative velocity according to Eq. (3.6), and $\Gamma$ is the local blade circulation. The classical Buckingham $\Pi$ Theorem states that there are N less dimensionless groupings than there are dimensional (physical) quantities; here N is the number of independent dimensions. In this case, the independent dimensions are length, time, and radians. Hence there are a total of $9 - 3 = 6$ dimensionless groupings, see Table 5.3.

**Table 5.2** Dimensional parameters relevant to stall delay.

| Blade geometry | Flow conditions | Solution dependent |
|---|---|---|
| $R \ r \ c \ \Delta\beta \ d\beta/dr$ | $V_0 \ \Omega$ | $V_{rel} \ \Gamma$ |

**Table 5.3** Stall-delay dimensionless groupings.

| Blade geometry | Flow conditions | Solution dependent |
|---|---|---|
| $B_1 = \dfrac{r}{R} \quad B_2 = \dfrac{c}{r} \quad B_3 = \dfrac{d\beta/dr}{\Delta\beta/R}$ | $\Phi_1 = \dfrac{\Omega R}{V_0}$ | $\Sigma_1 = \dfrac{\Gamma}{V_0 R} \quad \Sigma_2 = \dfrac{V_0}{V_{rel}}$ |

Note that the choice of dimensionless groupings is non-unique as any $\Pi$ – combination of given dimensionless groupings results in yet another (though not independent) dimensionless grouping. In Table 5.3, $B_1$ is the normalized local radial blade station, $B_2$ is the local chord ratio, $B_3$ is the ratio of the local-to-average twist rate, $\Phi_1$ is the tip speed ratio, $\Sigma_1$ is a dimensionless local circulation, and $\Sigma_2$ is a local velocity ratio. In Table 5.3, $\Sigma_1$ and $\Sigma_2$ are listed as "solution dependent" as neither the local circulation, $\Gamma$, nor the local relative velocity, $V_{rel}$, are known prior to the analysis. In a generalized sense, the 3-D corrected lift and drag coefficients can be written as

$$c_{l,3D}, c_{d,3D} = F(B_1^{b1}, B_2^{b2}, B_3^{b3}, \Phi_1^{f1}, \Sigma_1^{s1}, \Sigma_2^{s2}) \tag{5.40}$$

where the exponents $b_1, b_2, b_3, f_1, s_1,$ and $s_2$ are determined either empirically, from theoretical boundary-layer analyses, or a combination thereof. The local blade circulation, $\Gamma$, is readily available in vortex-based solvers (see Chapter 7) but can be also determined in BEM-type solvers at a local airfoil blade section from the K-J lift theorem, specifically Eq. (4.7), such that

$$\Gamma = \frac{1}{2} V_{rel} c_l c \tag{5.41}$$

which is dependent on $c$ and $V_{rel}$ but also on the actual local (3-D) lift coefficient, $c_l$, as it evolves over the iterative solution-dependent process. A methodology for a solution-dependent stall-delay model is outlined in Figure 5.10 and is the form currently implemented in the *XTurb* code ($STALLDELAY = 2$ in BEM mode).

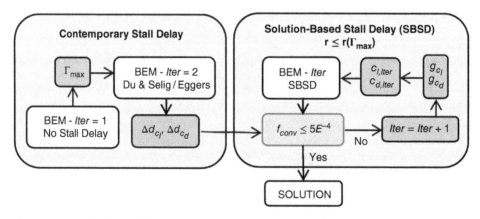

**Figure 5.10** Methodology of a solution-based stall-delay model.

At first, a baseline BEM analysis (*Iter* = 1) is being conducted that determines strength and radial location of the maximum circulation, $\Gamma_{max}$, which is taken as the outermost location on the blade where solution-dependent corrections are being applied. Next, airfoil data are corrected following the Du and Selig model for lift and that of Eggers et al. for the drag, as is done as a standard in, for example, NREL's *AirfoilPrep* worksheet (Hansen 2013), and a subsequent BEM analysis (*Iter* = 2) is performed. As a next step, the difference between computed lift and drag coefficients from the first two solution-dependent iterations are stored as $\Delta d_{c_l}$ and $\Delta d_{c_d}$, respectively. The following BEM analysis iteration (*Iter* = 3) now further corrects all airfoil tables inboard of the original $\Gamma_{max}$ according to

$$c_{l,Iter} = c_{l,Iter-1} + f_{conv} g_{c_l} \Delta d_{c_l} \tag{5.42}$$

$$c_{d,Iter} = c_{d,Iter-1} + f_{conv} g_{c_d} \Delta d_{c_d} \tag{5.43}$$

where $f_{conv}$ is a weighted correction factor normalized by $\Gamma_{max}$ according to

$$f_{conv} = \frac{\|\Gamma_{Iter-1} - \Gamma_{Iter-2}\|}{\Gamma_{max}} \tag{5.44}$$

The solution-dependent process is considered converged when $f_{conv} \leq 5.0 \times 10^{-4}$. The actual correction functions $g_{c_l}$ and $g_{c_d}$ based on dimensionless groupings in Table 5.3 are defined as:

$$g_{c_l} = \left(\frac{c}{r}\right)^2 \left(\frac{\Gamma}{V_{rel}R}\right)^{1/2} \left(\frac{2V_{rel}}{\Omega r}\right)^2 \tag{5.45}$$

$$g_{c_d} = \frac{1}{3}\left(\frac{r}{R}\right)\left(\frac{c}{r}\right)^{-1} \left(\frac{\frac{d\beta}{dr}}{\frac{\Delta\beta}{R}}\right)\left(\frac{2V_{rel}}{\Omega r}\right)^{-1} \tag{5.46}$$

Details on the actual choice of the exponents in Eqs. (5.45) and (5.46) can be found in Dowler (2013) and Dowler and Schmitz (2015), though noting here Snel's original $(c/r)^2$ and Lindenburg's $(2V_{rel}/\Omega r)^2$ scaling in the relation for $g_{c_l}$.

Figure 5.11. shows some examples of 3-D stall-delay corrected airfoil polars (Chaviaropoulos and Hansen, Dowler and Schmitz) for the NREL Phase VI rotor (Hand et al. 2001), which is equipped exclusively with the S809 airfoil. The XTurb input file *PhaseVI-Ch5-1-3.inp* in PREDICTION mode was used setting STALLDE-

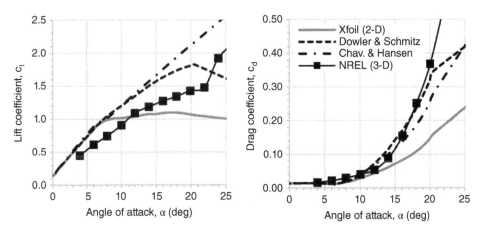

**Figure 5.11** Example of 3-D stall-delay corrected airfoil polar for the NREL Phase VI rotor at $r/R = 0.30$ (S809, $Re = 5 \times 10^5$).

LAY = 2 (Dowler and Schmitz). The 3-D corrected airfoil polars are output to *XTurb_Output_StallDelay.dat*.

---

**XTurb Example 5.1  NREL Phase VI Rotor – XTurb_Output_StallDelay.dat (PhaseVI-Ch5-1-3.inp)**

```
        *** Polar # 1***
          r/R      Original Polar #
          0.2500        1

        Number  AOA[deg]     CL        CD        CDP       CM
        _____  -

                ... ... ...
        *** Polar # 41***
          r/R      Original Polar #
          1.0000        1

        Number  AOA[deg]     CL        CD        CDP       CM
        _____  -

                ... ... ...
```

---

It can be seen in Figure 5.11 that 3-D stall-delay correction results in a notable lift increase at the inboard $r/R = 0.30$ station of the NREL Phase VI rotor. Quantitative comparisons against measured data follow in Section 5.2.4.

Note that the solution-dependent $g_{c_l}$ correction actually allows the lift-curve slope to exceed its respective 2-D value, as had already been observed in the original experiments by Himmelskamp, see Figure 5.6. It is of further note that $g_{c_d}$ is essentially a centrifugal pumping effect that is active only over the blade separated region (bounded by a pair of counter-rotating trailing vortices) between $\Gamma_{max}$ and an additional smaller maximum in the $\Gamma$ distribution inboard of $\Gamma_{max}$, both evaluated in the baseline BEM analysis (*Iter* = 1). Hence if the baseline BEM analysis does not suggest an additional $\Gamma$ maximum inboard of $\Gamma_{max}$, $g_{c_d} = 0$ throughout the solution-dependent iterative process, and the drag correction equals that suggested by Eggers et al. On the other hand, the solution-dependent process can reveal some robustness issues if there are multiple $\Gamma$ extrema inboard of $\Gamma_{max}$ (listed in *XTurb_Output_Circulation.dat*), a behavior the author hopes will be picked up by an interested reader. Indeed, the solution-dependent stall-delay model presented in this section should not be regarded as a final answer to rotational augmentation (as should probably not any contemporary stall-delay model) but more of a methodology useful to future research.

### 5.2.4  Scaling Rotational Augmentation from Small-Scale to Utility-Scale Turbines

It is intriguing (but misleading) to believe that rotational augmentation is a phenomenon limited to model-scale experimental turbines and small-scale wind turbines, simply because the rotational speed, $\Omega$, scales as $\sim 1/R$ for the same design tip speed ratio, $\lambda$. It is once more that a dimensionless parameter can elucidate the relative importance of rotational augmentation on a given wind turbine blade. The Coriolis force vector had been defined in Eq. (5.21); here we define similarly the centrifugal force vector as

$$\text{Centrifugal Force Vector}: \vec{F}_{Ce} = m\Omega^2 \vec{r} \tag{5.47}$$

and suggest that the ratio of magnitudes of the Coriolis and centrifugal force vectors provides a dimensionless grouping relevant to scaling rotational augmentation effects. We obtain the following rotational augmentation parameter:

$$\text{Rotational Augmentation Parameter}: \frac{F_{Co}}{F_{Ce}} = \frac{2v_{rel}}{\Omega r} \tag{5.48}$$

In Eq. (5.48), $v_{rel}$ is the magnitude of the total velocity vector (including spanwise/radial flow) of a fluid parcel relative to the rotating reference frame. In BEM theory, radial flow is not included. Therefore, a modified rotational augmentation parameter can be written in terms of the relative velocity, $V_{rel}$, from Eq. (3.6) as:

$$\text{Modified Rotational Augmentation Parameter}: \frac{2V_{rel}}{\Omega r} = 2B_1^{-1}\Phi_1^{-1}\Sigma_1^{-1} \tag{5.49}$$

Note that Eq. (5.49) is a dimensionless grouping written in terms of other dimensionless groupings in Table 5.3 and is indeed a solution-dependent parameter used in Eqs. (5.45) and (5.46). Values for the modified rotational augmentation parameter in Eq. (5.49) for a given wind turbine blade are a fine balance between the local entrainment speed, $\Omega r$, and the actual (solution-dependent) axial/angular induction factors, $a$ and $a'$, that determine $V_{rel}$ through Eq. (3.6). The interested reader is also referred to the work of Bangga et al. (2017) for a detailed computational investigation on a large utility-scale turbine that also considers inertial scaling effects by means of the Rossby number. Figure 5.12 shows radial distributions of the primary rotational augmentation parameters, $c/r$ and $2V_{rel}/\Omega r$, for the NREL Phase VI (Hand et al. 2001), MEXICO (Snel et al. 2014), and PSU 1.5-MW (Schmitz 2015) turbine rotors. It becomes clear that scaling rotational augmentation effects is in part governed by the blade geometry through the local chord ratio, $c/r$, and the modified rotational augmentation parameter, $2V_{rel}/\Omega r$.

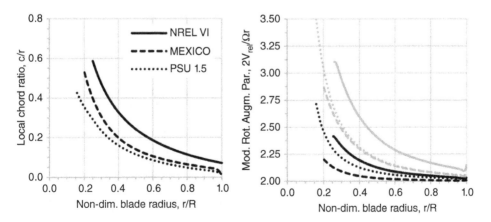

**Figure 5.12** Radial distributions of local chord ratio, $c/r$, and modified rotational augmentation parameter, $2V_{rel}/\Omega r$, for the NREL Phase VI ($V_0 = 7.13$ m s$^{-1}$), MEXICO ($V_0 = 10.24$ m s$^{-1}$), and PSU 1.5-MW ($V_0 = 8.14$ m s$^{-1}$) turbines.

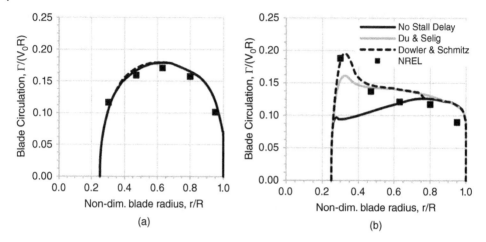

**Figure 5.13** NREL Phase VI rotor – Equivalent radial circulation distribution (a: $V_0 = 7$ m s$^{-1}$, b: $V_0 = 13$ m s$^{-1}$).

Note in Figure 5.12 that $2V_{rel}/\Omega r \to 2$ at the blade tip, though still being $> 2$, as $a' \to 0$ and $V_{rel}(a(r)) > \Omega r$. The gray lines for $2V_{rel}/\Omega r$ in Figure 5.12 denote a respective higher wind speed and lower $\lambda$. Figure 5.13 shows equivalent circulation distributions along the blade radius for the NREL Phase VI rotor (from *XTurb_Output_Circulation.dat*). Also shown are symbols denoting data from Tangler (2004) who derived measured radial circulation values from available sectional force coefficients. It can be seen that rotational augmentation effects are quite evident at the higher wind speed of $V_0 = 13$ m s$^{-1}$ and are captured quite well for the stall-delay models available in *XTurb*.

The circulation distribution scales to first order with the sectional airfoil lift coefficient, $c_l$, and the relative velocity, $V_{rel}$, see Eq. (5.41). However, the actual distribution of normal force [N m$^{-1}$] along the blade radius scales with $V_{rel}^2$ and hence determines rotor thrust and power, see Eqs. (3.16) and (3.17). Figure 5.14 plots *XTurb*

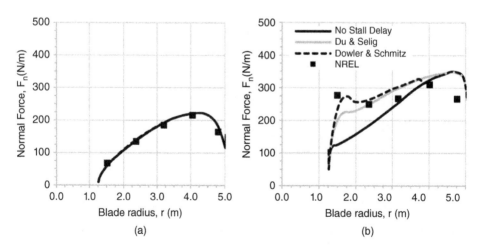

**Figure 5.14** NREL Phase VI rotor – Radial distribution of normal force (a: $V_0 = 7$ m s$^{-1}$, b: $V_0 = 13$ m s$^{-1}$).

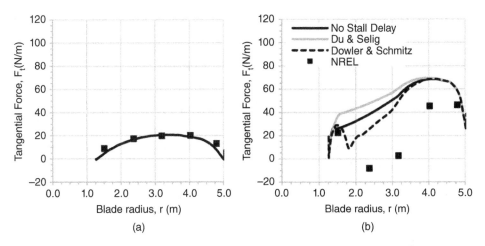

**Figure 5.15** NREL Phase VI rotor – Radial distribution of tangential force (a: $V_0 = 7$ m s$^{-1}$, b: $V_0 = 13$ m s$^{-1}$).

(from *XTurb_Output_PREDICTION.dat*) computed normal force distributions compared to measured NREL data (Hand et al. 2001; Simms et al. 2001) and illustrates the pronounced effect of rotational augmentation at a wind speed of $V_0 = 13$ m s$^{-1}$. Overall, the stall-delay model due to Du and Selig captures the trend quite well, with the iterative solution-based stall-delay model capturing the innermost part of the blade a bit better, which is likely due to the computed axial-/angular induction factors feeding back into the modified rotational augmentation parameter in Eq. (5.49).

The corresponding distribution of tangential force along the blade radius is shown in Figure 5.15. Note that sectional tangential forces are practically one order of magnitude smaller than the sectional normal forces. Again, the wind speed of $V_0 = 13$ m s$^{-1}$ stands out as exhibiting strong rotational augmentation effects on the NREL Phase VI rotor.

The trends are captured quite well, while the additional drop in the tangential force at the inboard part of the blade is predicted well with the solution-based $g_{c_d}$ correction in Eq. (5.46). At a wind speed of $V_0 = 13$ m s$^{-1}$, the (stall-controlled) NREL Phase VI rotor is practically separated/stalled along the entire blade and quite difficult to predict. Note that the NREL definition of the tangential force points to the airfoil leading edge, and not to the trailing edge (Hand et al. 2001), see also Figure 3.3.

### 5.2.5 Extraction of Rotational Augmentation Data from Computed Flow Fields

As of today, the capability of stall-delay models to predict rotational augmentation effects on rotating wind turbine blades is still fairly limited, the primary reason being that only a limited amount of comprehensive data is available on model-scale rotors and even less so on utility-scale wind turbines. The result is that current stall-delay models are designed to capture (or better model) existing measured data and do not necessarily predict similar effects on a different scale rotor. Tangler (2002) discusses this as the "Nebulous Art" of using limited wind-tunnel data to retrofit BEM predictions to actual measurements. In this context, one has to understand that what really matters for a good stall-delay model is to provide consistent predictions among a number of rotors; for example, quantified as a "delta" in performance comparisons. In the end, rotational augmentation effects do have a percentage contribution to rotor power, as can be, for example, quantified by Eq. (3.17), and this is definitely important to design optimization of future wind turbine blades to be used on large utility-scale machines.

The extraction of rotational augmentation data from computed high-fidelity flow fields sounds promising at first, though is challenged by an unambiguous way of determining the local angle of attack in a computed (or measured) flow field. Furthermore, given that CFD does (in general) still not capture an airfoil's maximum lift coefficient, $c_{l,max}$, at relevant Reynolds numbers of $O(10^6)$, establishing credibility in CFD computed 3-D airfoil characteristics is not an easy task. As for determining the local angle of attack in CFD, several methods have been proposed, for example Hansen et al. (1997), Sørensen et al. (2002), Johansen and Sørensen (2004), Schmitz and Chattot (2006), Shen et al. (2009), and Yang et al. (2011), which (among many others) have their own advantages/disadvantages based on the actual rotor they were applied to. In this context, it is worth mentioning an advanced stall-delay model proposed by Bak and Fuglsang (2004) and Bak et al. (2006), which is informed by actual CFD computed pressure distributions, an approach that could be further explored in the future using data from measured pressure taps and/or high-fidelity CFD analyses, see also Guntur and Sørensen (2013).

A recent contribution on determining the local blade angle of attack that is equally applicable to CFD analyses and measured particle-image velocimetry (PIV) fields is a new idea proposed by Herraez et al. (2018) who show theoretically that the flow angle measured/computed at the bisectrix line between rotor blades (in zero yaw) is equal to the flow angle at the respective blade section. Such innovative techniques applied to concurrent CFD analyses and PIV measurements have the potential to shed further light into the difficult topic of rotational augmentation.

## 5.3 Airfoil Characteristics at High Angles of Attack

In reference to Section 4.1.2.4, the region of separated flow over an upper airfoil surface grows with angle of attack, moving the separation point forward toward the leading edge. Hence the suction peak and, consequently, the adverse pressure gradient increase accordingly up to a point where the boundary-layer separation point moves abruptly to the leading edge. At this point, the airfoil enters the "deep-stall" region, and the corresponding angle of attack is referred to as the "deep-stall angle," $\alpha_{Stall}$. The airfoil nose radius is the primary parameter that determines the magnitude of the suction peak (and hence adverse pressure gradient at high angle of attack). Timmer (2010) derived a relation for the deep-stall angle, $\alpha_{Stall}$, by correlating measured deep-stall angles on Delft University of Technology (DUT) airfoils to the upper airfoil surface ordinate at 1.25% airfoil chord, $(y/c)_{0.0125c}$. The relation reads:

$$\text{Approximate Deep} - \text{Stall Angle}: \alpha_{Stall} = 0.1114(y/c)_{0.0125c} \tag{5.50}$$

Note that Eq. (5.50) was derived for DUT airfoils, however can be considered a good estimate for wind turbine airfoil families at utility-scale Reynolds numbers. One has to keep in mind here that $\alpha_{Stall}$ will change for airfoils operating at model-scale Reynolds numbers $Re = O(10^5)$ and also changes to some extent with turbulence intensity and leading-edge surface contamination.

Since not many airfoils have been tested for angles of attack $\alpha_{Stall} \le \alpha \le 90° \ldots 180°$ (also because of high wind-tunnel blockage) but data are needed for turbine startup and stop/brake analyses, models for lift and drag characteristics have been developed based on available data. Spera (2008) and Wood (2011) are excellent references for more in-depth documentation on designated experiments and model developments. In the following, we limit ourselves to the most common models used in wind turbine analysis codes.

### 5.3.1 Flat-Plate Correction

At high angles of attack, a rule-of-thumb is that a reasonably thin airfoil behaves similar to a 2-D flat plate with $c_l \to 0$ and $c_d \to 2$, see an early experiment by Fage and Johansen (1927). In reality, however, the aspect ratio, $AR$, of the model/blade plays a contributing rule as a result of induced edge effects. Hoerner (1965) suggests relations that resulted from approximating available experiments on flat-plate shapes as:

$$\text{Flat Plate Drag}: c_{d,max} = 1.111 + 0.018AR, \quad AR < 50 \tag{5.51}$$

$$c_{d,\max} = 2.01, \quad AR \ge 50 \tag{5.52}$$

In the context of wind turbine blades, $AR$ is defined as the blade radius, $R$, divided by the blade chord at the 75% radial blade station, that is, $AR = R/c_{0.75R}$. Here it is worth noting that practically all wind turbine blades have $AR < 50$, which can be important when analyzing startup and stop/brake conditions.

### 5.3.2 Viterna–Corrigan Correction

Based on Eqs. (5.51) and (5.52), Viterna and Corrigan (1981) proposed airfoil lift and drag characteristics at high angles of attack for $\alpha > \alpha_s$ where $\alpha_s$ is a reference angle of attack somewhere in the stall region between the angle of attack at $c_{l,max}$ and the deep-stall angle, $\alpha_{Stall}$. The Viterna–Corrigan correction is the most commonly used high AoA method today. For the drag coefficient, $c_d$, the following relation is proposed

$$\text{High-}\alpha \text{ Drag Coefficient}: c_d = c_{d,max}\sin^2\alpha + K_1 \cos\alpha \tag{5.53}$$

where $K_1 = [c_{d,s} - c_{d,max}\sin^2\alpha_s]/\cos\alpha_s$ and the index $s$ refers to airfoil properties at the reference angle of attack, $\alpha_s$. The high AoA lift coefficient, $c_l$, is computed from

$$\text{High-}\alpha \text{ Lift Coefficient}: c_l = \frac{1}{2}c_{d,max} \sin 2\alpha + K_2 \frac{\cos^2\alpha}{\sin\alpha} \tag{5.54}$$

where $K_2 = (c_{l,s} - c_{d,max} \sin\alpha_s \cos\alpha_s) \sin\alpha_s/\cos^2\alpha_s$ and referring to the same reference $\alpha_s$ and corresponding $c_{l,s}$. Figure 5.16 shows an example of the Viterna–Corrigan correction applied to the S809 airfoil at $Re = 5 \times 10^5$ as used in the *XTurb* example *PhaseVI-Ch5-1.inp* (Xfoil 2-D extrap: STALLDELAY=0)

Note that the Viterna–Corrigan correction can be switched on by setting VITERNA = 1 in the BLADE input list. The corrected airfoil polar is output to *XTurb_gnuplot_Airfoils_Data_CASE_1.plt*.

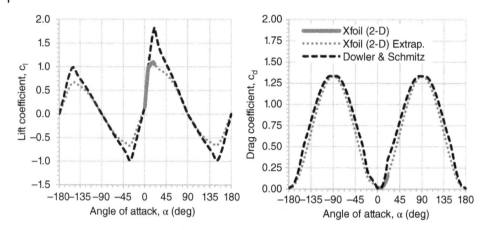

**Figure 5.16** Example of high AoA corrected airfoil polar for the NREL Phase VI rotor (S809, $Re = 5 \times 10^5$).

### 5.3.3  Comments on High Angle-of-Attack Corrections

Note that the widespread Viterna–Corrigan correction is practically a global method that has proven itself to predict high AoA rotor performance quite well. One can debate about whether or not the correction method should be applied to existing (2-D) airfoil data before/after applying a stall-delay correction. It is surmised that the original correction was probably meant to be a total correction (i.e. including all 3-D corrections) as actual stall-delay models in use today were developed well after the original paper by Viterna and Corrigan (1981). Today, the Viterna-Corrigan correction is usually referred to as an extrapolation method of (otherwise 3-D corrected or not) airfoil data to $\alpha_{Stall} \leq \alpha \leq 90° \ldots 180°$, which is helpful to both startup and stop/brake analyses and also during the iterative BEM solution process. The choice of $\alpha_s$ is in itself somewhat rotor dependent. For example, Tangler and Kocurek (2005) concluded that $\alpha_s = \alpha_{Stall}$ gave the best predictions for the NREL Phase VI rotor, while $\alpha_s = \alpha_{c_{l,max}}$ might be better suited for other rotors. Other high AoA correction models are those suggested by Lindenburg (2003) and Spera (2008) that resulted in slightly different values for $c_{d,max}$ based on more airfoil-like shapes and wedges found in the literature. Another aspect that has to be mentioned is the inherent unsteady (and 3-D) nature of high AoA airfoil flows, and one can conjecture that all $c_{d,max}$ variations based on various data sets are being "washed away" by the uncertainties associated with both models and experiments. In this context, CFD analyses are both expensive and subject to their own uncertainties associated with turbulence-model parameters that were not calibrated for these flow conditions.

## References

Aguasvivas, S., Lavely, A., Vijayakumar, G., Schmitz, S., Brasseur, J., and Duque, E. P. N. (2015) Nonsteady wind turbine loading response to passage of daytime atmospheric eddies. *68th Annual Meeting of the American Physical Society (APS) Division of Fluid*

*Dynamics*, Gallery of Fluid Motion. http://gfm.aps.org/meetings/dfd-2015/ 55f598c069702d060d7e0300. Last accessed December 30, 2018.

Bak, C. and Fuglsang, P. (2004) A method for deriving 3D airfoil characteristics for a wind turbine. *42$^{nd}$ AIAA Aerospace Sciences Meeting and Exhibition* 2004, Reno, NV, USA.

Bak, C., Madsen, H., and Johansen, J. (2001). Influence from blade-tower interaction on fatigue loads and dynamics. In: *European Wind Energy Conference and Exhibition (EWEC '01), Copenhagen, Denmark* (ed. P. Helm and A. Zervos), 394–397. München: WIP Renewable Energies.

Bak, C., Johansen, J., and Andersen, P. B. (2006) Three-dimensional corrections of airfoil characteristics based on pressure distributions. *Proceedings of the European Wind Energy Conference* 2006, Athens (Greece).

Bangga, G., Lutz, T., Jost, E., and Krämer, E. (2017). CFD studies on rotational augmentation at the inboard sections of a 10 MW wind turbine rotor. *Journal of Renewable and Sustainable Energy* 9 (2): 023304. https://doi.org/10.1063/1.4978681.

Banks, W. and Gadd, G. (1963). Delaying effect of rotation on laminar separation. *AIAA Journal* 1: 941–942.

Branlard, E. (2017). *Wind Turbine Aerodynamics and Vorticity-Based Methods – Fundamentals and Recent Applications* (Chapter 3.2.3). London: Springer.

Breton, S. P. (2008) *Study of the Stall Delay Phenomenon and of Wind Turbine Blade Dynamics using Numerical Approaches and NREL's Wind Tunnel Tests*. Ph.D. Dissertation, Norwegian University of Science and Technology, Trondheim, Norway.

Breton, S.P., Coton, F., and Moe, G. (2008). A study on rotational effects and different stall delay models using a prescribed wake vortex scheme and NREL phase VI experiment data. *Wind Energy* 11 (5): 459–482.

Carcangiu, C., Sørensen, J., Cambuli, F., and Mandas, N. (2007). CFD-RANS analysis of the rotational effects on the boundary layer of wind turbine blades. *Journal of Physics: Conference Series* 75: 012031.

Chattot, J. J. (2006) Extension of a helicoidal vortex model to account for blade flexibility and tower interference. *44th AIAA Aerospace Sciences Meeting and Exhibit, AIAA-2006-0391*.

Chaviaropoulos, P.K. and Hansen, M.O.L. (2000). Investigating three-dimensional and rotational effects on wind turbine blades by means of a quasi-3D Navier–Stokes solver. *Journal of Fluids Engineering* 122: 330–336.

Coriolis, G.G. (1835). Sur les équations du mouvement relatif des systèmes de corps. *Journal de l'École polytechnique* 15: 144–154.

Corrigan, J. J. and Schillings, J. J. (1994) Empirical model for stall delay due to rotation. *American Helicopter Society Aeromechanics Specialists Conference*, San Francisco, CA, USA.

Corten, G. P. (2001) *Flow Separation on Wind Turbine Blades*. Ph.D. Dissertation, University of Utrecht, Rotterdam, Netherlands.

Dowler, J. L. (2013) *A Solution-Based Stall Delay Model for Horizontal-Axis Wind Turbines*. M.S. Thesis. The Pennsylvania State University

Dowler, J.L. and Schmitz, S. (2015). A solution-based stall delay model for horizontal-axis wind turbines. *Wind Energy* 18 (10): 1793–1813.

Du, Z. and Selig, M. S. (1998) 3-D stall-delay model for horizontal axis wind turbine performance prediction. *1998 ASME Wind Energy Symposium, AIAA-1998-0021*.

Du, Z. and Selig, M.S. (2000). The effect of rotation on the boundary layer of a wind turbine blade. *Renewable Energy* 20: 167–181.

Dumitrescu, H., Cardos, V., and Dumitrache, A. (2007). Modelling of inboard stall delay due to rotation. *Journal of Physics: Conference Series* 75: 012022.

Dumitrescu, H., Frunzulică, F., and Cardoş, V. (2013). Improved stall-delay model for horizontal-axis wind turbines. *AIAA Journal of Aircraft* 50 (1): 315–319.

Eggers, A. J., Chaney, K., and Digumarthi, R. (2003) An assessment of approximate modeling of aerodynamic loads on the UAE rotor. *$41^{st}$ Aerospace Sciences Meeting and Exhibit, AIAA-2003-0868.*

Fage, A.R. and Johansen, F.C. (1927). On the flow of air behind an inclined flat plate of infinite span. *Proceedings of the Royal Society of London Series A* 166: 170–197.

Galbraith, R. A. M., Gracey, M. W., and Leitch, E. (1992) Summary of Pressure Data for thirteen Aerofoils on the University of Glasgow Aerofoil Database. Glasgow University Aero Report 9221.

Gerber, B., Tangler, J.L., Duque, E., and Kocurek, J.D. (2005). Peak and post-peak power aerodynamics from phase VI NASA Ames wind turbine data. *ASME Journal of Solar Energy Engineering* 127: 192–199.

Gonzalez, A. and Munduate, X. (2008). Three-dimensional and rotational aerodynamics on the NREL phase VI wind turbine blade. *ASME Journal of Solar Energy Engineering* 130 (3): https://doi.org/10.1115/1.2931506.

Guntur, S. K. and Sørensen, N. N. (2013) *A Detailed Study of the Rotational Augmentation and Dynamic Stall Phenomena for Wind Turbines.* DTU Wind Energy Ph.D., No. 0022(EN).

Hand, M. M., Simms, D. A., Fingersh, L. J., Jager, D. W., Cotrell, J. R., Schreck, S., and Larwood, S. M. (2001) *Unsteady Aerodynamics Experiment Phase VI: Wind Tunnel Test Configurations and Available Data Campaigns.* National Renewable Energy Laboratory, NREL/TP-500-29955.

Hansen, A. C. (2013) NWTC Computer-Aided Engineering Tools (AirfoilPrep by Dr. Craig Hansen). 2012. URL: http://www.nrel.gov/designcodes/preprocessors/airfoilprep, Last accessed February 14, 2013.

Hansen, M., Sørensen, N., Sørensen, J., and Michelsen, J. (1997). Extraction of lift, drag and angle of attack from computed 3D viscous flow around a rotating blade. In: *Scientific Proceedings of the European Wind Energy Conference, EWEC'97*, 499–501. Dublin (Ireland): EWEC.

Hansen, M. H., Gaunaa, M., and Madsen, H. A. (2004) A Beddoes–Leishman type dynamic stall model in state-space and indicial formulations. Risoe-R-1354, Roskilde, Denmark.

Herraez, I., Daniele, E., and Schepers, J.G. (2018). Extraction of the wake induction and angle of attack on rotating wind turbine blades from PIV and CFD results. *Wind Energy Science* 3: 1–9.

Himmelskamp, H. (1947) Profile Investigations on a Rotating Airscrew. *MAP Volkenrode*; September 1, 1947.

Hoerner, S.F. (1965). *Fluid-Dynamic Drag.* New Jersey: S.F Hoerner.

Jha, P.K. and Schmitz, S. (2016). Blade load unsteadiness and turbulence statistics in an actuator-line computed turbine–turbine interaction problem. *ASME Journal of Solar Energy Engineering* 138 (3): 031002. https://doi.org/10.1115/1.4032545.

Johansen, J. and Sørensen, N.N. (2004). Aerofoil characteristics from 3D CFD rotor computations. *Wind Energy* 7 (4): 283–294.

Johansen, J., Sørensen, N.N., Michelsen, J., and Schreck, S. (2002). Detached-eddy simulation of flow around the NREL Phase-VI rotor. *Wind Energy* 5 (2–3): 185–197.

Jonkman, J., Butterfield, S., Musial, W., and Scott, G. (2009) Definition of a 5-MW Reference Turbine for Offshore System Development. Technical Report NREL/TP-500-38060, National Renewable Energy Laboratory.

von Kármán, T. and Sears, W.R. (1938). Airfoil theory for non-uniform motion. *Journal of the Aeronautical Sciences* 5 (10): 379–390.

Küssner, H.G. (1935). Zusammenfassender Bericht über den instationären Auftrieb von Flügeln. *Luftfahrtforschung* 13 (12): 410–424.

Leishman, J.G. (2002). Challenges in modelling the unsteady aerodynamics of wind turbines. *Wind Energy* 5: 85–132.

Leishman, J.G. (2006). *Principles of Helicopter Aerodynamics* (Chapter 8),, 2e. Cambridge: Cambridge University Press.

Leishman, J.G. and Beddoes, T.S. (1989). A semi-empirical model for dynamic stall. *Journal of the American Helicopter Society* 34 (3): 3–17.

Lindenburg, C. (2003) *Investigation into Rotor Blade Aerodynamics*. Energy Research Centre of the Netherlands, ECN-C--03-025 2003, Petten, Netherlands.

Lindenburg, C. (2004) Modelling of rotational augmentation based on engineering considerations and measurements. *2004 European Wind Energy Conference Proceedings*, London, UK.

Madsen, H. and Christensen, H. (1990). On the relative importance of rotational, unsteady and three-dimensional effects on the HAWT rotor aerodynamics. *Wind Engineering* 14 (6): 405–415.

Madsen, H. A. and Rasmussen, F. (1988) Derivation of three-dimensional airfoil data on the basis of experiments and theory. *Proceedings of Windpower* 1988, Honolulu, Hawaii.

McCroskey, W. J. (1971) *Measurements of Boundary Layer Transition, Separation and Streamline Direction on Rotating Blades*. National Aeronautics and Space Administration, NASA TN-D-6321.

McCroskey, W. and Yaggy, P. (1968). Laminar boundary layers on helicopter rotors in forward flight. *AIAA Journal* 6 (10): 1919–1926.

Nandi, T.N., Herrig, A., and Brasseur, J.G. (2017). Non-steady wind turbine response to daytime atmospheric turbulence. *Philosophical Transactions of the Royal Society of London, Series A: Mathematical, Physical and Engineering Sciences* 375: https://doi.org/10.1098/rsta.2016.0103.

Øye, S. (1991) Dynamic stall, simulated as a time lag of separation. *Proceedings of the 4th IEA Symposium on the Aerodynamics of Wind Turbines*, ETSU-N-118, Harwell Laboratory, Harwell, United Kingdom.

Ronsten, G. (1992). Static pressure measurements on a rotating and a non-rotating 2.375 m wind turbine blade. Comparison with 2D calculations. *Journal of Wind Engineering and Industrial Aerodynamics* 39: 105–118.

Sant, T., van Kuik, G., and van Bussel, G.J.W. (2006). Estimating the angle of attack from blade pressure measurements on the NREL phase VI rotor using a free wake vortex model: axial conditions. *Wind Energy* 9: 549–577.

Schmitz, S. (2015) *PSU Generic 1.5-MW Turbine*. DOI: 10.13140/RG.2.2.22492.18567.

Schmitz, S. and Chattot, J.J. (2006). Characterization of three-dimensional effects for the rotating and parked NREL phase VI wind turbine. *ASME Journal of Solar Energy Engineering* 128: 445–454.

Schreck, S. and Robinson, M. (2002). Rotational augmentation of horizontal axis wind turbine blade aerodynamic response. *Wind Energy* 5: 133–150.

Schreck, S. and Robinson, M. (2007). Horizontal axis wind turbine blade aerodynamics in experiments and modeling. *IEEE Transactions on Energy Conversion* 22 (1): 61–70.

Schreck, S., Sørensen, N.N., and Robinson, M. (2007). Aerodynamic structures and processes in rotationally augmented flow fields. *Wind Energy* 10: 159–178.

Schreck, S., Sant, T., and Micallef, D. (2010) *Rotational augmentation disparities in the MEXICO and UAE phase VI experiments*. National Renewable Energy Laboratory, NREL/CP-500-47759.

Schreck, S., Fingersh, L., Siegel, K., Singh, M., and Medina, P. (2013) Rotational augmentation on a 2.3-MW rotor blade with thick flatback airfoil cross sections. *Proceedings of the 51ˢᵗ AIAA Aerospace Sciences Meeting, AIAA 2013-0915*.

Shen, W.Z., Hansen, M.O.L., and Sørensen, J.N. (2009). Determination of the angle of attack on rotor blades. *Wind Energy* 12: 91–98.

Sheng, W., Galbraith, R. A. M., Coton, F. N., and Gilmour, R. (2006a) The Collected Data for Tests on an S809 Aerofoil. Volume I: Pressure Data From Static, Ramp and Triangular Wave Tests. Glasgow University Aero Report 0606.

Sheng, W., Galbraith, R. A. M., Coton, F. N., and Gilmour, R. (2006b) The Collected Data for Tests on an S809 Aerofoil, Volume II: Pressure Data From Static and Oscillatory Tests. Glasgow University Aero Report 0607.

Sheng, W., Galbraith, R.A.M., and Coton, F.N. (2006c). A new stall-onset criterion for low speed dynamic-stall. *ASME Journal of Solar Energy Engineering* 128 (4): 461–471.

Sheng, W., Galbraith, R.A.M., and Coton, F.N. (2008). A modified dynamic stall model for low Mach numbers. *ASME Journal of Solar Energy Engineering* 130 (3): https://doi.org/10.1115/1.2931509.

Simms, D., Schreck, S., Hand, M., and Fingersh, L. J. (2001) *NREL Unsteady Aerodynamics Experiment in the NASA-Ames Wind Tunnel: A Comparison of Predictions to Measurements*. National Renewable Energy Laboratory, NREL/TP-500-29494.

Snel, H. (1991) Scaling laws for the boundary layer flow on rotating wind turbine blades. *4th IEA Symposium Aerodynamics of Wind Turbines*, ETSU-N-118.

Snel, H., Houwink, R., and Piers, W. (1992) Sectional prediction of 3D effects for separated flow on rotating blades. *18th European Rotorcraft Forum, B10-1–B10-18*.

Snel, H., Houwink, R., van Bussel, G.J.W., and Bruining, A. (1993a). Sectional prediction of 3D effects for stalled flow on rotating blades and comparison with measurements. In: *Proceedings of the European Community Wind Energy Conference*, 395–399. Lübeck-Travemünde (Germany).

Snel, H., Houwink, R., and Bosschers, J. (1993b) *Sectional Prediction of Lift Coefficients on Rotating Wind Turbine Blades in Stall*. Netherlands Energy Research Foundation, ECN-C-93-052.

Snel, H., Schepers, J., and Montgomerie, B. (2014). The MEXICO project (model experiments in controlled conditions): the database and first results of data processing and interpretation. *The Science of Making Torque from Wind Conference*, . *Journal of Physics: Conference Series* 75: 012014.

Sørensen, N. N. (2000) Evaluation of 3D effects from 3D CFD computations. *Proceedings of the 14ᵗʰ IEA Symposium on the Aerodynamic Effects of Wind Turbines*, Boulder CO, USA.

Sørensen, N.N., Michelson, J.A., and Schreck, S. (2002). Navier–Stokes predictions of the NREL phase VI rotor in the NASA Ames 80 ft × 120 ft wind tunnel. *Wind Energy* 5: 151–169.

Spera, A. D. (2008) *Models of lift and drag coefficients of stalled and unstalled airfoils in wind turbines and wind tunnels*. National Aeronautics and Space Administration, NASA/CR-2008-215434.

Tangler, J. L. (2002) *The Nebulous Art of Using Wind-Tunnel Airfoil Data to Predicting Rotor Performance*. Published as: National Renewable Energy Laboratory, NREL/CP-500-31243.

Tangler, J.L. (2004). Insight into wind turbine stall and post-stall aerodynamics. *Wind Energy* 7: 247–260.

Tangler, J. L. and Kocurek, J. D. (2005) *Wind turbine post-stall airfoil performance characteristics guidelines for blade-element momentum methods*. National Renewable Energy Laboratory, NREL/CP-500-38456.

Tangler, J. and Selig, M. (1997) *An evaluation of an empirical model for stall delay due to rotation for HAWTs*. National Renewable Energy Laboratory, NREL/CP 440–23258.

Theodorsen, T. (1935) General Theory of Aerodynamic Instability and the Mechanism of Flutter. National Aeronautics and Space Administration, NACA Report 496.

Timmer, W. A. (2010) Aerodynamic characteristics of wind turbine blade airfoils at high angles of attack. *Conference Proceedings Torque 2010 – The Science of Making Torque from Wind*, Heraklion (Greece), pp. 71–97.

Viterna, L. A. and Corrigan, R. D. (1981) Fixed pitch rotor performance of large horizontal axis wind turbines. *DOE/NASA Workshop on Large Horizontal Axis Wind Turbines*, Cleveland OH.

Wagner, H. (1925). Über die Entstehung des dynamischen Auftriebes von Tragflügeln. *Zeitschrift für Angewandte Mathematic und Mechanik* 5 (1): 17–35.

Wilcox, D.C. (2010). *Basic Fluid Mechanics* (Chapter 10), 4e. DCW Industries.

Wood, D. (2011). Aerofoils: lift, drag, and circulation. In: *Small Wind Turbines – Analysis, Design, and Application* (Chapter 4), 57–75. London: Springer.

Yang, H., Shen, W.Z., Sørensen, J.N., and Zhu, W.J. (2011). Extraction of airfoil data using PIV and pressure measurements. *Wind Energy* 14 (4): 539–556.

## Further Reading

Branlard, E. (2017). *Wind Turbine Aerodynamics and Vorticity-Based Methods – Fundamentals and Recent Applications* (Chapter 3.2.3). London: Springer.

Leishman, J.G. (2000). *Principles of Helicopter Aerodynamics*. Cambridge: Cambridge University Press.

Timmer, W.A. (2009). Aerodynamic characteristics of wind turbine blade airfoils. In. In: *Wind Turbine Technology – Fundamental Concepts of Wind Turbine Engineering* (Chapter 4), vol. 2 (ed. D.A. Spera). New York: ASME Press.

Wood, D. (2011). *Small Wind Turbines – Analysis, Design, and Application*. London: Springer.

# 6

# Vortex Wake Methods

*The Laws of Prandtl and the Laws of Nature*
                                        – David Bloor, *The Enigma of the Aerofoil*

## 6.1  Fundamentals of Prandtl Lifting-Line Theory

Prandtl developed a simple and elegant theory for calculating the effect of tip vortices on the aerodynamic lift and drag characteristics of wings of finite span (Prandtl 1918, 1923). As one of the most influential aerodynamicists of all time, Prandtl developed boundary-layer theory and lifting-line theory in the early twentieth century, both of which have lost nothing of their influence on the design of aircraft. In this context, Anderson's work (1978, 2001) is a good reference for the interested reader.

### 6.1.1  Vortex Sheet and Horseshoe Vortices

Prandtl's conceptual ideas of a bound vortex and a trailing vortex sheet thought of as a superposition of elemental horseshoe vortices are rooted in satisfying two of Helmholtz's vortex theorems:

- The strength of a vortex filament is constant along its length.
- A vortex filament cannot end in a fluid; it must either close on itself or extend to the fluid boundary.

Figure 6.1 illustrates Prandtl's lifting-line concept. At first, the fluid flow associated with a traversing wing is represented as a quadrilateral vortex filament whose left side (A–B) represents the wing lifting line and is typically referred to as the "bound vortex"; the right side (D–C) of the quadrilateral vortex filament is the so-called "starting vortex" of equal-and-opposite circulation than the bound vortex, a simple flow model from inviscid circulation theory. In order to satisfy Helmholtz's vortex theorems that a vortex filament cannot end in a fluid, bound, and starting vortices have to be connected by piecewise vortex filaments (A–D, B–C) of equal strength, representing the tip vortices trailing from the bound lifting line. This fundamentally implies that any lifting body must have a pair of tip vortices trailing into the wake. By the time startup processes have decayed and the starting vortex has convected sufficiently far downstream, the quadrilateral vortex filament appears as a horseshoe vortex whose tip filaments extend to the downstream fluid boundary.

*Aerodynamics of Wind Turbines: A Physical Basis for Analysis and Design,* First Edition. Sven Schmitz.
© 2020 John Wiley & Sons Ltd. Published 2020 by John Wiley & Sons Ltd.
Companion website: www.wiley.com/go/schmitz/wind-turbines

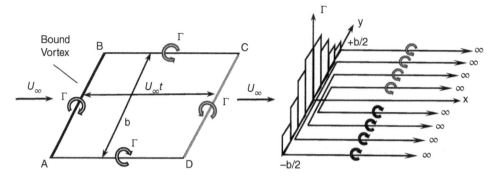

**Figure 6.1** Lifting-line concept and horseshoe vortices.

For a wing of finite span, the flow physics postulates that the bound circulation, $\Gamma(y)$, has to be zero at the wing tips due to the pressure balance immediately adjacent to the wing tips. Hence, the bound vortex has to be of variable strength and would thus violate one of Helmholtz's vortex theorems, unless the bound vortex (or lifting line) is indeed a superposition of the bases of multiple elemental horseshoe vortices, see Figure 6.1 (Right). Consequently, a continuous vortex sheet trails from the lifting line and into the far wake to the fluid boundary. Note that the strengths of the trailing vortex filaments appear equal-and-opposite for a wing with symmetric loading and that an integral over a wake plane results in zero net vortex strength.

Before proceeding with the mathematical description and solution of the lifting-line problem, let us first describe some geometrical parameters of a finite wing:

$$\text{Wing Aspect Ratio:} \quad AR = \frac{b^2}{S} \tag{6.1}$$

Here, $b$ is the wing span and $S$ is the wing area. In general, wings/blades of high aspect ratio, that is, $AR \geq 7$, have negligible influence of the spanwise direction compared to the streamwise and normal directions. In other words, each spanwise ($y = const.$) wing section can be considered as a two-dimensional airfoil cross-section, a simple and elegant way of including (airfoil) profile drag in the wing performance evaluation. The wing area, $S$, itself is computed from:

$$\text{Wing Area:} \quad S = \int_{-b/2}^{+b/2} c(y)\,dy \tag{6.2}$$

In Eq. (6.2), $c(y)$ is the spanwise distribution of wing chord. An associated mean aerodynamic chord, $\bar{c}$, is defined as:

$$\text{Mean Aerodynamic Chord:} \quad \bar{c} = \frac{1}{S}\int_{-b/2}^{+b/2} c^2(y)\,dy \tag{6.3}$$

A simple example is that of a rectangular wing where $\bar{c} = c$ and $AR = b/c$. Figure 6.2 shows the conceptual lifting-line model and trailing vortex sheet, with the bases of all horseshoe vortices fixed to the wing (lifting line) and trailing vortex filaments extending to infinity. In this context, we define the "Trefftz" plane as the downstream wake plane

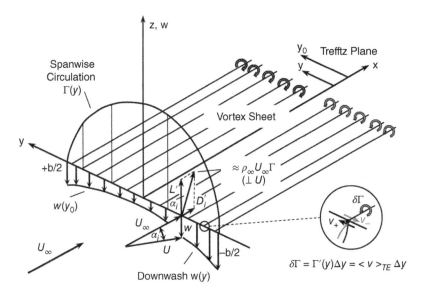

**Figure 6.2** Trailing vortex sheet behind lifting line.

where streamwise gradients vanish. The circulation distribution along the lifting line in Figure 6.2 is the result of the superposition of a series of elemental horseshoe vortices, with the maximum circulation, $\Gamma_{max}$, occurring at mid-span, that is, $y = 0$.

Both the streamwise and vertical velocities, $u$ and $w$, are continuous at the wing trailing edge (TE) to satisfy the Kutta condition, see Eq. (4.10). (Note that the airfoil cross-section is reduced to the concept of the lifting line, and hence the TE is located on the lifting line itself.) Nevertheless, intersecting streamlines at the TE result in a "jump" in the spanwise velocity, $v$, just above (upper surface) and below (lower surface) the lifting line, which is the physical mechanism to generate the trailing vortex filaments and hence trailing vortex sheet. Denoting with $\Gamma(y)$ the spanwise distribution of bound circulation, its spanwise gradient, $\Gamma'(y)$, is a direct consequence of the jump in the spanwise velocity component above/below the lifting line:

$$\text{Trailing Vorticity/Circulation:} \quad \Gamma'(y) = \frac{d\Gamma}{dy} = \langle v \rangle_{TE} \tag{6.4}$$

In Eq. (6.4), $\langle v \rangle_{TE} = v_+ - v_-$ is the jump in spanwise (perturbation) velocity at the TE and essentially determines the strength of the vortical trailer emanating from that particular location along the wing span. The total wing lift is obtained by integrating the section KJ theorem from Eq. (4.8) along the wing span to obtain:

$$\text{Wing Lift (KJ Theorem:)} \quad L = \rho_\infty U_\infty \int_{-b/2}^{+b/2} \Gamma(y) dy \tag{6.5}$$

Note that at the wing tips, $\Gamma(\pm b/2) = 0$ due to the pressure equalization. In the Trefftz plane, the trailing vortex filaments extend between $\pm\infty$ and their induced vertical (downwash) velocity along a straight line perpendicular to the vortex filaments and in the vortex-sheet plane can be computed from two-dimensional potential theory as:

$$\text{Downwash (Trefftz Plane):} \quad w_T(y) = -\frac{1}{2\pi} \int_{-b/2}^{+b/2} \frac{\Gamma'(\eta)}{y-\eta} d\eta \tag{6.6}$$

$$\text{Wing Downwash (Lifting Line):} \quad w_W(y) = -\frac{1}{4\pi} \int_{-b/2}^{+b/2} \frac{\Gamma'(\eta)}{y-\eta} d\eta \tag{6.7}$$

Here Eqs. (6.6) and (6.7) are principal-value integrals, and we note the factor of 1/2 in induced downwash from the Trefftz plane (T) to the wing lifting line (W), a result from simple geometric considerations. Furthermore, it is important to understand that the planar vortex-sheet concept with a straight lifting line does not induce any $u$ (or $v$) velocities on the lifting line itself. The effect of the wing downwash, $w_W(y)$, however is shown schematically in Figure 6.2 (denoted as w) as an effective change in direction of the blade section velocity. This results in an effective tilt of the lift vector in the stream-wise direction, thus generating an "induced drag" associated with the flow induction due to the wing vortex sheet. The induced drag, $D_i$, can also be thought of a result of depositing kinetic energy, $\frac{1}{2}\rho_\infty(v^2 + w^2)$, in the Trefftz plane:

$$\text{Wing Induced Drag:} \quad D_i = +\frac{1}{2}\rho_\infty \iint_T (v^2 + w^2) dA$$

$$= -\frac{1}{2}\rho_\infty \int_{-b/2}^{+b/2} \Gamma(y)\, w_T(y) dy \tag{6.8}$$

Here we assume that $|w_W(y)| \ll V_\infty$ and hence small angles such that an "induced angle of attack," $\alpha_i(y)$, can be written as:

$$\text{Induced Angle of Attack:} \quad \alpha_i(y) = \frac{w_W(y)}{u_\infty} = -\frac{1}{4\pi U_\infty} \int_{-b/2}^{+b/2} \frac{\Gamma'(\eta)}{y-\eta} d\eta \tag{6.9}$$

In essence, an induced angle of attack, $\alpha_i(y) < 0$, reduces the geometric wing setting angle, $\alpha_g(y)$, of the corresponding wing section (on a wind turbine blade, $\alpha_g(y)$ corresponds to the local blade pitch angle, $\beta$), thus reducing both the section lift coefficient, $c_l$, and generating induced drag.

### 6.1.2 Inviscid Flow: Lifting-Line Theory

The task in lifting-line theory is to find the spanwise distribution of the effective angle of attack, $\alpha_e(y) = \alpha(y) - \alpha_0(y)$, from a thin-airfoil theory perspective where $\alpha_0(y)$ is the zero-lift angle of the corresponding airfoil polar ($= -2d/c$ for a thin airfoil with a parabolic camber line, see Eq. (4.26)). Using the thin-airfoil section lift coefficient in Eq. (4.7) in conjunction with the K–J theorem, we write $\alpha_e(y)$ in the following form:

$$\text{Effective Angle of Attack:} \quad \alpha_e = \frac{1}{2\pi} \frac{2\Gamma}{U_\infty c} = \alpha_g + \alpha_i \tag{6.10}$$

For a given wing setting angle $\alpha_g(y)$, we can use Eq. (6.9) and rearrange (6.10) to obtain the following relation:

$$\text{Fundamental Equation:} \quad \alpha_g(y) = \frac{\Gamma(y)}{\pi U_\infty c(y)} + \frac{1}{4\pi U_\infty} \int_{-b/2}^{+b/2} \frac{\Gamma'(\eta)}{y-\eta} d\eta \tag{6.11}$$

Equation (6.11) is also referred to as the integro-differential equation of Prandtl. In the following, a transformation similar to the one used in thin-airfoil theory (see Section 4.4.1) is used as:

$$y(\Theta) = -\frac{b}{2}\cos\Theta, \quad 0 \le \Theta \le \pi \tag{6.12}$$

In the transformed space, a solution approach is formulated as an infinite series:

$$\text{Bound Circulation:} \quad \Gamma[y(\Theta)] = 2U_\infty b \sum_{n=1}^{\infty} A_n \sin n\Theta \tag{6.13}$$

Other useful transformations (at least for the first three modes) when integrating the fundamental Eq. (6.11) are:

$$\text{Series Transformations:} \quad \cos\theta = -\frac{2y}{b}, \quad -\frac{b}{2} \le y \le \frac{b}{2}$$

$$\sin\theta = \sqrt{1 - \cos^2\theta} = \sqrt{1 - \left(\frac{2y}{b}\right)^2}$$

$$\sin 2\theta = 2\sin\theta\cos\theta = -\frac{4y}{b}\sqrt{1 - \left(\frac{2y}{b}\right)^2} \tag{6.14}$$

$$\sin 3\theta = \sin\theta[4\cos^2\theta - 1] = \sqrt{1 - \left(\frac{2y}{b}\right)^2}\left[4\left(\frac{2y}{b}\right)^2 - 1\right]$$

Dividing Eq. (6.5) by $\frac{1}{2}\rho_\infty U_\infty^2 S$ results in the wing lift coefficient, $c_L$, with:

$$\text{Wing Lift Coefficient:} \quad c_L = \frac{2}{U_\infty S}\int_{-b/2}^{+b/2}\Gamma(y)dy = \pi \cdot AR \cdot A_1 \tag{6.15}$$

Note that Eq. (6.15) depends only on the first mode. In the transformed plane with $0 \le \Theta \le \pi$, it can be shown that all modes $n > 1$ integrate exactly to zero. Normalizing Eq. (6.8) by $\frac{1}{2}\rho_\infty U_\infty^2 S$ results in the wing induced-drag coefficient, $c_{D_i}$, written in the following form:

$$\text{WingInduced Drag Coefficient:} \quad c_{D_i} = -\frac{1}{U_\infty^2}\int_{-b/2}^{+b/2}\Gamma(y)\, w_T(y)dy$$

$$= \pi b^2 \sum_{n=1}^{\infty} nA_n^2 \tag{6.16}$$

It is apparent from Eq. (6.16) that for a lifting wing (i.e. $A_1 > 0$), the induced drag, $c_{D_i}$ can be minimized for $A_2, A_3, \ldots, A_n = 0$. In this special case, Eqs. (6.15) and (6.16) can be combined such that the minimum induced drag becomes:

$$\text{"Minimum" Induced-Drag coefficient:} \quad c_{D_i,min} = \frac{c_L^2}{\pi\,AR} \tag{6.17}$$

It is important to realize that the minimum induced-drag condition can be only achieved, if the bound circulation in Eq. (6.13) consists of only the first mode in the transformed $\theta$-space, corresponding to an elliptical circulation (and lift) distribution in the physical $y$-space, see the example given in Section 6.1.2.1. A general induced-drag coefficient is written as:

$$\text{General Induced-Drag coefficient:} \quad c_{D_i} = \frac{c_L^2}{\pi\,AR}(1+\delta) = \frac{c_L^2}{\pi\,AR\,e} \tag{6.18}$$

Equation (6.18) describes a wing polar and is often referred to as the "drag-due-to-lift" relationship. In general, $c_{D_i}$ is written either in terms of a relative drag increase, $\delta$, or the "Oswald efficiency factor," $e$, defined as:

$$\text{Relative Drag Increase:} \quad \delta = \sum_{n=2}^{\infty} n\left(\frac{A_n}{A_1}\right)^2 \tag{6.19}$$

$$\text{Oswald Efficiency Factor:} \quad e = \frac{1}{1+\delta} = \frac{1}{\pi\,AR}\frac{c_L^2}{c_{D_i}} \tag{6.20}$$

In the special case of minimum induced drag with an elliptic spanwise circulation distribution, $\delta = 0$ and $e = 1$, respectively. As for the wing lift curve, $c_L(\alpha_W)$, where $\alpha_W$ is the wing setting angle at mid-span with $\alpha_W = \alpha_g(y=0)$, the Prandtl integro-differential Eq. (6.11) can be manipulated, see for example, Chattot and Hafez (2015), such that the wing lift slope, $dc_L/d\alpha$, becomes:

$$\text{Wing Lift Slope:} \quad \frac{dC_L}{d\alpha} = \frac{2\pi}{1 + \frac{2(1+\tau)}{AR}} \approx \frac{2\pi}{1 + \frac{2}{AR}} \tag{6.21}$$

In Eq. (6.21), the parameter $\tau$ ranges between 0.05 and 0.25, see also Glauert (1926). It is interesting and important to note that the wing lift slope approaches the $2\pi$ result obtained from thin-airfoil theory, see Section 4.1.1.2, for a wing aspect ratio $AR \to \infty$.

Some exact wing analyses integrating the Prandtl integro-differential Eq. (6.11) can be, for example, found in Moran (1984). The general case can be also solved numerically. The following subsections provide some *XTurb* examples. The numerical algorithm in *XTurb* is based on the numerical solution procedure in Section 6.7.8 of Chattot and Hafez (2015).

### 6.1.2.1 Elliptic Loading (Inviscid Airfoil Polar)

Let us consider a simple design problem for an ideal wing with minimum induced drag at a desired lift coefficient, $C_L$, and aspect ratio, $AR$. Hence the only remaining Fourier coefficient is $A_1 = C_L/\pi AR$. The design task is then to find wing chord, $c(y)$, and wing twist $\alpha_g(y)$. In the ideal case, the spanwise circulation distribution, $\Gamma(y)$, is of the following form:

$$\text{EllipticLoading:} \quad \Gamma(y) = \Gamma_0\sqrt{1 - \left(\frac{2y}{b}\right)^2}$$

$$= \pi U_\infty c(y)[\alpha_g(y) + \alpha_i(y)] \tag{6.22}$$

For the special case of an elliptic spanwise circulation distribution, it can be shown from Eq. (6.9) that the induced angle of attack $\alpha_i(y) = -C_L/\pi AR = const$, which is indeed a minimum energy condition for a trailing vortex sheet. Also, the mid-span circulation becomes $\Gamma_0 = \tfrac{1}{2}U_\infty c_L c_0$ from the KJ theorem. Consequently, the two simplest designs are either for (i) an elliptical $c(y)$ with $\alpha_g(y) = const$ or (ii) a constant

$c(y)$ and a shifted elliptical $\alpha_g(y)$. In this context, it is instrumental to understand that a given spanwise circulation distribution, whether it be optimal or not, can be realized by an infinite number of wing/blade designs; however, the beauty of design option (i) lies in the fact that the spanwise circulation distribution remains being elliptic if $\alpha_g(y)$ is collectively changed to obtain a different wing $c_L$. In that particular case, the wing chord, $c(y)$, becomes:

$$\text{Elliptical Wing Chord:} \quad c(y) = c_0\sqrt{1 - \left(\frac{2y}{b}\right)^2} \tag{6.23}$$

For an elliptical wing planform such as in Eq. (6.23), the aspect ratio is $AR = 4b/\pi c_0$ and the mid-span (or root) chord, $c_0$, can be computed as:

$$c_0 = \frac{4b}{\pi\,AR} \tag{6.24}$$

Substituting all findings thus far and Eqs. (6.23) and (6.24) into (6.22) yields an expression for the wing twist, $\alpha_g(y)$, as:

$$\alpha_g(y) = \frac{c_L}{2\pi}\left(1 + \frac{2}{AR}\right) = const \tag{6.25}$$

Equation (6.25) means nothing else than that the elliptical planform wing is untwisted and set at an angle [rad] being a factor $1 + 2/AR$ higher than for the 2-D limit case of infinite aspect ratio.

We proceed with a *XTurb* example (Schmitz 2012) of a wing with an elliptical planform and an aspect ratio of $AR = 10$. The corresponding *XTurb* input file is *EllipticWing-AR10-Ch6-1-2.inp*. The non-rotating (or parked) blade analysis can be realized in *XTurb* by setting negative TSR or RPM values. In addition, the Helicoidal Vortex Model (HVM) model (*METHOD* = 2) has to be activated (Note: A wing analysis cannot be performed with the standard Blade-Element Momentum Theory (BEMT) model with *METHOD* = 1). The used airfoil polar is *ThinSymmAirf-Inv.polar*. Results obtained from *XTurb_Output_PARKED.dat* are shown next:

**XTurb Example 6.1  Elliptic Wing – XTurb_Output_PARKED.dat (EllipticWing-AR10-Ch6-1-2.inp)**

```
6.1.2 - EllWingAR10                **** XTurb V1.9    -   OUTPUT PARKED ****
  Blade Number        BN =  1

                              + Prescribed-Wake Method [Chattot, Schmitz] +

Solidity  Aspect Ratio
 0.0318   10.0114

 Drag Coefficient       CD = CTD/SOLIDITY
 Lift Coefficient       CL = CPL/SOLIDITY

 Number   TSR      PITCH [deg]    CTD      CPL       CD       CL
     1  -2.0000   94.0000       0.0001    0.0116   0.0042    0.3657
     2  -2.0000   86.0000       0.0001   -0.0116   0.0042   -0.3656
     3  -2.0000   82.0000       0.0005   -0.0233   0.0170   -0.7313
     4  -2.0000   78.0000       0.0012   -0.0349   0.0383   -1.0973
```

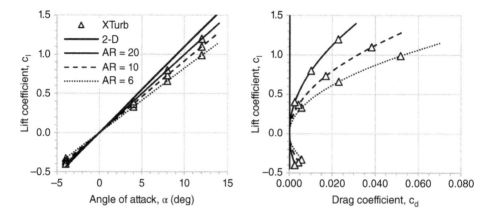

**Figure 6.3** Lift curves and drag polars for elliptic-planform wings of varying aspect ratio, AR.

Note that the force convention is different for a non-rotating (or parked) case. In reference to Figures 3.3 and 6.2, a non-rotating (or parked) blade is pitched at $90° - \alpha_g$ in the *XTurb* input file (see previous). The *XTurb* computed thrust coefficient, $C_T$, for example, is related to the blade/wing drag coefficient, $c_D$, by the ratio of wing/blade and rotor disk areas (or solidity), $\sigma = S/A$, via:

$$\text{Force Coefficients:} \quad c_L = \frac{C_{TD}}{\sigma} \quad c_D = \frac{C_{PL}}{\sigma} \tag{6.26}$$

Figure 6.3 shows the lift curves and drag polars of three elliptical wings with varying aspect ratio. Results obtained by *XTurb* agree very well with exact theory outlined before (within one drag count of $\Delta c_d = 10^{-4}$!). It is apparent that the two-dimensional limit case of $2\pi$ lift-curve slope and zero induced drag, $c_{D_i}$, is obtained for $AR \to \infty$.

The corresponding circulation, $\Gamma(y)$, and downwash, $w_W(y)$, distributions at wing setting angles of $\alpha_W = 4°$, $12°$ are shown in Figure 6.4, with data taken from *XTurb-Output_Circulation.dat* and *XTurb_Output_Method.dat*. There is very close agreement between results obtained from exact theory and the *XTurb* code.

Note that using a cosine discretization in the spanwise (or radial) direction (COS-DISTR = 1) in the SOLVER input list results in higher accuracy, which is directly related to the transformation in Eq. (6.12) that enables the infinite-series solution approach.

### 6.1.2.2 Parked NREL Phase VI Rotor (Viscous Airfoil Polar)

The NREL Phase VI rotor can be analyzed in parked mode using the *XTurb* code in parked HVM mode. Figure 6.5 shows the parked NREL Phase VI rotor tested in the National Aeronautics and Space Administration (NASA) Ames (80′ × 120′) wind tunnel, see also description of Sequences L and O in Hand et al. (2001). Note that a multi-bladed rotor in parked conditions is essentially a set of interacting wings, and the mutual effect of each wing/blade vortex sheet and bound vortex on one another has to be taken into account. This is beyond classical lifting-line theory and is a special case of a formulation of interacting vortex sheets using the Biot–Savart law, see Section 6.2.

In order to quantitatively compare measured and computed normal and tangential force coefficients versus the local geometric angle of attack, $\alpha_g(y)$, one has to be aware

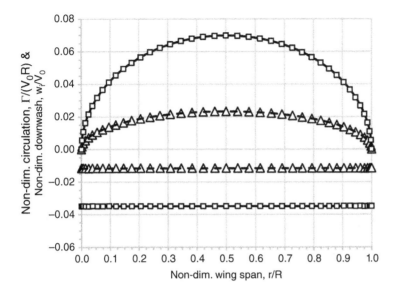

**Figure 6.4** Spanwise circulation and downwash distribution ($AR = 10$) for $\alpha_w = 4°$ (triangles), 12° (squares) - lines = exact solution; symbols = XTurb results.

**Figure 6.5** Parked (sequences L and O) NREL phase VI rotor in the NASA Ames (80′ × 120′) wind tunnel. Source: Courtesy of Lee Jay Fingersh, NREL 55063.

of the relative pitch between spanwise (or radial) blade sections. For the parked NREL Phase VI rotor, this was defined based on the 47% radial station. Table 6.1 summarizes the relative pitch angles in reference to $r/R = 0.47$ where measured data are available.

Next, let us consider the *XTurb* example with input file *PhaseVI-Parked-Ch6-1-2.inp*. The parked NREL Phase VI rotor is analyzed in ANALYSIS mode using the HVM model

**Table 6.1** Relative pitch angles due to blade twist (NREL phase VI rotor).

| $\alpha_{30}$ | $\alpha_{47}$ | $\alpha_{63}$ | $\alpha_{80}$ | $\alpha_{95}$ |
|---|---|---|---|---|
| $-9.58°$ | $0°$ | $+3.60°$ | $+5.10°$ | $+6.12°$ |

($METHOD = 2$) along with negative RPM values to force a parked case. The used airfoil polar file is *S809_Re700E5.polar*. The corresponding *XTurb_Output_PARKED.dat* is shown next:

---

**XTurb Example 6.2  NREL Phase VI Rotor – XTurb_Output_PARKED.dat (PhaseVI-Parked-Ch6-1-2.inp)**

```
 6.1.2 - NRELVI-Parked              **** XTurb V1.9    -  OUTPUT PARKED ****
        Blade Number        BN =  2

                             + Prescribed-Wake Method [Chattot, Schmitz] +
     Solidity  Aspect Ratio
      0.0519   12.2775

     Drag Coefficient         CD = CTD/SOLIDITY
     Lift Coefficient         CL = CPL/SOLIDITY

      Number   TSR     PITCH [deg]        CTD      CPL       CD        CL
        1   -7.5850   80.0300          0.0008  -0.0127   0.0156  -0.2451
```

---

Results computed by *XTurb* are compared to measured NREL normal and tangential force coefficients, $c_n$ and $c_t$, in Figure 6.6. It can be seen that measured and computed

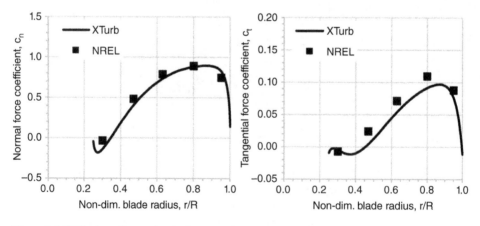

**Figure 6.6** NREL phase VI rotor (parked) – normal and tangential force coefficients (L – sequence, $V_0 = 20.1$ m s$^{-1}$, $\alpha_{47} = 3.53°$).

XTurb Example 6.3 NREL 5-MW – XTurb_Output.dat (NREL-5MW-Parked-Ch6-1-6.inp)

```
6.1.2 - NREL5MW-Parked        ***** XTurb V1.9     -     OUTPUT *****
    Blade Number      BN =  3

                      + Prescribed-Wake Method [Chattot, Schmitz] +

PREDICTION
Blade Radius      BRADIUS  =   63. [m]
Air Density       RHOAIR   =   1.225 [kg/m**3]
Air Dyn. Visc.    MUAIR    =   0.000018 [kg/(m*s)]
Number of Cases   NPRE     =   10

Thrust          T  = 0.5*RHOAIR*VWIND**2.*(pi*BRADIUS**2.)*CT
Power           P  = 0.5*RHOAIR*VWIND**3.*(pi*BRADIUS**2.)*CP
Torque          TO = P / RPM
Bending Moment  BE = 0.5*RHOAIR*VWIND**2.*(pi*BRADIUS**2.)*BRADIUS*CB
```

| Number | VWIND[m/s] | RPM[1/min] | TSR | PITCH[deg] | ... | T[N] | ... | TO[Nm] | BE[Nm] |
|---|---|---|---|---|---|---|---|---|---|
| 1 | 3.0000 | -12.1000 | -26.6093 | 90.0000 | ... | 380.2899 | ... | 10470.6447 | -1273.3488 |
| 2 | 3.0000 | -12.1000 | -26.6093 | 80.0000 | ... | 356.8660 | ... | -81244.1520 | -1528.8750 |
| 3 | 3.0000 | -12.1000 | -26.6093 | 70.0000 | ... | 616.0489 | ... | -125919.6618 | -5006.0575 |
| 4 | 3.0000 | -12.1000 | -26.6093 | 60.0000 | ... | 1202.9516 | ... | -116978.3345 | -11609.5457 |
| 5 | 3.0000 | -12.1000 | -26.6093 | 50.0000 | ... | 2065.1712 | ... | -99284.5089 | -20302.2225 |
| 6 | 25.0000 | -12.1000 | -3.1931 | 95.0000 | ... | 33488.2958 | ... | 3694473.0456 | -146010.1155 |
| 7 | 25.0000 | -12.1000 | -3.1931 | 90.0000 | ... | 26409.0241 | ... | 727128.1041 | -88427.0019 |
| 8 | 25.0000 | -12.1000 | -3.1931 | 85.0000 | ... | 24579.4062 | ... | -2378032.7096 | -83512.8824 |
| 9 | 25.0000 | -12.1000 | -3.1931 | 80.0000 | ... | 24782.3645 | ... | -5641954.9978 | -106171.8738 |
| 10 | 25.0000 | -12.1000 | -3.1931 | 75.0000 | ... | 30356.1355 | ... | -8003244.9852 | -185143.6713 |

results agree quite well along the blade span. A more detailed comparison over a range of blade pitch angles can be found in Schmitz (2006) and Schmitz and Chattot (2006).

The interested reader is encouraged to investigate the effect of taking away one blade in the *XTurb* input file and hence the mutual induction. One may surmise that this has a small effect on computed normal and tangential force coefficients, $c_n$ and $c_t$, at inboard blade stations.

### 6.1.2.3  Parked NREL 5-MW Turbine – Optimum Blade Pitch in Low-/High Winds

A useful application (XTurb Example 6.3) for analyzing parked utility-scale wind turbines is finding optimum blade tip pitch settings in low-/high winds. In low winds, the task is often to find the optimum pitch angle that maximizes the turbine startup torque; in high winds, the task may well be to minimize root-flap bending moment. Here we investigate both by considering the NREL 5-MW turbine (Jonkman et al. 2009) in *XTurb* and the PREDICTION mode. The input file is *NREL-5MW-Parked-Ch6-1-6.inp*, see associated *XTurb_Output.dat*.

From the data shown here, it can be suggested that the total startup torque, TO [Nm] for three blades, for the NREL 5-MW turbine is maximized in low winds somewhere around a blade tip pitch angle of 70°, while root-flap bending moment, BE [Nm] for one blade, is minimized in high winds in the vicinity of 85° blade tip pitch angle. Note that one can easily further refine the search for the optimum pitch angles.

## 6.2  Prescribed-Wake Methods

Prescribed-wake methods are the next logical step between blade-element momentum (BEM) methods and high-fidelity approaches that involve computational fluid dynamics (CFD). In essence, vortex wake methods are an extension of Prandtl lifting-line theory to the flow past a rotor. In this context, the Biot–Savart formula (see Section 6.2.3) is used to compute the induced wake flow over the rotor disk and, in particular, along the blade lifting lines. This is also referred to as the Goldstein model (Goldstein 1929). It is important to note, though, that the blade-element assumption still holds in vortex wake methods, that is, each blade section is considered to operate as an airfoil (two-dimensional). In addition, we continue to consider a constant incoming wind speed, $V_0$, at all times (steady) and no radial flow gradients along the blade lifting line (no radial flow).

The assumptions associated with prescribed-wake methods are:

- Two-dimensional (blade aerodynamics)
- Steady
- No radial flow
- Fixed wake structure (No vortex/sheet rollup)

Note that the BEM assumption of root-/tip losses (see Section 3.1) has been replaced by a fixed wake structure whose mutual flow induction of multiple blades accounts for both a downwash $w$ (angular induction) and streamwise $u$ (axial induction) along the blade lifting line, thus eliminating the need for root-/tip corrections with all their limitations discussed in Section 3.4.3. The relationship between induced velocities and axial-/angular induction factors is derived in Section 6.2.5. While any prescribed-wake method has some degree of freedom with respect to for example, helix pitch and wake expansion, the trailing vortex sheets are assumed to be rigid as done in classical lifting-line theory. In other words, the rollup process of root-/tip vortices and the inboard vortex sheet are not accounted for, a limitation that has some implications on the prediction of blade tip loads as is discussed later in this chapter. In comparison to BEM, on the other hand, the induced velocity field computed by any prescribed-wake method does not require model adjustments in the "turbulent wake state" as is the case for BEM solution methods (see Section 3.5.3).

### 6.2.1 Helicoidal Vortex Filaments

Let us proceed by introducing some basic kinematic description of a vortical wake sheet (Chattot 2006). For simplicity, we consider first a single vortical trailer emanating from the blade lifting line, with the blade chord and hence TE essentially compressed onto the lifting line, see Figure 6.7. The starting point for vortical trailer $j$ then becomes:

$$\text{Vortical Trailer (Start):} \quad \vec{r}_j' = (x_{TE,j}, y_{TE,j}, z_{TE,j})^T \tag{6.27}$$

Note that this applies to all radial stations $j = 1 \dots jx$ to form a complete trailing vortex sheet. The wake position vector, $\vec{r}_j = (x, y, z)^T$, has the following components:

Vortical Trailer (Position): $x = x_{TE,j} + x_i$

$$y = r_j \cos\left(\frac{x}{adv_i} + \phi_j\right) \tag{6.28}$$

$$z = r_j \sin\left(\frac{x}{adv_i} + \phi_j\right)$$

In Eq. (6.28), $r_j = y_{TE,j}$ is the radius of filament $j$ emanating from the blade TE at phase angle $\tan^{-1}(z_{TE,j}/y_{TE,j})$ of the initial position vector, $\vec{r}_j'$. Furthermore, $x_i$ is the axial distance from any point on the vortical trailer to its origin on the lifting line. As far as $y$ and $z$ directions are concerned, each vortical trailer performs revolutions of varying radius $r_j$ about the $x$ axis.

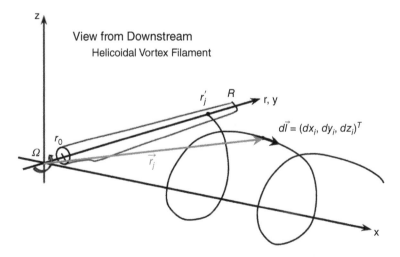

**Figure 6.7** Kinematic description of helicoidal vortex filament.

The helical pitch is described by a local advance ratio (inverse of tip speed ratio) according to the following relation:

$$\text{Sheet Advance Ratio:} \quad adv_i = \frac{V_0 + u(x_i)}{\Omega R} \tag{6.29}$$

Hence the helix pitch changes with the axial wake velocity, $u(x_i)$, defined as the average axial velocity inside the streamtube at downstream location $x_i$, with the index $i$ running from $i = 1 \dots ix$ along each wake trailer.

### 6.2.2 Vortex-Sheet Geometry

At the blade lifting line, $adv_B = (V_0 + u)/\Omega R$ and far downstream in the Trefftz plane $adv_T = (V_0 + u_1)/\Omega R$. Here $u = u(x_i = 0)$ and $u_1$ are related to the average axial induction factor, $a$, via Eq. (2.12). This means in particular that a prescribed-wake structure has to either assume average values for $u$ and $u_1$ or iterate on them during the solution process, see Section 6.2.6, using either the integrated thrust or power coefficient, that is, $C_T$ or $C_P$. The *XTurb* code follows the method described in Chattot (2006) and iteratively adjusts the helix pitch by solving Eq. (2.16) for an average axial induction factor, $a$, and then determining $u$ and $u_1$ from Eq. (2.12). Figure 6.8 gives an example of the equilibrium $adv$ distribution in the wake of the NREL Phase VI rotor at an operating tip speed ratio of $\lambda = 3.77$.

As the helicoidal vortex structure is essentially described as a sequence of piecewise straight vortex filaments such that the induced flow can be computed using the

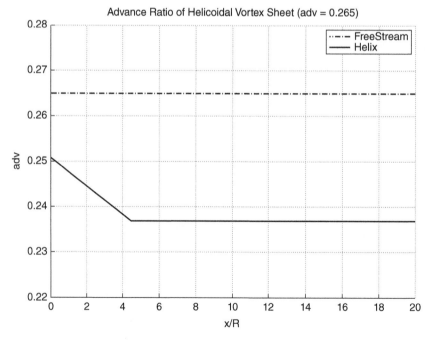

**Figure 6.8** Example of helix advance ratio vs. axial distance traveled into the wake (NREL phase VI rotor, $\lambda = 3.77$) from Schmitz (2006). Source: Reproduced with permission by S. Schmitz.

Biot–Savart formula (see Section 6.2.3), the numerical discretization includes a few parameters that have proven to give consistent results for integrated performance coefficients. An example involves the *XTurb* input file *PhaseVI-Ch6-2-2.inp* run in HVM (METHOD = 2) mode. The discretization of the vortex structure, see next, involves the number of trailing filaments (JX-1), the initial axial spacing along the helix in radii (dx0), the end location of the helix-mesh stretching in radii (XSTR) along with the corresponding number of helix sectors (NSEC) within one sheet revolution, and the location of the Trefftz plane (XTREFFTZ) in radii.

---

**XTurb Example 6.4  NREL Phase VI Rotor – Screen Output (PhaseVI-Ch6-2-2.inp)**

```
  ...

***** HVM *****

 +++++ Vortex Structure - HELIX +++++
         # Filaments jx-1 = 40
 End Mesh Stretching xstr = 1.
             Set xtrefftz = 20.
    Initial Mesh Step dx0 = 0.0001
   Number of Sectors nsec = 30

 +++++ Vortex Structure - BOUND +++++
         # Filaments jx-1 = 40
            # Segments ib = 2
                          = (jx-1)*(ib+1) = 120   Points

 +++++ Vortex Structure - Wake Expans./Contr. +++++
 Wake Expansion/Contraction    WAKEEXP = 0

 +++++ Influence Coefficients +++++

  ...
```

---

The input settings shown here are typical when using the HVM mode, and the reader is encouraged to perform her/his own studies on the effect of parameter variations on integrated thrust and power coefficients. Figure 6.9 shows the corresponding helicoidal vortex structure for the two-bladed NREL Phase VI rotor plotted from *XTurb_gnuplot_helix_data_CASE_1.plt*. More detailed investigations on the accuracy of straight-line vortex segmentation can be found in, for example, Bhagwat and Leishman (2001a,b) and Gupta and Leishman (2005).

As average axial velocity decreases in the wake, the streamtube (see Section 2.1.5) and hence the helicoidal vortex sheet(s) should expand to satisfy continuity according to the following:

$$\text{Wake Expansion:} \quad \frac{r_{W,j}}{r_j} = \sqrt{\frac{adv_i}{adv_B}} \qquad (6.30)$$

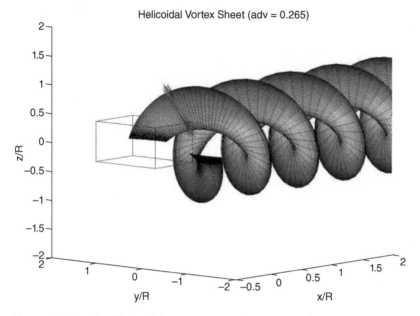

Helicoidal Vortex Sheet (adv = 0.265)

**Figure 6.9** Example of helicoidal vortex structure downstream of a two-bladed rotor (NREL phase VI rotor, $\lambda = 3.77$) from Schmitz (2006). Source: Reproduced with permission by S. Schmitz.

Here, $r_{W,j}(x_i)$ replaces $r_j$ in Eq. (6.28). The effect is in general quite small for low thrust coefficients, $C_T$, but can become notable for higher $C_T$ and included in *XTurb* computations via the parameter WAKEEXP $= 1$.

### 6.2.3 Biot–Savart Law

In order to compute the induced velocities at the blade lifting line from a given vortical wake structure, the Biot–Savart formula is used as:

$$\text{Biot–Savart Law:} \quad \vec{V}_C = (u_C, v_C, w_C)^T = \int \frac{\Gamma}{4\pi} \cdot \frac{\vec{dl} \times \vec{r}}{r^3} \tag{6.31}$$

Here the subscript $C$ denotes a general point $\vec{C} = (x_C, y_C, z_C)^T$ in the flowfield, $\Gamma$ is the circulation strength of a piecewise straight vortex filament of length $\vec{dl}$, and $\vec{r}$ is the distance vector between point $\vec{C}$ and the center of the vortex filament. An illustration is given in Figure 6.10 where both the bound vortex (left) on the lifting line and a trailing vortex filament (right) are shown.

As both the bound vortex and each trailing vortex filament have different circulation strengths, the total induced velocity, $\vec{V}_C$, at point $\vec{C}$ reads in discrete form:

Biot–Savart Law (Discrete):

$$\vec{V}_C = \sum_j \left[ \frac{\Gamma_j}{4\pi} \sum_k \frac{\vec{\Delta l}_{Bo} \times \vec{r}_{Bo}}{r_{Bo}^3} + \frac{\delta\Gamma_j}{4\pi} \sum_i \frac{\vec{\Delta l}_{He} \times \vec{r}_{He}}{r_{He}^3} \right] \tag{6.32}$$

In Eq. (6.32), the subscript *Bo* stands for the "bound" vortex and *He* for the "helicoidal" vortex sheet. Note that the summation is along the blade lifting line $(j)$ with $j = 1 \dots jx$.

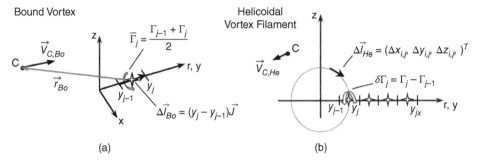

**Figure 6.10** Bound vortex element (a) and trailing vortex filament (b).

Furthermore, each bound vortex element of strength $\overline{\Gamma}_j = \frac{1}{2}(\Gamma_j + \Gamma_{j-1})$ is broken into $kx$ pieces (not necessary if element is straight); similarly, there are a total of $jx - 1$ helicoidal vortex filaments of circulation strength $\delta\Gamma_j = \Gamma_j - \Gamma_{j-1}$ originating between the bound vortex elements, with each helicoidal vortex filament consisting of $ix$ piecewise straight segments (see also Figure 6.9). Investigating Eq. (6.32), it is important to realize that a straight bound vortex does not induce any velocity on itself as $\Delta\vec{l}_{Bo}$ and $\vec{r}_{Bo}$ are collinear. In this context, it is interesting to realize that for a two-bladed rotor with straight blades (i.e. no sweep or dihedral) such as the NREL Phase VI rotor, the bound vortices of both blades do not induce velocities on one another; however, this is different for a three-bladed rotor.

### 6.2.4 Induced Velocities and Influence Coefficients

Let us continue by assuming steady conditions and a rigid (or prescribed) vortex structure and write Eq. (6.32) in a more compact form as:

Induced Velocities: $\quad u_C = \dfrac{1}{4\pi} \displaystyle\sum_{j=1}^{jx} \overline{\Gamma}_j a_{j,Bo} + \delta\Gamma_j (a_{j,He} + a_{j,He}^{rem})$

$$v_C = \frac{1}{4\pi} \sum_{j=1}^{jx} \overline{\Gamma}_j b_{j,Bo} + \delta\Gamma_j (b_{j,He} + b_{j,He}^{rem})$$

$$w_C = \frac{1}{4\pi} \sum_{j=1}^{jx} \overline{\Gamma}_j c_{j,Bo} + \delta\Gamma_j (c_{j,He} + c_{j,He}^{rem}) \tag{6.33}$$

Here, we define $a$, $b$ and $c$ as "influence coefficients" that are essentially scaled geometric relations that when multiplied by a constant circulation strength along its spatial extent, provide induced velocities at an arbitrary point $\vec{C}$:

Influence Coefficients (Bound):

$$a_{j,Bo} = \sum_{k=2}^{kx} \frac{(z_C - z_{k,j}) \cdot \Delta y_{k,j} - (y_C - y_{k,j}) \cdot \Delta z_{k,j}}{[(x_C - x_{k,j})^2 + (y_C - y_{k,j})^2 + (z_C - z_{k,j})^2]^{3/2}} \bigg|_{Blades}$$

$$b_{j,Bo} = \sum_{k=2}^{kx} \frac{(x_C - x_{k,j}) \cdot \Delta z_{k,j} - (z_C - z_{k,j}) \cdot \Delta x_{k,j}}{[(x_C - x_{k,j})^2 + (y_C - y_{k,j})^2 + (z_C - z_{k,j})^2]^{3/2}} \bigg|_{Blades} \tag{6.34}$$

$$c_{j,Bo} = \sum_{k=2}^{kx} \frac{(y_C - y_{k,j}) \cdot \Delta x_{k,j} - (x_C - x_{k,j}) \cdot \Delta y_{k,j}}{[(x_C - x_{k,j})^2 + (y_C - y_{k,j})^2 + (z_C - z_{k,j})^2]^{3/2}} \Bigg|_{Blades}$$

Note that the summations also occur over all blades, assuming steady conditions and hence the same blade loading for all blades. For the helicoidal vortex sheet, we find the following:

Influence Coefficients (Helix):

$$a_{j,He} = \sum_{i=2}^{ix} \frac{(z_C - z_{i,j}) \cdot \Delta y_{i,j} - (y_C - y_{i,j}) \cdot \Delta z_{i,j}}{|\vec{r}_C - \vec{r}_{i,j}|^3} \Bigg|_{Blades}$$

$$b_{j,He} = \sum_{i=2}^{ix} \frac{(x_C - x_{i,j}) \cdot \Delta z_{i,j} - (z_C - z_{i,j}) \cdot \Delta x_{i,j}}{|\vec{r}_C - \vec{r}_{i,j}|^3} \Bigg|_{Blades} \qquad (6.35)$$

$$c_{j,He} = \sum_{i=2}^{ix} \frac{(y_C - y_{i,j}) \cdot \Delta x_{i,j} - (x_C - x_{i,j}) \cdot \Delta y_{i,j}}{|\vec{r}_C - \vec{r}_{i,j}|^3} \Bigg|_{Blades}$$

Equation (6.35) includes all helicoidal vortex filaments originating on the lifting line and up to the location of the Trefftz plane, $x_{ix,j}$. Chattot (2006) derived asymptotic remainder terms as:

Asymptotic Remainder Terms (Helix):

$$a_{j,He}^{rem} = \int_{x_{ix,j}}^{\infty} \frac{(z_C - z_{i,j}) \cdot dy_{i,j} - (y_C - y_{i,j}) \cdot dz_{i,j}}{x_{i,j}^3} \Bigg|_{Blades} \cong \frac{r_j^2}{adv_T \cdot x_{ix,j}^2}$$

$$b_{j,He}^{rem} = \int_{x_{ix,j}}^{\infty} \frac{(x_C - x_{i,j}) \cdot dz_{i,j} - (z_C - z_{i,j}) \cdot dx_{i,j}}{x_{i,j}^3} \Bigg|_{Blades} \cong \frac{z_C - 2r_j \sin(x_{ix,j}/adv_T)}{x_{ix,j}^2}$$

$$\qquad (6.36)$$

$$c_{j,He}^{rem} = \int_{x_{ix,j}}^{\infty} \frac{(y_C - y_{i,j}) \cdot dx_{i,j} - (x_C - x_{i,j}) \cdot dy_{i,j}}{x_{i,j}^3} \Bigg|_{Blades} \cong \frac{y_C - 2r_j \cos(x_{ix,j}/adv_T)}{x_{ix,j}^2}$$

It is apparent from Eq. (6.36) that the remainder terms behave as $O(1/x_{T,j}^2)$ and are quite small if the Trefftz plane is located far downstream, for example, XTREFFTZ = 20. In Eq. (6.36), only the leading terms have been retained.

### 6.2.5 Relationship Between Vortex Theory and Blade-Element Theory

In the following, it is instructive to understand the relationship between vortex wake methods and BEM theory as derived earlier in Section 3.2. From a conceptual method point-of-view, one can say that vortex theory replaces the momentum-theory component in BEM by replacing the "M" with a vortex method "VM," thus essentially becoming a Blade-Element Vortex Method (BEVM). As a consequence, the BEM system of two equations for two unknowns $a$ and $a'$ (see Section 3.5.1) is replaced by a methodology of solving for the radial blade circulation, $\Gamma(r)$, and associated induced axial-/angular velocities, $u_i$ and $w_i$, at the blade lifting line. For a point $\vec{C}$ being on the blade lifting line, we write axial and angular induced velocities as $u_i$ and $w_i$, while keeping in mind that they vary along the blade radius. Here the subscript $i$ refers to "induced" in the present context. These vortex-induced velocities can be written in terms of $a$ and $a'$ as:

$$\text{Lifting Line (Induced Velocities):} \quad u_i = u_C, \ w_i = w_C \tag{6.37}$$

$$\text{Axial-/Angular Induction:} \quad a = -u_i/V_0, \ a' = +w_i/(V_0\lambda_r) \tag{6.38}$$

Note that Eq. (6.38) is given in the header of every *XTurb* output file *XTurb_Output_Method.dat* as:

---

**XTurb Example 6.5  Induced Velocities (Sample XTurb-Output_Method.dat)**

```
...

  "Axial   Induction Factor"   a       = -ui
  "Angular Induction Factor"   a_prime = +wi/(r/R*TSR)

  r/R  Chord/R  Twist[deg]  AOA[deg]  PHI[deg]  CL  CD  CL/CD  FR  FT  ui  wi  a  a_prime

...
```

---

As far as blade-element aerodynamics is concerned, the trigonometric relations (3.7)–(3.9) derived from the velocity triangle in Figure 3.3 can now be written in terms of $u_i$ and $w_i$ as:

$$\text{Velocity Triangle:} \quad \tan\phi(r) = \left(1+\frac{u_i(r)}{V_0}\right) \Big/ \left(\lambda_r + \frac{w_i(r)}{V_0}\right) \tag{6.39}$$

$$\sin\phi(r) = \frac{V_0}{V_{rel}(r)}\left(1+\frac{u_i(r)}{V_0}\right) \tag{6.40}$$

$$\cos\phi(r) = \frac{\Omega r}{V_{rel}(r)}\left(1+\frac{w_i(r)}{V_0\lambda_r}\right) \tag{6.41}$$

The basic angle relation (3.5) on the velocity triangle remains unaltered with:

$$\text{Angle Relation:} \quad \phi(r) = \alpha(r) + \beta(r) \tag{6.42}$$

The KJ lift theorem locally along the lifting line combines $V_{rel}$, $c_l$, and $c$ to a single solution quantity and is given by:

$$\text{KJ Theorem (Local, } r): \quad \Gamma = \frac{1}{2}V_{rel}c_l c \tag{6.43}$$

Equation (6.43) is essential to an iterative solution methodology for vortex wake methods. It is further interesting to note that in anticipation of optimization schemes in Chapter 8, an optimum circulation distribution can be rewritten in terms of the BEM optimum blade parameter $\sigma' c_l$ defined in Section 3.3.1.

### 6.2.5.1  Sectional Thrust and Torque in Vortex Theory
Incremental lift and drag forces are based on a local blade element according to Eqs. (3.12)–(3.13). Here we apply the KJ theorem in Eq. (6.43) to these relations and obtain:

$$\text{Incremental lift (One blade):} \quad dL = \rho V_{rel}\Gamma\, dr \tag{6.44}$$

$$\text{Incremental drag (One blade):} \quad dD = c_d/c_l \,\rho V_{rel}\Gamma\, dr \tag{6.45}$$

Note that (numerically) one has to be aware of the special case $\Gamma = 0$ for $c_l = 0$. Similarly, we can rewrite Eqs. (3.14) and (3.15) for incremental thrust and torque for $B$ blades and find the following using Eqs. (6.37)–(6.41):

**Vortex Theory ($B$ blades):**

Incremental thrust:
$$dT = B\frac{1}{2}\rho V_0^2 R^2 \frac{2\Gamma}{V_0 R}\left[\left(\lambda_r + \frac{w_i}{V_0}\right) + \frac{c_d}{c_l}\left(1 + \frac{u_i}{V_0}\right)\right]d\frac{r}{R} \quad (6.46)$$

Incremental torque:
$$dQ = B\frac{1}{2}\rho V_0^2 R^3 \frac{2\Gamma}{V_0 R}\left[\left(1 + \frac{u_i}{V_0}\right) - \frac{c_d}{c_l}\left(\lambda_r + \frac{w_i}{V_0}\right)\right]\frac{r}{R}d\frac{r}{R} \quad (6.47)$$

Equations (6.46)–(6.47) are now exclusively in terms of the blade circulation, $\Gamma$, as all other variables $u_i$, $w_i$, $c_l$, and $c_d$ are essentially functions of the former as derived in this section.

### 6.2.5.2 Rotor Thrust and Power in Vortex Theory

For known quantities in Eqs. (6.46) and (6.47), one can integrate for the respective rotor thrust and power using the following relations:

Thrust:
$$T = B\frac{1}{2}\rho V_0^2 R^2 \int_{r_{Root}}^{R} \frac{2\Gamma}{V_0 R}\left[\left(\lambda_r + \frac{w_i}{V_0}\right) + \frac{c_d}{c_l}\left(1 + \frac{u_i}{V_0}\right)\right]dr \quad (6.48)$$

Power:
$$P = B\frac{1}{2}\rho V_0^3 R^2 \lambda \int_{r_{Root}}^{R} \frac{2\Gamma}{V_0 R}\left[\left(1 + \frac{u_i}{V_0}\right) - \frac{c_d}{c_l}\left(\lambda_r + \frac{w_i}{V_0}\right)\right]\frac{r}{R}dr \quad (6.49)$$

Here, $r_{Root}$ is the root cut-out where the inboard blade stations merge with the hub of the rotor. The rotor performance coefficients can now be written as:

Thrust Coefficient:

$$C_T = \frac{T}{\frac{1}{2}\rho V_0^2 A} = \frac{B}{\pi}\int_{r_{Root}/R}^{1} \frac{2\Gamma}{V_0 R}\left[\left(\lambda_r + \frac{w_i}{V_0}\right) + \frac{c_d}{c_l}\left(1 + \frac{u_i}{V_0}\right)\right]d\frac{r}{R} \quad (6.50)$$

Power Coefficient:

$$C_P = \frac{P}{\frac{1}{2}\rho V_0^3 A} = \frac{B}{\pi}\lambda\int_{r_{Root}/R}^{1} \frac{2\Gamma}{V_0 R}\left[\left(1 + \frac{u_i}{V_0}\right) - \frac{c_d}{c_l}\left(\lambda_r + \frac{w_i}{V_0}\right)\right]\frac{r}{R}d\frac{r}{R} \quad (6.51)$$

We note once more that the only unknown in the preceding equations is the radial circulation, $\Gamma$, as all other variables can be derived from the former. The task is now to defining a solution methodology for $\Gamma$, which is the topic of the next section.

## 6.2.6 Iterative Prescribed-Wake Solution Methodology

In the following, let us define an iterative prescribed-wake solution methodology. Note again that there are probably as many algorithms as there are codes available; however, we focus here on a general strategy based on the preceding analysis and highlight its differences to BEM solution strategies described in Section 3.5.2. At first, the *blade analysis* problem is defined as:

**Given: Rotor Operating Conditions and Blade Planform**

Tip Speed Ratio, $\lambda$
Tip Pitch Angle, $\beta_{Tip} = \beta(r/R = 1)$
Blade Number, $B$
Root Cut-Out, $r_{Root}/R = 0.10$
Blade Pitch, $\beta = \beta(r/R)$
Non-Dimensional Blade Chord, $\frac{c}{R} = \frac{c}{R}(r/R)$
Airfoil Tables, $c_l(\alpha)$ and $c_d(\alpha)$

The task is then to compute integrated rotor thrust and power coefficients, that is, $C_T$ and $C_P$. In prescribed-wake theory, a solution strategy consists of an outer loop over an initially assumed vortex structure, which is adjusted based on integrated rotor performance coefficients, $C_T$ and $C_P$, after completing the inner iteration cycle. The purpose of the inner iteration scheme is to find an equilibrium radial circulation distribution, $\Gamma(r)$, on a given vortex structure, with the goal of converging the flow state at each blade element, see next.

**Find: Rotor Performance Coefficients, $C_T$ and $C_P$**

**For a given vortex structure, …**

$$
\text{Iterative Solution,} \left\{
\begin{array}{c}
\textbf{Find equilibrium Circulation:} \\
\text{(Loop over entire blade)} \\
\text{Induced Velocities, } u \text{ and } w \text{ Eqs. (6.33), (6.37)} \\
\text{Velocity Ratio, } V_{rel}/V_0 \text{ Eqs. (3.6), (6.38)} \\
\text{Blade Flow Angle, } \phi \text{ Eq. (6.39)} \\
\text{Airfoil Data, } c_l \text{ and } c_d
\end{array}
\right.
$$

Radial Circulation, $\Gamma/V_0$ Eq. (6.43)

**… until radial blade circulation $\Gamma$ is converged!**
**Integrate over blade span, …**
Thrust Coefficient, $C_T$ Eq. (6.50)
Power Coefficient, $C_P$ Eq. (6.51)

**… adjust vortex structure based on $C_P$**

Note that while BEM methods iterate separately over each blade element, the radial circulation distribution, $\Gamma(r)$, is converged over the entire blade length, simply because the induced velocities at each radial station depend on flow induction from all vortical trailers, see Eq. (6.32). Figure 6.11 shows a flowchart of the prescribed-wake solution methodology.

Induced velocities are computed for a given vortex structure and radial circulation distribution. Next, the local flow angles are computed to find the respective angles of attack and hence blade-element $c_l(r)$ and $c_d(r)$. Following, the radial blade circulation, $\Gamma(r)$ is updated using the KJ theorem. This process is repeated in the inner iteration loop until $\Gamma(r)$ is converged. The primary difference compared to any BEM-type solution algorithm is that $\Gamma(r)$ is converged over the entire blade and not separately at each blade element. Once $\Gamma(r)$ is converged, integrated rotor performance coefficients, $C_T$ and $C_P$, define an updated vortex structure (see Section 6.2.2), and the inner iteration process

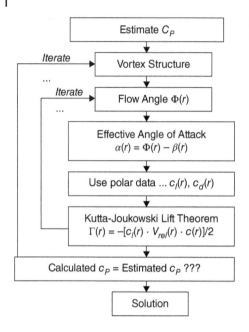

Estimate $C_P$

Iterate — Vortex Structure

...

Iterate — Flow Angle $\Phi(r)$

...

Effective Angle of Attack
$\alpha(r) = \Phi(r) - \beta(r)$

Use polar data ... $c_l(r)$, $c_d(r)$

Kutta-Joukowski Lift Theorem
$\Gamma(r) = -[c_l(r) \cdot V_{rel}(r) \cdot c(r)]/2$

Calculated $c_P$ = Estimated $c_P$ ???

Solution

**Figure 6.11** Flowchart of prescribed-wake solution algorithm.

is repeated using the converged $\Gamma(r)$ from the previous vortex structure as an initial condition. In general, a converged radial circulation and vortex structure are obtained in about five outer iterations, see the *XTurb* example next, for typical operating tip speed ratios, $\lambda$. More iterations are usually required at higher $\lambda$ and/or high thrust coefficient, $C_T$, when the helix pitch becomes quite small and trailing vortex sheets are relatively close to one another.

---

**XTurb Example 6.6 NREL Phase VI Rotor – XTurb_Convergence.dat (PhaseVI-Ch6-2-2.inp)**

```
PhaseVI-Ch6-2-2                                ***** XTurb-PSU  -  CONVERGENCE *****

...


   HVM  :  Lists convergence of CP and/or GAM and will say whether or
           not convergence was achieved.


   Prescribed-Wake Method (HVM)


   Number    TSR      PITCH [deg]
        1   5.4179    3.0000
IT CPOLD CP DCP  2   0.474074095 -0.382995696   0.857069792
IT CPOLD CP DCP  3  -0.382995696 -0.349939696   0.0330560005
IT CPOLD CP DCP  4  -0.349939696 -0.353219005   0.00327930921
IT CPOLD CP DCP  5  -0.353219005 -0.352938608   0.000280397166

Convergence was achieved in 4 iterations over vortex structure.
Change in Power Coefficient DCP =   0.000280397166 < 0.00100000005
```

For parked blades (or wings), the vortex structure is a planar vortex sheet emanating from the lifting line. In this case, only one outer iteration is required to converge the radial (or in this case spanwise) circulation distribution.

### 6.2.6.1 Krogstad Turbine – Prescribed-Wake versus BEM Solution Method

Let us conduct a numerical experiment and analyze a given turbine using both the BEMT (METHOD $= 1$) and prescribed-wake, aka HVM (METHOD $= 2$), solution methodologies. We run *XTurb* in PREDICTION mode applied to the Krogstad turbine (Krogstad and Lund 2012) and compare $C_T$, $C_P$ versus $\lambda$ distributions between methods and against data obtained in the 2.7 × 2.0 m wind-tunnel test section at the Norwegian University of Science and Technology. The Krogstad turbine is a three-bladed 0.9 m diameter rotor that is exclusively equipped with the S826 airfoil (Somers 2005). One corresponding *XTurb* input file is *Krogstad-Ch6-2-6-TSR6-8.inp*. (Note that the actual sectional Reynolds numbers can be found in the corresponding *XTurb_Output1.dat*, and the user may find that those vary significantly over the operating range of tip speed ratios, $\lambda$. Hence the corresponding *XTurb* input files are separated into smaller $\lambda$ ranges such that the used airfoil polar files correspond to the Reynolds range at these conditions.)

Figure 6.12 shows computed rotor thrust and power coefficients, $C_T$ and $C_P$, versus tip speed ratio, $\lambda$, for both BEM and HVM compared to measured data. It is interesting to see in Figure 6.12 that the prescribed-wake method (HVM) and BEM results agree well with each other and with measured data in the low-$\lambda$ range and up to about $C_{P, max}$ at $\lambda = 6.6$. The low-$\lambda$ range is characterized by high angles of attack along most of the blade span (see *XTurb* output files), and the interested reader is encouraged to investigate the effect of using a stall-delay model (STALLDELAY $= 1$) to witness a small effect of rotational augmentation (see Chapter 5) on the power coefficient, $C_P$. For $\lambda > 6.6$, HVM and BEM solutions begin to differentiate themselves from one another. While HVM results are very close to measured $C_P$ values, BEM results compare very closely with measured $C_T$ values. In this context, it is important to understand that HVM does not include any particular treatment in the "turbulent wake state" ($C_T > 0.4$), while the BEM mode in *XTurb* uses the method described in Section 3.5.3.3.

The results shown in Figure 6.12 are representative when compared to an analysis of blind-comparison runs (Krogstad and Eriksen 2013). The original paper by Krogstad and Lund (2012) emphasized the uncertainty of computed $C_T$ and $C_P$ with respect to the airfoil data used. Indeed, the author intentionally chose the example of Figure 6.12 to demonstrate the "dilemma" faced oftentimes in wind turbine aerodynamics when comparing computational results to small-scale experiments, that is, that the uncertainty and sensitivity in airfoil data for $Re < 2x10^5$ has a notable effect on wind turbine performance prediction. Once more, the reader is encouraged to conduct independent research by replacing the sample airfoil files with airfoil polars obtained by Xfoil (Drela 1989) or higher-fidelity CFD analyses with variations in transition and turbulence models.

Figure 6.13 compares converged radial axial-/angular (equivalent) induction factors, $a$ and $a'$, computed by HVM and BEMT for the Krogstad turbine at $\lambda = 6.1$. Differences between the HVM and BEMT modes are apparent starting at about mid-span and are a result of practically half the blade operating in the far "turbulent wake state." Indeed,

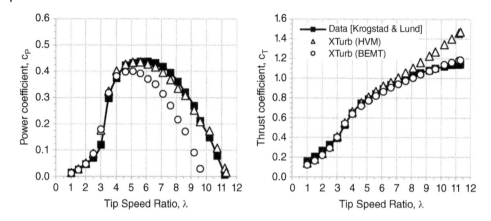

**Figure 6.12** Rotor power coefficient, $C_p$, and thrust coefficient, $C_T$, versus tip speed ratio, $\lambda$ (Krogstad turbine).

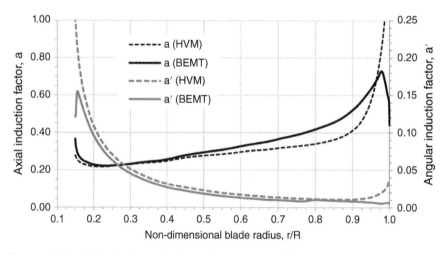

**Figure 6.13** Radial distributions of induction factors $a$ and $a'$ obtained by *XTurb* in HVM and BEMT modes (Krogstad turbine, $\lambda = 6.1$).

Figure 6.13 reveals that HVM results close to the tip are actually in the "propeller brake state," see also Figure 2.6. In the present context, this indicates that differences between vortex methods and BEMT approaches reveal themselves, if the integrated rotor thrust coefficient becomes larger than one, that is, $C_T > 1$. While this is not as prevalent for more lightly-loaded rotors, for example, the NREL Phase VI rotor, it may have an effect on the design of large utility-scale wind turbine blades and associated scaled rotor experiments, see Chapter 8.

### 6.2.7 Limitations of Prescribed-Wake Methods

It is correct to say that prescribed-wake methods do have their place as an intermediate-fidelity tool between BEMT and free-wake methods as they do not require any tip correction and neither an empirical model in the "turbulent wake state."

In addition, their heritage of being an extended lifting-line theory is useful in analyzing parked (non-rotating) cases, as seen in Sections 6.1.2.2–6.1.2.3, and proves itself also being useful toward finding optimal radial circulation distributions in Chapter 8. In the end, however, the major shortcoming of a prescribed-wake method is that it does not include vortex/sheet rollup effects, particularly at the blade tip, some compelling examples of which are provided in Section 6.3.7. In this respect, it is important to understand that vortex rollup processes are not only limited to high thrust coefficients (close to the design point of modern utility-scale wind turbine blades) but can also occur at lower thrust coefficients depending on the slope of the radial circulation distribution (or wake vorticity), $\Gamma'(r)$, in the tip region (Schmitz and Maniaci 2017).

## 6.3 Free-Wake Methods

A natural extension to prescribed-wake methods is to have the wake structure distort and roll up freely downstream of the rotor. In that respect, free-wake methods are a generalized vortex wake method, with blades/wings being represented by lifting lines (or surfaces), and sectional aerodynamic forces determined by local inflow and airfoil lookup tables. As such free-wake methods are rooted in incompressible and irrotational potential flow. Similar to prescribed-wake methods, the Biot–Savart formula (see Section 6.2.3) computes the induced flowfield of all piecewise straight vortex filaments at the lifting lines, thus eliminating the need for tip corrections as used in BEMT methods.

The assumptions associated with free-wake methods are:

- Two-dimensional (blade aerodynamics)
- Time-varying
- No radial flow
- Free-wake structure (vortex/sheet rollup)

It is important to understand that for free-wake methods, the Biot–Savart formula is also used to compute advection speeds for all vortical wake markers (see Section 6.3.2). Hence free-wake methods are time-varying in nature as the wake structure deforms according to the computed marker advection speeds. Consequently, the runtime per iteration increases substantially compared to prescribed-wake methods as new influence coefficients need to be computed at each time step and for all elements of the vortex structure. The result is that the computational cost of free-wake methods increases as $\sim N^3$ where $N$ is the number of wake particles/filaments (as opposed to $\sim N^2$ for prescribed-wake methods). Further note that at least five full wake rotations need to be computed to reach a periodic solution and that time steps are on the order of 1° blade azimuth. In addition to the increased computational cost, experience has shown that free-wake methods are prone to being numerically unstable. The tendency toward numerical instability is due to vortex filaments becoming disorganized quickly in the wake combined with singular induced flow velocities resulting from filaments becoming very close to one another (see Section 6.3.4). Advanced methods (see Section 6.3.5) have been developed that tackle specifically the numerical issue described previously. Free-wake methods have seen more than one renaissance over the last generation, and they are increasingly confronted with competition from advanced computational methods (see Chapter 7) that do not share the same numerical difficulties but resolve the same and additional flow physics.

Nevertheless, free-wake methods are and will continue to be an important "bridging tool" as their foundational relation to lifting-line theory provides opportunities to extract flow physics, for example, tip vortex rollup and skewed wake effects, for model adjustments in prescribed-wake or BEMT methods. Here, the reader is referred to Section 6.3.6 where the capability of a free-wake method to predict blade tip loads is demonstrated and to the theses of Sant (2007) and Kloosterman (2009) for a more detailed introduction to free-wake methods in general and their use in improving BEMT-type methodologies. In this context, the work of Sebastian and Lackner (2011, 2013) is of further note in their application of a free-wake method to investigate blade aerodynamics and its coupled interaction with floating offshore platforms.

In the following subsections, the primary theory and ingredients of free-wake methods are introduced, without necessarily referring to one (out of the many) model variations currently used in the wind energy community.

### 6.3.1 Trailing Vortices versus Shed Vortices

To this end, we have learned about bound circulation along the lifting line and trailing circulation/vorticity emanating from the lifting line and extending far into the wake up to the Trefftz plane. A third type of circulation is the so-called "shed" circulation/vorticity, which is the result of a time-varying bound circulation, $\Gamma = \Gamma(r, t)$. Note that in a steady problem, $\Gamma = \Gamma(r)$ along the bound vortex and hence both bound vortex and vortical trailers do not change their respective strengths over time. In a time-varying problem, however, the bound circulation becomes a function of time, either because of a startup process, yaw misalignment, or time-varying inflow conditions. Consequently, vortex filaments trailing from the lifting line (or bound vortex) into the wake are of strength $(\partial\Gamma/\partial r)_t$, that is, depending on the radial circulation gradient at a particular time, $t$, and this time-varying strength is propagated along the respective vortical trailer with each subsequent time step. In order to satisfy Helmholtz's vortex theorems, a "shed" vortex element of strength $(\partial\Gamma/\partial t)_r$ parallel to the lifting and in between two adjacent trailing vortex filaments has to be released from the lifting line. An illustration is given in Figure 6.14.

Conceptually, a "shed" vortex filament can be interpreted as an elemental starting vortex that is shed into the wake as a reaction to a temporal change in the circulation strength at the respective radial blade element. Thus, the wake vorticity sheet downstream of a lifting line becomes a matrix of discrete circulation values, $\Gamma_{i,j}$, which increases significantly storage and computational cost of the respective analysis method as the induced flow of the shed circulation/vorticity has to be computed in addition to the trailing vorticity, see also for example, Leishman (2006) for a more detailed discussion. Also shown in Figure 6.14 is the rollup process of the blade tip vortex where vortical trailers originating outboard of $\Gamma_{max}(r)$ progressively bundle up into a single concentrated tip vortex, a process important for the accurate prediction of blade tip loads, though accompanied by numerical stability issues as discussed in Section 6.3.4.

### 6.3.2 Lagrangian Markers and Blade Model

In free-wake methods, Lagrangian markers connect piecewise straight vortex filaments. The Lagrangian markers are allowed to move freely in space at a time-varying advection

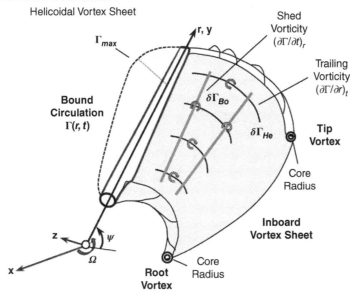

View from Downstream
Helicoidal Vortex Sheet

$\Gamma_{max}$

r, y

Shed
Vorticity
$(\partial\Gamma/\partial t)_r$

Trailing
Vorticity
$(\partial\Gamma/\partial r)_t$

Bound
Circulation
$\Gamma(r, t)$

$\delta\Gamma_{Bo}$

$\delta\Gamma_{He}$

Tip
Vortex

Core
Radius

Inboard
Vortex Sheet

z

$\psi$

$\Omega$

x

Core
Radius

Root
Vortex

**Figure 6.14** Blade trailing versus shed wake vorticity.

speed computed by the induced velocities (Biot–Savart formula) of all vortex filaments (bound, trailing, and shed). An advection equation for the Lagrangian markers reads:

$$\text{Advection Velocity 1:} \quad \frac{d\vec{r}}{dt} = \vec{V}(\vec{r}, t), \vec{r}(t_0) = \vec{r}_0 \tag{6.52}$$

In Eq. (6.52), $\vec{r}_0$ is the initial marker position vector, $t_0$ is the time at which the respective marker was first released/formed, and $\vec{V}$ is the marker advection speed consisting of a superposition of various contributors in linear potential-flow theory. We proceed by writing the time-rate-of-change of the marker position vector, $\vec{r}$, in terms of an azimuth angle, $\psi$, and wake age, $\zeta$, as:

$$\text{Advection Velocity 2 :} \quad \frac{d\vec{r}(\psi, \zeta)}{dt} = \vec{V}(\vec{r}(\psi, \zeta)), \vec{r}(\psi_0, \zeta_0) = \vec{r}_0 \tag{6.53}$$

Here $\psi_0 = \Omega t_0$ is the blade azimuthal position at time, $t_0$, when the marker was first released at the corresponding marker wake age of $\zeta_0 = 0$. This means that in general $\psi = \Omega t$ and $\zeta = \Omega(t - t_0)$ as pseudo spatial and temporal coordinates (Bagai and Leishman 1995). We can then write the time-rate-of-change of the marker position vector as:

$$\text{Marker Position:} \quad \frac{d\vec{r}(\psi, \zeta)}{dt} = \frac{\partial\vec{r}}{\partial\psi}\frac{d\psi}{dt} + \frac{\partial\vec{r}}{\partial\zeta}\frac{d\zeta}{dt} = \Omega\left(\frac{\partial\vec{r}}{\partial\psi} + \frac{\partial\vec{r}}{\partial\zeta}\right) \tag{6.54}$$

Using Eq. (6.52), we can rearrange (6.54) to the following marker advection equation:

$$\text{Advection Equation:} \quad \frac{\partial\vec{r}}{\partial\psi} + \frac{\partial\vec{r}}{\partial\zeta} = \frac{\vec{V}(\vec{r})}{\Omega} \tag{6.55}$$

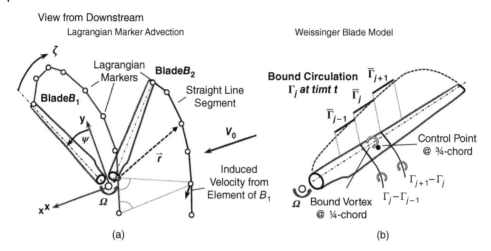

**Figure 6.15** Lagrangian wake markers (a) and Weissinger blade model (b).

Figure 6.15 illustrates the marker advection process for the tip vortex filament. For a numerical solution procedure, Eq. (6.55) is written in terms of difference operators as:

$$\text{Advection Equation (Discrete):} \quad (D_\psi + D_\zeta)\vec{r} = \frac{\vec{V}(\vec{r})}{\Omega} \quad (6.56)$$

Here $D_\psi$ and $D_\zeta$ correspond to spatial and temporal difference operators, respectively. Next, we consider further the marker advection velocity, $\vec{V}$, which in linear potential-flow theory can be written as the superposition of the wind speed, $\vec{V}_0$, the local rotor entrainment speed, $\vec{V}_{rot} = \vec{\Omega} \times \vec{r}_0$, and the total induced velocity, $\vec{V}_{ind}$, as:

$$\text{Advection Velocity 3:} \quad \vec{V} = \vec{V}_0 + \vec{V}_{rot} + \vec{V}_{ind} \quad (6.57)$$

Note that $\vec{V}_{ind}$ has to be computed at every time step and for every Lagrangian marker using repeated application of discrete forms of the Biot–Savart formula. The total induced velocity, $\vec{V}_{ind}$, itself is a superposition of contributions from the bound vortex, $\vec{V}_{Bound}$, from all blades and the trailing/shed wake filaments, $\vec{V}_{Wake} = \vec{V}_{Trailed} + \vec{V}_{Shed}$, such that $\vec{V}_{ind}$ becomes:

$$\text{Marker Induced Velocity:} \quad \vec{V}_{ind} = \vec{V}_{Bound} + \vec{V}_{Trailed} + \vec{V}_{Shed} \quad (6.58)$$

Hence it comes as no surprise that free-wake methods have a higher computational cost (0.5 hours to 1 day) when compared to typical prescribed-wake methods (0.5–2 min).

As far as the blade model is concerned, free-wake methods can be used with the classical lifting-line approach for the bound vortex as described in preceding sections. An improved, though simple and computationally efficient, blade model is that by Weissinger (1947), shown in Figure 6.15, which is essentially a vortex panel method with only one chordwise panel. Here a control point is introduced at the ¾-chord location where flow tangency is required to be satisfied during the iterative solution process for the radial blade circulation, $\Gamma(r, t)$. In this context, the radial control points can be considered markers that are fixed to the blades. A primary difference compared to a straight lifting line is that the control points also see induced velocities from all bound vortex elements. Satisfying a tangency condition at the local three-quarter-chord location derives from thin-airfoil theory where the chordwise vorticity distribution is such that flow tangency is satisfied exactly at that location if the camber line is parabolic.

### 6.3.3  Iterative Free-Wake Solution Methodology

We refer to the *blade analysis* problem defined in Section 6.2.6 and continue by describing a (simplified) general solution methodology for free-wake methods, accounting for Lagrangian marker advection in space and time as well as subiterations to satisfy flow tangency at each time step. An illustration is given in Figure 6.16 with the intent of providing an objective overview diagram of the many variations in use today.

A step-by-step description is given next where the focus of the algorithm is to obtain a periodic solution, which can be the case in yawed flow conditions, considering blade-tower interactions (see Chapter 5), or simply a steady-state solution:

$\psi = 0°$. Start with blades at initial position.

*Solve N blades.* Compute (or guess initially) the blade loading given the local velocity vector and resulting angles of attack along the blade radius.

*Blade circulation, $\Gamma_j^{B,n}$.* Determine the radial blade circulation for all blades.

*Flow tangency/adjust $\Gamma$ distribution.* Use the Weissinger blade model to check flow tangency; iterate on circulation with a suitable algorithm. Proceed when converged.

*Marker advection.* Determine the influence coefficients of the current vortex structure and use new circulation distributions from previous step to compute induced velocities at each Lagrangian marker location. Advect markers to the next time step to obtain an updated vortex structure.

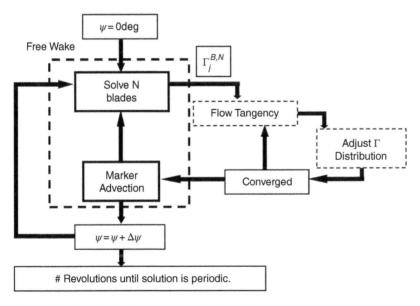

**Figure 6.16** Flowchart of free-wake solution algorithm.

$\psi = \psi + \Delta\psi$. Move blades to next azimuthal position and generate new Lagrangian markers.

*# Revolutions until Solution is Periodic.* Repeat the previous loop for several revolutions. In general, a free-wake method is considered converged if the blade loading becomes periodic; that is, reoccurs with each subsequent rotor revolution.

### 6.3.4 Handling Singularities – Viscous Core Models

The potential-flow nature of the Biot–Savart formula in Eq. (6.31) has one particular property that is exclusively responsible for the fundamental numerical stability issues associated with free-wake methods. Because induced velocities are inversely proportional to the projected normal distance of a given location $\vec{C}$ to a vortex filament, that is, $\sim 1/r^3$, this can result in very high induced velocities in the case of freely moving vortex filaments becoming very close to one another. While the actual singularity for $r = 0$ can be avoided easily in a numerical procedure by specifying a small tolerance in conditional statements, the actual problem is the gradient (and curvature) of induced velocities if markers become close to any vortex filament, the reason being that all markers are advected by the same time step. If too large a time step is chosen, a given marker at location $\vec{C}$ might get advected far further than its immediate neighbors, thus distorting the wake and leading to numerical instabilities. This can be limited by reducing the time step, though resulting in further increased computational cost and just delaying the problem to a later advection time. The solution to this fundamental issue is to "desingularize" the solution, which can be achieved by either (i) defining a cut-off distance below which induced velocities are set to zero, (ii) introducing viscous vortex cores (see next) of radius $r_c$ around the filament, or (iii) eliminating the singularity all together by formulating the problem as a "singularity-free" wake (see Section 6.3.5).

#### 6.3.4.1 Vortex Stretching

Vortex filaments with discrete markers at their end points are subject to vortex stretching (either elongation or compression) as both the filament length and orientation change during the marker advection process. This is simply because the end markers are advecting at different velocities. However, the filament continues to have the same circulation strength, independent of whether it originated as a shed or trailing filament. Nevertheless, the vorticity inside a "stretched" vortex core has to obey Stokes' theorem, see Eq. (4.6), and hence has to change accordingly to integrate to the same circulation strength. An illustration is given in Figure 6.17. A more detailed investigation can be found in Ananthan and Leishman (2004), for example.

Let us assume a vortex filament of length, $l$, and (viscous) core radius, $r_c$. At a given time, the filament is stretched by $\varepsilon = \Delta l / l$ over a time step $\Delta t = \Delta \zeta / \Omega$. In order to conserve the total strength of the vortex tube, the following relation must be satisfied:

$$\pi r_c^2 l = \pi (r_c - \Delta r_c)^2 (l + \Delta l) \tag{6.59}$$

Subsequent solution for the physically-relevant solution for the change in core radius, $\Delta r_c$, yields:

$$\text{Vortex Stretching:} \quad \Delta r_c = r_c \left(1 - \frac{1}{\sqrt{1 + \varepsilon}}\right) \tag{6.60}$$

Original Filament

"Strained" Filament

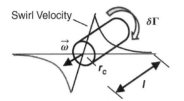
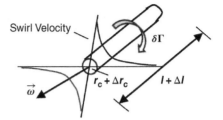

**Figure 6.17** Schematic of stretched vortex filament.

Note that in Eq. (6.60), $\varepsilon > -1$ has to be satisfied, meaning that the filament cannot be compressed to zero length. On the other hand, the more a filament is stretched/elongated, the smaller the adjusted core radius, $r_c^* = r_c - \Delta r_c$. Here it is important to realize that this results in a true "dilemma" of classical free-wake methods as (viscous) vortex cores are introduced for numerical stability, but the stabilizing effect vanishes as filaments get progressively stretched. Again, the practical way of counteracting this effect is to decrease the time step, thus increasing computational cost and ultimately only delaying the problem.

### 6.3.4.2 Rankine Vortex
A commonly used core model is that of a Rankine vortex, see Lamb (1932), that assumes a finite core with solid-body rotation inside the core. The azimuthally induced velocity, $V_\Theta$, reads:

$$\text{Rankine Vortex:} \quad V_\Theta(\bar{r}) = \begin{cases} \left(\dfrac{\Gamma}{2\pi r_c}\right)\bar{r} & 0 < \bar{r} < 1 \\ \left(\dfrac{\Gamma}{2\pi r_c}\right)\dfrac{1}{\bar{r}} & \bar{r} \geq 1 \end{cases} \tag{6.61}$$

In Eq. (6.61), $\bar{r} = r/r_c$ is the core-radius normalized distance between a point $\vec{C}$ and the vortex filament. Also note that the slope $dV_\Theta/d\bar{r}$ is not continuous at $\bar{r} = 1$.

### 6.3.4.3 Lamb–Oseen Vortex
The Lamb–Oseen vortex is an actual exact solution to the Navier–Stokes equations, see Lamb (1932) and Oseen (1911), with a continuous azimuthally induced (or swirl) velocity, $V_\Theta$, defined as:

$$\text{Lamb–Oseen Vortex 1:} \quad V_\Theta(\bar{r}) = \frac{\Gamma}{2\pi r_c}\frac{1 - e^{-\alpha\bar{r}^2}}{\bar{r}} \tag{6.62}$$

In Eq. (6.62), $\bar{r} = r/r_c$ is again the core-radius normalized distance between a point $\vec{C}$ and the vortex filament, see, for example, Batchelor (1967); Saffman et al. (1992); and Wu et al. (2006), and $\alpha$ is a constant. Note that the Lamb–Oseen vortex decays over time due to the action of viscosity. As such the core radius, $r_c$, grows in time and reads:

$$\text{Core Radius:} \quad r_c = \sqrt{4\alpha\nu t} \tag{6.63}$$

In Eq. (6.63), $t$ is the physical time and $v$ is the fluid kinematic viscosity. In this respect, the growing core radius due to the action of viscosity somewhat counteracts the respective reduction due to vortex stretching. Using a simple series expansion (Scully 1975) and further generalization (Vatistas et al. 1991), $V_\Theta$ can be written as a function of the dimensional $r$ as:

$$\text{Lamb–Oseen Vortex 2}: \quad V_\Theta(r) = \frac{\Gamma}{2\pi r_c} \frac{\bar{r}}{(1 + \bar{r}^{2n})^{1/n}} \tag{6.64}$$

Here, $n$ is a free parameter ($n = 1$ for the "Scully Vortex") that can be chosen based on empirical data in viscous flow where the core growth is a function of turbulence and the net rate of diffusion. In practice, only $n = 1, 2$ are physically meaningful.

### 6.3.4.4 Difficulties of Viscous Core Models

The inherent difficulty of all viscous core models is that an inviscid vortex model/method is augmented to take into the account the action of viscosity on vortex stretching and diffusion. Remaining free parameters are chosen as a "tweak" to gain numerical stability and do not necessarily represent the correct physics. There is a myriad of other "tricks" available in the wind energy and rotorcraft communities that are truly too exhaustive to list. One commonly used strategy though is to simply limit the number of vortical trailers in the wake, thus enabling the vortical wakes to develop for a number of revolutions before stretched filaments become close to one another and local marker advection speeds are subject to non-physical artificial core effects. This is actually a valid approach if a sparse distribution of vortical trailers is released from a blade region where the circulation distribution is close to linear; however, this is not the general case in the tip region where curvature (second derivative) in the radial circulation distribution governs the subtle details of the immediate vortex rollup process and has a notable effect on computed blade tip loads. And here we are back at the aforementioned dilemma that the vortex/sheet rollup is the physical effect that free-wake methods are (in theory) capable of capturing and is their primary advantage compared to prescribed-wake and BEMT methods. Another alternative is to simply define a "cut-off" distance around each filament such that the marker advection speed is set to zero within that distance to the singularity; however, this approach has been known to result in unphysical solutions that are very sensitive to the cut-off distance chosen for the analysis.

### 6.3.5 Singularity-Free-Wake – Distributed Vorticity Elements (DVEs)

As shown in the preceding sections, conventional free-wake methods use a series of discrete vortex filaments or vortex particles to transport shed and trailing vorticity into the wake. Numerical singularities that destabilize the solution are avoided by employing a cut-off distance from the filament singularity or by using vortex core models, though at the expense of physical accuracy. An alternative is the DVE method, first developed by Bramesfeld and Maughmer (2008) for fixed-wing applications. The conceptual idea is rooted in the multi-lifting-line method of Horstmann (1987) and Horstmann et al. (2007) where discrete vortex filaments are replaced by a continuous vortex sheet composed of second-order-accurate sheet elements to represent the radial circulation distribution.

### 6.3.5.1 The Multi-Lifting-Line Method of Horstmann

In general, vortex filaments emanating from the blade lifting line are of constant strength. Consequently, model accuracy is controlled by the number of spanwise elements. Horstmann et al. (2007) developed a multi-lifting-line method using vortex elements with parabolic distributions instead of constant-strength filaments to determine the induced drag of nonplanar wing configurations. The parabolic vortex elements of Horstmann form a continuous-shed vortex sheet on the lifting surface and wake that by design eliminates discrete vortex filaments and all associated issues described earlier in Section 6.3. The parabolic circulation distributions of the Horstmann elements result in sheets of shed vorticity with linear $\gamma$ (vorticity) distributions that are continuous in the magnitude of vorticity, thus requiring fewer spanwise elements for a desired accuracy than obtained using constant-strength vortex elements. Due to the continuity in shed vorticity, the velocity fields modeled by each wake element are finite, and the tangentially induced velocity in the plane of the sheet is undetermined, see Horstmann (1987) for more details.

### 6.3.5.2 The Singularity-Free-Wake Method of Bramesfeld and Maughmer

Bramesfeld and Maughmer (2008) used the accuracy of Hortstmann's parabolic vortex elements to develop a free-wake method that eliminates numerical stability issues caused by the wakes of standard free-wake methods. The original method was developed for fixed-wing applications and later extended to inviscid wind turbine analysis by Basom and Maughmer (2011) (see also the thesis by Basom 2010) and viscous effects as well as winglets by Maniaci and Maughmer (2012) (see also the thesis by Maniaci 2013). By adding many DVEs, the blade and wake can be modeled with the ability to relax and deform to a force-free-wake shape. An example of a fully-relaxed wake modeled using DVEs is shown in Figure 6.18. The computational cost of the DVE method is similar to standard free-wake methods due to the $\sim N^3$ behavior with the number of

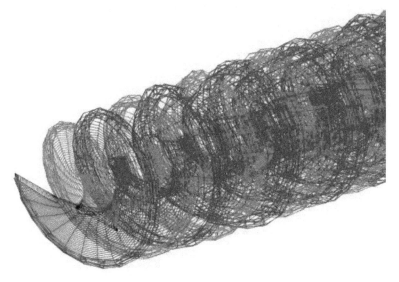

**Figure 6.18** Example of DVE-computed wake (NREL phase VI rotor S-sequence, $V_0 = 7$ m s$^{-1}$) from Schmitz and Maniaci (2016).

wake sheets/particles, $N$. Nevertheless, the DVE method can be used in general with larger time steps, and a baseline of typically 41 radial elements is used to compute the wake rollup process such as illustrated in Figure 6.18.

### 6.3.6 Prediction of Blade Tip Loads – Free-Wake versus Prescribed-Wake/BEM Methods

From a blade aerodynamics standpoint, the primary capability of free-wake methods as a mid-fidelity tool is to compute details of the tip vortex rollup process and its subsequent effect on predicting blade tip loads, thus bridging efficient BEMT (and prescribed-wake) methodologies and computationally expensive high-fidelity CFD methods.

A compelling example is shown in Figure 6.19 for the NREL Phase VI rotor in attached flow and at a medium thrust coefficient. Measured normal force coefficients (Hand et al. 2001) are compared against various results computed by *XTurb* and the *WindDVE* code (Maniaci 2013). It can be seen in Figure 6.19 (a) that both the standard BEMT (METHOD = 1) and prescribed-wake (HVM) method (METHOD = 2) in *XTurb* over-predict the measured normal force coefficient at the $r/R = 0.95$ station. This behavior is very typical of practically all BEMT-type methods and a consequence of the limitations associated with classical tip corrections, see Section 3.4.3. Note that the HVM mode in *XTurb* leads to very similar results compared to the respective BEMT analysis, thus supporting the hypothesis that the remaining discrepancies between measured data and model predictions in the tip region might be a consequence of not accounting for the subtleties associated with the rollup process of the tip vortex.

On the other hand, analyzing the rotor using the (singularity-free) free-wake *Wind-DVE* code, with data taken from Schmitz and Maniaci (2017), shows some notable improvement in predicting blade tip loads, see Figure 6.19 (b), thus providing evidence that vortex/sheet rollup does affect computed blade loads in the tip region. Indeed, applying a general $g$ function derived from a suite of free-wake analyses as proposed by Schmitz and Maniaci (2017), see also Section 3.4.4.4, within *XTurb* (METHOD = 1, TIPLOSS = 2) does result in improved load predictions at $r/R = 0.95$ and compares

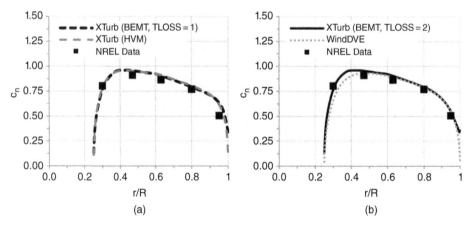

**Figure 6.19** NREL phase VI rotor – normal force coefficient (S – sequence, $V_0 = 7$ m s$^{-1}$) – (a) BEMT, TLOSS = 1 versus HVM; (b) BEMT, TLOSS = 2 versus WindDVE.

very closely to *WindDVE* results, as intended by a general *g* function. Note that the *XTurb* results were obtained in only a fraction of a second, while the *WindDVE* analysis requires about 1 hour of time. In the wind energy community, the idea of improving BEMT analyses by vortex theory has seen significant improvements over the past decade. In this context, the interested reader is referred to recent accomplishments by Sant (2007), Branlard et al. (2013), Branlard and Gaunaa (2014), Wood et al. (2016), and Wood (2018) for somewhat different approaches, though with equally compelling results.

### 6.3.7 Limitations of Free-Wake Methods

The primary issue of most free-wake methods remains numerical stability, though this has been resolved in practical terms by the DVE method described in Section 6.3.5. Nevertheless, the primary advantage of free-wake methods compared to BEMT analyses is the computation of vortex rollup effects. The required computational time is reasonable, ranging from tens of minutes to several hours, though notably higher than HVM and particularly BEMT methods and therefore somewhat limiting their use for analyzing a large number of design cases. In this context, the TIPLOSS = 2 option in *XTurb* BEMT mode has shown promise in capturing vortex rollup effects at lower computational cost. Furthermore, free-wake methods are still rooted in lifting-line theory, thus relying on quality (possibly 3-D corrected) airfoil tables and therefore not capable of capturing real 3-D effects associated with blade separation/stall in off-design conditions.

In the end, it becomes a matter of return-on-investment in the sense that one has to critically evaluate the potential benefit(s) of a higher computational cost compared to predictive capability and uncertainty of the analysis method. In this respect, free-wake methods in general, whether it be in the wind energy, rotorcraft, and even fixed-wing communities, have seen more than one renaissance over time.

## References

Ananthan, S. and Leishman, J.G. (2004). The role of filament stretching in the free-vortex modeling of rotor wakes. *Journal of the American Helicopter Society* 49: 176–191.

Anderson, J.D. (1978). *Introduction to Flight*. McGraw Hill.

Anderson, J.D. (2001). *Fundamentals of Aerodynamics*, McGraw Hill Series in Aeronautical and Aerospace Engineering, 3e. New York: McGraw Hill.

Bagai, A. and Leishman, J.G. (1995). Rotor free-wake modeling using a pseudo-implicit relaxation algorithm. *AIAA Journal of Aircraft* 32 (6): 1276–1285.

Basom, B. J. (2010) Inviscid Wind-Turbine Analysis Using Distributed Vorticity Elements. M.S. Thesis, The Pennsylvania State University.

Basom, B. J. and Maughmer, M. D. (2011) Inviscid Analysis of Horizontal-Axis Wind Turbines Using Distributed Vorticity Elements. *49th AIAA Aerospace Sciences Meeting including the New Horizons Forum and Aerospace Exposition*, AIAA-2011-0539.

Batchelor, G.K. (1967). *An Introduction to Fluid Dynamics*. Cambridge University Press.

Bhagwat, M.J. and Leishman, J.G. (2001a). Stability, consistency and convergence of time-marching free-vortex rotor wake algorithms. *Journal of the American Helicopter Society* 46 (1): 59–71.

Bhagwat, M.J. and Leishman, J.G. (2001b). Accuracy of straight-line segmentation applied to curvilinear vortex filaments. *Journal of the American Helicopter Society* 46 (2): 166–169.

Bramesfeld, G. and Maughmer, M.D. (2008). Relaxed-wake vortex-lattice method using distributed vorticity elements. *AIAA Journal of Aircraft* 45 (2): 560–568.

Branlard, E. and Gaunaa, M. (2014). *Development of new tip-loss corrections based on vortex theory and vortex methods. Journal of Physics: Conference Series (Online)* 555: 1–8.

Branlard, E., Dixon, K., and Gaunaa, M. (2013). *Vortex methods to answer the need for improved understanding and modelling of tip-loss factors. IET Renewable Energy Generation* 7 (4): 311–320.

Chattot, J.J. (2006). Helicoidal vortex model for steady and unsteady flows. *Computers & Fluids* 35: 733–741.

Drela, M. (1989). X-Foil: an analysis and design system for low Reynolds number airfoils. In: *Low Reynolds Number Aerodynamics*, vol. 54 (ed. T.J. Mueller), 1–12. New York: Lecture Notes in Engineering. Springer-Verlag.

Glauert, H. (1926). *The Elements of Aerofoil and Airscrew Theory*. Cambridge, UK: Cambridge University Press.

Goldstein, S. (1929). On the vortex theory of screw propellers. *Proceedings of the Royal Society of London. Series A* 123: 440–465.

Gupta, S. and Leishman, J.G. (2005). Accuracy of the induced velocity from helicoidal vortices using straight-line segmentation. *AIAA Journal* 43 (1): 29–40.

Hand, M.M., Simms, D.A., Fingersh, L.J. et al. (2001). *Unsteady Aerodynamics Experiment Phase VI: Wind Tunnel Test Configurations and Available Data Campaigns*. National Renewable Energy Laboratory, NREL/TP-500-29955.

Horstmann, K.H. (1987). *Ein Mehrfach-Traglinienverfahren und seine Verwendung für Entwurf und Nachrechnung nichtplanarer Flügelanordnungen*, 87–51. Braunschweig, Germany, DFVLR-FB: DVFLR, Institut für Entwurfsaerodynamik.

Horstmann, K.H., Engelbrecht, T., and Liersch, C. (2007). *LIFTING _ LINE Version 2.2 Handbook. German Aerospace Center (DLR)*. Braunschweig, Germany: http://www.dlr.de/as/en/desktopdefault.aspx/tabid-188/ 379_read-625/ (accessed: April 17, 2019).

Jonkman, J., Butterfield, S., Musial, W., and Scott, G. (2009) *Definition of a 5-MW Reference Turbine for Offshore System Development*. Technical Report NREL/TP-500-38060, National Renewable Energy Laboratory.

Kloosterman, M. H. M. (2009) Development of the Near Wake behind a Horizontal Axis Wind Turbine. M.S. Thesis, Delft University of Technology, The Netherlands.

Krogstad, P.A. and Eriksen, P.E. (2013). "Blind test" calculations of the performance and wake development for a model turbine. *Renewable Energy* 50: 325–333.

Krogstad, P.A. and Lund, J.A. (2012). An experimental and numerical study of the performance of a model turbine. *Wind Energy* 15: 443–457.

Lamb, H. (1932). *Hydrodynamics*. Cambridge, UK: Cambridge University Press pp. 592–593, 668–669.

Leishman, J.G. (2006). *Principles of Helicopter Aerodynamics*, 2e. Chapter 10.7. Cambridge: Cambridge University Press.

Maniaci, D. C. (2013) Wind Turbine Design using a Free-Wake Vortex Method with Winglet Application. Ph.D. Dissertation, The Pennsylvania State University.

Maniaci, D. C. and Maughmer, M. D. (2012) Winglet design for wind turbines using a free-wake vortex analysis method. 50th AIAA Aerospace Sciences Meeting, AIAA-2012-1158.

Moran, J. (1984). *An Introduction to Theoretical and Computational Aerodynamics*. New York: Wiley.

Oseen, C.W. (1911). Über Wirbelbewegung in einer reibenden Flüssigkeit. *Ark. Mat. Astron. Fys.* 7: 14–21.

Prandtl, L. (1918) Tragflügeltheorie I. und II. *Mitteilung. Nachr. von der Kgl. Gesellschaft der Wissenschaften, Math.-phys. Klasse*, p. 451.

Prandtl, L. (1923). *Applications of Modern Hydrodynamics to Aeronautics*. National Advisory Committee for Aeronautics, NASA-TR-116.

Saffman, P.G., Ablowitz, M.J., Hinch, E.J. et al. (1992). *Vortex Dynamics*, 253. Cambridge: Cambridge University Press. ISBN: 0-521-47739-5.

Sant, T. (2007) Improving BEM-based Aerodynamic Models in Wind Turbine Design Codes. Ph.D. Dissertation, Delft University of Technology, The Netherlands.

Schmitz, S. (2006) Coupling of Navier–Stokes Solver With Helicoidal Vortex Model for the Computational Study of Horizontal Axis Wind Turbines. Ph. D. Dissertation, University of California Davis.

Schmitz, S. (2012). *XTurb-PSU: A Wind Turbine Design and Analysis Tool*. University Park, PA: Pennsylvania State University.

Schmitz, S. and Chattot, J.J. (2006). Characterization of three-dimensional effects for the rotating and parked NREL phase VI wind turbine. *ASME Journal of Solar Energy Engineering* 128: 445–454.

Schmitz, S. and Maniaci, D. (2016) *Analytical Method to Determine a Tip Loss Factor for Highly-Loaded Wind Turbine Rotors*. AIAA-2016-0752.

Schmitz, S. and Maniaci, D. (2017). *Methodology to determine a tip-loss factor for highly loaded wind turbine. AIAA Journal* 55 (2): 341–351. https://doi.org/10.2514/1.J055112.

Scully, M. (1975) Computation of Helicopter Rotor Wake Geometry and its Influence on Rotor Harmonic Airloads. Ph.D. Dissertation, Massachusetts Institute of Technology.

Sebastian, T. and Lackner, M. (2011). Development of a free vortex wake model code for offshore floating wind turbines. *Renewable Energy* 46: 269–275.

Sebastian, T. and Lackner, M. (2013). Characterization of the unsteady aerodynamics of offshore floating wind turbines. *Wind Energy* 16 (3): 339–352.

Somers, D.M. (2005). *The S825 and S826 Airfoils*. State College, PA: Airfoils Inc. Published as: National Renewable Energy Laboratory, NREL/SR-500-36344.

Vatistas, G.H., Kozel, V., and Mih, W.C. (1991). A simpler model for concentrated vortices. *Experiments in Fluids* 11 (1): 73–76.

Weissinger, J. (1947). *The Lift Distribution of Swept-Back Wings*. National Advisory Committee for Aeronautics, NACA-TM-1120.

Wood, D.H. (2018). *Application of extended vortex theory for blade element analysis of horizontal-axis wind turbines. Renewable Energy* 121: 188–194.

Wood, D.H., Okulov, V.L., and Bhattacharjee, D. (2016). *Direct calculation of wind turbine tip loss. Renewable Energy* 95: 269–276.

Wu, J.Z., Ma, H.Y., and Zhou, M.D. (2006). *Vorticity and Vortex Dynamics*, 262. Berlin: Springer-Verlag. ISBN: 3-540-29027-3.

## Further Reading

Chattot, J.J. and Hafez, M.M. (2015). *Theoretical and Applied Aerodynamics – and Related Numerical Methods*. Chapter 6.7. Dordrecht: Springer.

Leishman, J.G. (2000). *Principles of Helicopter Aerodynamics*. Cambridge: Cambridge University Press.

Stepniewski, W.Z. and Keys, C.N. (1984). *Rotary-Wing Aerodynamics*. New York: Dover.

# 7

# Advanced Computational Methods

*To CFD or not to CFD, that's the question.*

— S. Schmitz

## 7.1 High-Fidelity Blade-Resolved CFD Solutions

The Navier–Stokes equations are derived from mass- and momentum principles in differential form. They present the most general form of equations in incompressible fluid mechanics (in compressible flow, the energy equation is needed in addition for closure). Numerical discretization is realized through either finite-difference, finite-element, or finite-volume methods, with the latter having become the most prominent method of discretization. The basic ingredients of any computational fluid dynamics (CFD) method are "The Grid, The Scheme, The Solver." Grid generation around wind turbine blades, nacelles, and the tower requires user experience, skill, and a significant amount of time; here either structured or unstructured grids can be generated depending on the solver. The numerical scheme(s) to solve the governing set of partial differential equations depend primarily on the discretization method used to solve the equations. The order-of-accuracy is of importance in realizing accurate solutions whose (numerical) discretization error is proportional to the respective order of the grid spacing. The actual solver describes the iterative strategy to solve the discretized set of partial differential equations. A sufficient condition for an adequate numerical solver is to behave numerically stable and to converge the (flow) solution to a specified solution residual; in this context, the residual is obtained by applying the iterative flow solution to the numerical scheme in order to quantify how well the solution satisfies the discretized equations.

The "Verification and Validation" (V&V) of CFD solvers is an integral part of continued grid, scheme, and solver developments. Here, "verification" typically refers to ensuring that numerical schemes and solver algorithm do indeed recover test functions (or exact solutions) to the expected order-of-accuracy and solution residual to the limit of round-off errors. As for "validation" of CFD solvers, this is typically a step-by-step process in which experimental data obtained from model problems are used to compare against computed solutions. These model problems do not necessarily include the complexity of the full problem (e.g. turbine blades + nacelle + tower in time-varying inflow) but a subset of smaller problems with reduced complexity (e.g. wind tunnel

*Aerodynamics of Wind Turbines: A Physical Basis for Analysis and Design,* First Edition. Sven Schmitz.
© 2020 John Wiley & Sons Ltd. Published 2020 by John Wiley & Sons Ltd.
Companion website: www.wiley.com/go/schmitz/wind-turbines

tests of airfoils, model rotors, etc. in uniform inflow) to properly validate the CFD solvers.

Today, CFD solvers continue to be further validated against newly available data, and they serve increasingly as a linking element between measured data and reduced-order methods such as Blade Element Momentum (BEM) and vortex models. Due to the time/cost of grid generation and numerical solution (on the order of "days" for $O(10^6)$ grid points), CFD is not expected to become the standard analysis method to evaluate large numbers of turbine blade designs in the foreseeable future. On the contrary, CFD is used to evaluate a selected number of final design candidates in practice and provides insightful data on 3-D flow effects and separation, tip vortex rollup, and wake development of use in further improving reduced-order BEM and vortex models.

### 7.1.1 Unsteady Reynolds-Averaged Navier–Stokes Equations

The steady and incompressible Navier–Stokes equations have been introduced in Chapter 4 and Eqs. (4.32) and (4.33), respectively, when introducing the airfoil (blade) boundary-layer equations. In this context, it is important to understand that unlike atmospheric boundary-layer (ABL) flow, blade airfoils do (in general) experience at least some portion of laminar flow before transitioning to turbulence occurs. In addition, the viscous region is confined to a thin layer (boundary layer) for high-$Re$ airfoil flows. This allows to average quantities in that region and "model," instead of resolving, all turbulent scales in the airfoil boundary layer and their effect on surface shear stress and hence skin-friction and profile drag, $c_d$. Here we write every flow quantity, $f$, as the sum of a time-averaged value, $\bar{f}$, and a fluctuating component, $f'$, as:

$$f = \bar{f} + f' \tag{7.1}$$

$$\text{Time Average: } \bar{f} = \frac{1}{T} \int_t^{t+T} f(x, y, z, t)dt \tag{7.2}$$

Substituting Eqs. (7.1) and (7.2) into the Navier–Stokes equations and noting that $\overline{f'} = 0$ and hence $\overline{f_1 + f_2} = \bar{f_1} + \bar{f_2}$ and similarly for respective partial derivatives, yields the Unsteady Reynolds-Averaged Navier–Stokes (URANS) equations in tensor notation:

$$\text{URANS Equations: } \frac{\partial \bar{u}_j}{\partial x_j} = 0 \tag{7.3}$$

$$\frac{\partial \bar{u}_i}{\partial t} + \bar{u}_j \frac{\partial \bar{u}_i}{\partial x_j} = -\frac{1}{\rho} \frac{\partial \bar{p}}{\partial x_i} + \nu \frac{\partial^2 \bar{u}_i}{\partial x_j^2} - \frac{\partial \overline{u_i' u_j'}}{\partial x_j} \tag{7.4}$$

Here, the subscript $i$ refers to the respective Cartesian component and $j$ denotes summation. In Eq. (7.4), the kinematic viscosity is $\nu = \mu/\rho$ where $\mu$ is the fluid dynamic viscosity. The last term $-\partial \overline{u_i' u_j'}/\partial x_j$ in Eq. (7.4) represents the Reynolds stress terms, adding a total of six unknowns to the URANS equations as a result of a symmetric Reynolds stress tensor. In order to achieve closure, the Reynolds stress terms must be modeled. A common approach in turbulence modeling is to perform order-of-magnitude analyses for the respective Reynolds stress terms and essentially cast them into an eddy-viscosity term adding to the laminar stress term, $\nu \partial^2 \bar{u}_i/\partial x_j^2$.

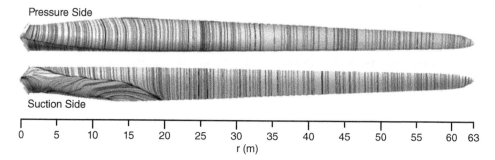

Pressure Side

Suction Side

0  5  10  15  20  25  30  35  40  45  50  55  60  63
r (m)

**Figure 7.1** Limit surface streaklines over surface pressure contours (NREL 5-MW turbine, $V_0 = 8$ m s$^{-1}$). Source: Reproduced with permission from (Chow and Van Dam 2012).

An example of a full URANS solution (Chow and Van Dam 2012) of the National Renewable Energy Laboratory (NREL) 5-MW turbine is shown in Figure 7.1 where surface streamlines are approximated by skin-friction lines on the blade surface. Full URANS solutions such as depicted in Figure 7.1 are still the state-of-the-art in blade-resolved simulations. Here it is interesting to realize that the blade flow is indeed 2-D for most of the blade. The root region, in particular, experiences 3-D flow effects whose detailed analysis can provide informative data for improved BEM-type models such as discussed in Chapter 5.

The effect of the blade root region on inboard blade loads had also been found for the NREL Phase VI rotor, see, for example, comprehensive works of Sørensen et al. (2002), Le Pape and Lecanu (2004), Le Pape and Gleize (2006), and Duque et al. (2003).

### 7.1.2 Turbulence Modeling

Turbulence models provide closure to the URANS equations by means of model expressions for the Reynolds stresses. In an eddy-viscosity model, one assumes that turbulence is constituted of small eddies, which continuously form and dissipate. One classical example is that of the well-known mixing-length hypothesis (or zero-equation model) due to Prandtl (Schlichting 1979) where Reynolds stresses are modeled as proportional to an eddy viscosity and the mean velocity gradient. Today, primarily one-equation and two-equation turbulence models are used in the CFD community where "one" or "two" refers to the number of additional transport equations to be solved. The wind energy science community has almost exclusively focused on two-equation turbulence models as they offer a good compromise between solution accuracy and computational cost. Most often the turbulent kinetic energy, $k$, is the primary transport variable. Common choices for the second transport variable are either the turbulent energy dissipation rate, $\varepsilon$, or the turbulent energy frequency, $\omega$, with the former relating to dissipation of turbulent energy and the latter to the time scale associated with turbulent energy.

#### 7.1.2.1 $k$-$\varepsilon$ Turbulence Model

Today, the $k$-$\varepsilon$ two-equation turbulence model is an industry standard used in a multitude of applications, with proven numerical accuracy and stability. Here, $k$ [m$^2$ s$^{-2}$] is the (specific) turbulent kinetic energy defined as the variance of (turbulent) velocity fluctuations, while $\varepsilon$ [m$^2$ s$^{-3}$] is the (specific) turbulent energy dissipation rate, which is a measure of the rate at which turbulent velocity fluctuations are being dissipated. The original version of the $k$-$\varepsilon$ turbulence model is due to Launder and Spalding (1974) where an effective (dynamic) viscosity is used as $\mu_{eff} = \mu + \mu_t$, where $\mu$ is the fluid (dynamic) viscosity and $\mu_t$ is a turbulent (dynamic) viscosity defined as:

$$\text{Turbulent (Dynamic) Viscosity: } \mu_t = \rho C_\mu \frac{k^2}{\varepsilon} \tag{7.5}$$

In Eq. (7.5), $C_\mu = 0.09$ is a model constant. The transport equations for $k$ and $\varepsilon$ become:

$$\text{Transport Eq. } k: \frac{\partial(\rho k)}{\partial t} + \frac{\partial(\rho k \overline{u}_i)}{\partial x_i} = \frac{\partial}{\partial x_j}\left[\frac{\mu_t}{\sigma_k}\frac{\partial k}{\partial x_j}\right] + P_k - \rho\varepsilon \tag{7.6}$$

$$\text{Transport Eq. } \varepsilon: \frac{\partial(\rho\varepsilon)}{\partial t} + \frac{\partial(\rho\varepsilon\overline{u}_i)}{\partial x_i} = \frac{\partial}{\partial x_j}\left[\frac{\mu_t}{\sigma_\varepsilon}\frac{\partial\varepsilon}{\partial x_j}\right] + C_{1\varepsilon}\frac{\varepsilon}{k}P_k - C_{2\varepsilon}\rho\frac{\varepsilon^2}{k} \tag{7.7}$$

Here, $P_k$ is the turbulent production rate and essentially a function of the Reynolds stresses, while $\sigma_k = 1.00$, $\sigma_\varepsilon = 1.30$, $C_{1\varepsilon} = 1.44$, and $C_{2\varepsilon} = 1.92$ are empirical model constants found for a wide range of turbulent flows. Note that the convective transport of the respective modeled turbulent quantity occurs on the left-hand side of Eqs. (7.6) and (7.7), while the right-hand side balances the transport part by respective diffusion, production, and destruction/dissipation terms.

### 7.1.2.2 $k$-$\omega$ Turbulence Model

It is generally agreed that turbulence models based on the $\varepsilon$ equation result in delayed predictions of the onset of separation for airfoils and hence underpredict the amount of flow separation at high angles of attack. The Wilcox $k$-$\omega$ turbulence model (Wilcox 1988, 1998), on the other hand, was developed to alleviate exactly this problem. In this model, a turbulent (kinematic) viscosity is defined as the ratio of the (specific) turbulent kinetic energy, $k$, and a turbulent energy frequency, $\omega$ [1/s], as:

$$\text{Turbulent (Kinematic) Viscosity: } v_t = \frac{k}{\omega} \tag{7.8}$$

The associated transport equations for $k$ and $\omega$ become:

$$\text{Transport Eq. } k: \frac{\partial k}{\partial t} + \overline{u}_j\frac{\partial k}{\partial x_j} = \tau_{ij}\frac{\partial\overline{u}_i}{\partial x_j} - \beta^* k\omega + \frac{\partial}{\partial x_j}\left[(v + \sigma^* v_t)\frac{\partial k}{\partial x_j}\right] \tag{7.9}$$

$$\text{Transport Eq. } \omega: \frac{\partial\omega}{\partial t} + \overline{u}_j\frac{\partial\omega}{\partial x_j} = \alpha\frac{\omega}{k}\tau_{ij}\frac{\partial\overline{u}_i}{\partial x_j} - \beta\omega^2 + \frac{\partial}{\partial x_j}\left[(v + \sigma v_t)\frac{\partial\omega}{\partial x_j}\right] \tag{7.10}$$

In Eqs. (7.9) and (7.10), the model constants are $\alpha = 0.556$, $\beta = 0.075$, $\beta^* = 0.090$, and $\sigma$, $\sigma^* = 0.500$. Note that the transport equations are again cast in a form where convective transport is handled on the left-hand side, while diffusion, production, and destruction/dissipation terms are comprised on the right-hand side. The $k$-$\omega$ turbulence model does require some limiters, for example, for the production rate (Menter 1994), and has shown some sensitivity to freestream conditions for $\omega$.

### 7.1.2.3 Shear-Stress Transport (SST) $k$-$\omega$-Based Turbulence Model

In essence, the SST formulation of the $k$-$\omega$-based turbulence model combines the best of the two models described previously. Here, the $k$-$\omega$ formulation in the lower parts of the boundary layer enables us to use the model also for low-$Re$ without the need for additional damping functions. The SST formulation then switches to a $k$-$\varepsilon$ type model in the freestream and thus avoids the sensitivity of $k$-$\omega$ type models to freestream conditions. The resulting SST $k$-$\omega$ based turbulence model is known for its robust behavior in adverse pressure gradients and separated flow. The primary weakness of both the $k$-$\varepsilon$ and $k$-$\omega$ turbulence models is that both assume zero SST, thereby over-predicting the amount of turbulent (kinematic) eddy viscosity, $v_t$. Nevertheless, $v_t$ can be limited following:

$$\text{Turbulent (Kinematic) Viscosity: } v_t = \frac{\alpha k}{\max(\alpha \omega, SF_2)} \tag{7.11}$$

Here $S$ is a measure of the strain rate and $F_2$ is a blending function that restricts the limiter to the wall boundary layer. The respective transport equations for $k$ and $\omega$ now read:

$$\text{Transport Eq. } k: \frac{\partial k}{\partial t} + \bar{u}_j \frac{\partial k}{\partial x_j} = P_k - \beta^* k\omega + \frac{\partial}{\partial x_j}\left[(v + \sigma_k v_t)\frac{\partial k}{\partial x_j}\right] \tag{7.12}$$

$$\text{Transport Eq. } \omega: \frac{\partial \omega}{\partial t} + \bar{u}_j \frac{\partial \omega}{\partial x_j} = \alpha S^2 - \beta\omega^2 + \frac{\partial}{\partial x_j}\left[(v + \sigma_\omega v_t)\frac{\partial \omega}{\partial x_j}\right]$$

$$+ 2(1 - F_1)\sigma_{\omega 2}\frac{1}{\omega}\frac{\partial k}{\partial x_i}\frac{\partial \omega}{\partial x_i} \tag{7.13}$$

In Eqs. (7.12) and (7.13), $S^2 = \frac{\omega}{k}\tau_{ij}\frac{\partial \bar{u}_i}{\partial x_j}$ and all remaining model constants and additional limiter relations can be found in, for example, Menter (1993, 1994). The SST $k$-$\omega$ model does produce a bit too large turbulence levels in regions with large normal strain, for example, regions of stagnation and those with strong acceleration. This tendency is much less pronounced than with a normal $k$-$\varepsilon$ model though. The SST $k$-$\omega$ based turbulence model is the most widely used two-equation turbulence model in the wind energy science community.

### 7.1.3 Effect of Laminar-/Turbulent Transition on CFD Predictions

The different physical mechanisms for laminar-turbulent transition and general effects on airfoil drag polars have been introduced in Section 4.1.2.6. As far as CFD modeling is concerned, fully turbulent simulations (i.e. a turbulence model is used throughout the domain) best represent, though not being exactly the same, as "forced transition" located at the (blade) leading edge. Transitional CFD predictions focusing on "natural transition" have only become possible since the development of the Langtry–Menter correlation-based transition model (Langtry and Menter 2009), which is a (natural) transition model built around the SST $k$-$\omega$ based turbulence model that uses one additional transport equation for the intermittency and one for a transition onset criterion based on the momentum-thickness Reynolds number. A modeled approach to "bypass transition" on rotating wind turbine blades has not been completed in the wind energy science community. In general, the lower drag coefficient associated with a partial run of laminar flow along the blade chord results in higher performance of wind turbine blades due to an accompanying higher lift-to-drag ratio, $c_l/c_d$, see also Figure 3.14. The effect of laminar-/turbulent transition is important for both the prediction of scaled wind turbine experiments and the performance of utility-scale turbines.

As for model-scale wind turbine rotors and associated CFD validation, a focus has been on studying the effect of laminar-turbulent transition on the onset of separation, stall, and possibly dynamic-stall behavior. Indeed, for chord-based Reynolds numbers $Re_c \leq 1 \times 10^6$, significant discrepancies in airfoil data have been documented among various experiments and CFD predictions, thus making it very difficult to predict scaled turbine data. In this context, an example of transitional delayed detached-eddy simulations (DDESs) showed that the freestream turbulence intensity does have a measurable effect on transition location and hence the onset of stall on the NREL Phase VI rotor (Sørensen and Schreck 2014). This again has implications to both the effect of atmospheric turbulence intensity on wind turbine blade transitional effects and on model-predictive accuracy in a scaled experimental setting. A more recent CFD study applied to the new MEXICO data (Sørensen et al. 2016) also revealed the effects of laminar-turbulent transition on the prediction of blade loads in the tip region, without necessarily improving the predictive capability of computationally expensive CFD predictions when natural transition effects are accounted for. It appears that the wind energy science community will continue to be challenged with dissecting the subtleties of scaled wind turbine aerodynamics.

As far as utility-scale wind turbines and laminar-/turbulent transition effects are concerned, researchers are facing a dilemma in the sense that validation data does not exist in practice, that is, there is no comprehensive data set for CFD (transition) model validation of large rotating wind turbine blades with chord-based $Re_c \geq 3 \times 10^6$. It is also somewhat a paradox that even if CFD methods were truly validated against scaled experimental data, the effects of laminar-/turbulent transition are very sensitive to the actual airfoils and particularly to the chord-based Reynolds number, and one could reasonably question the approach of validating transitional CFD methods against scaled turbine experiments when they are actually intended to predict utility-scale performance. Indeed, for utility-scale wind turbines, the implications of laminar-turbulent transition are primarily focused around performance. An example of a "free" versus "fixed" transition airfoil polar is given in Figure 4.12 where it can be seen that the maximum lift-to-drag-ratio, $(c_l/c_d)_{max}$, reduces notably in case of fixed transition at the blade leading edge. The task then seems to be to predict the radial transition line along the blade such that accurate predictions of the turbine power coefficient, $C_p$, can be achieved. In this context, accounting for cross-flow instabilities in the boundary-layer receptivity seems paramount in the tip region, thus adding another challenge to an already difficult problem. Furthermore, the associated design lift coefficient (e.g. $c_l$ at max. lift-to-drag ratio) may be lower/higher in transitional flow depending on the airfoil and chord-based Reynolds number. This in turn may have an effect on the optimum blade parameter, $\sigma' c_l$, introduced in Chapter 3, as it affects the local blade solidity, $\sigma'$, and hence the blade design for optimum performance. In addition, blade surface contamination such as dirt, insects, and blade erosion damage affect the transition location and hence performance, all of which are not easily accounted for in high-fidelity CFD analyses.

### 7.1.4 Coupling of Navier–Stokes Solver with Helicoidal Vortex Model

The chief advantage of blade-resolved high-fidelity CFD analyses is that three-dimensional flow effects in the root-/tip regions and other (3-D) radial flow phenomena such as stall delay (see Chapter 5) and flow/separation associated with centrifugal pumping are resolved within the capabilities of the computational grids and analysis method. Nevertheless, the computational cost of high-fidelity CFD is on the order of hours (mostly days) on large parallel computing systems, thus prohibiting CFD methods for routine design/analysis applications. In practice, advanced CFD methods are used primarily to evaluate a selected number of final design contestants.

On the other hand, the primary strength of vortex wake methods (see Chapter 6) is a dissipation-free advection of vortical wake structures at very low computational expense compared to solving dense computational CFD wake grids. This gave rise to the conceptual idea of "hybrid CFD" solvers that combine the strengths of both CFD and vortex methods, that is (i) resolve the 3-D blade flow by CFD using a computational domain limited to the immediate vicinity of the blade and (ii) model wake advection by means of a vortex method. This is the basis for "hybrid CFD" or "Coupled Navier–Stokes/Vortex-Panel" methods. One of the first application of a hybrid CFD/Vortex method to a wind turbine rotor was performed by Xu and Sankar (2002) and Benjanirat and Sankar (2003) who modified a hybrid CFD solver for helicopter flows and applied it to the NREL Phase VI rotor (Hand et al. 2001; Simms et al. 2001). An example of an alternative (steady) coupled CFD/Vortex methodology is shown in Figure 7.2 where the CFD domain is confined to approximately two mean blade chord lengths around a single wind turbine blade and where the wake grid is aligned with the expected advection path of the trailing vortex sheet at that particular tip speed ratio (Schmitz 2006). The close proximity of the CFD domain boundary to the turbine blade requires physical boundary conditions that include the full rotor inflow everywhere along the CFD boundary. The idea of the hybrid methodology is that the (off-blade) vortex method provides the induced velocity and pressure fields as boundary conditions to the CFD domain; however, the vortex method requires the

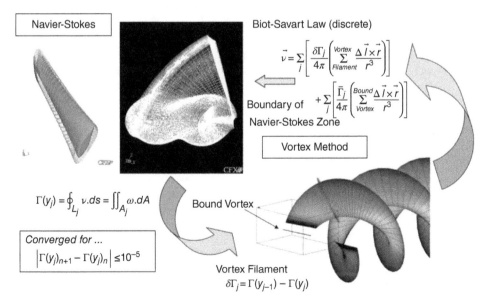

**Figure 7.2** Coupled Navier–Stokes/Vortex-Panel Solver (Schmitz and Chattot 2005a). Source: Published with permission from ASME Journals.

radial circulation distribution, $\Gamma(r)$, in order to compute induced velocities from both bound and trailing vortex filaments of all blades. For this reason, $\Gamma(r)$ is determined inside the CFD domain by means of a suitable set of closed-contour line integrals along the blade radius, see again Figure 7.2, following Stokes' theorem in Eq. (4.6) to include all sources of vorticity in the blade boundary layer up to the trailing-edge line (see also Section 4.1).

The actual coupling sequence is as follows: Initially, the CFD domain is solved with uniform inflow along its boundary. After the CFD solution has converged, the radial circulation distribution, $\Gamma(r)$, is determined inside the CFD domain and passed on to a prescribed wake method (see Section 6.2), with a wake structure chosen to balance the power extracted from the flow. Following, the prescribed wake method computes induced velocity and pressure (using Bernoulli's equation) along the CFD domain boundary, thus completing the first coupled iteration. Next, the CFD solver is restarted with an updated set of boundary conditions and solved to convergence; subsequently, $\Gamma(r)$ is recomputed and passed on to the prescribed wake method, which updates its geometry and the associated induced velocity and pressure fields to the CFD boundary. This coupling process is typically repeated 3–5 times until $\Gamma(r)$ has converged to a specified criterion (Schmitz and Chattot 2005a; Schmitz 2006).

Figure 7.3 shows an example of converged radial circulation distributions for the NREL Phase VI rotor at a wind speed of $V_0 = 7m/s$ (Schmitz and Chattot 2006b). Here the stand-alone prescribed wake method (Section 6.2) is referred to as VLM (Vortex Line Method) and the hybrid CFD methodology as PCS (Parallelized Coupled Solver). Both viscous and inviscid runs (zero drag in VLM, free-slip wall in PCS) are compared. It is interesting to see that for a test case with attached flow along the entire blade, both methods agree quite well, thus supporting credibility in a lifting-line type of approach in attached flow. Of further note are some discrepancies at inboard blade sections that are attributed to three-dimensional effects resolved by PCS that cannot be captured by VLM (run without any 3-D airfoil corrections in this case), including the effect of the Coriolis force in the CFD momentum equation.

Note that $\Gamma(r)$ is available in *XTurb* (Schmitz 2012) output files for this purpose, see also Schmitz and Chattot (2007a,b) for a more detailed investigation. As we

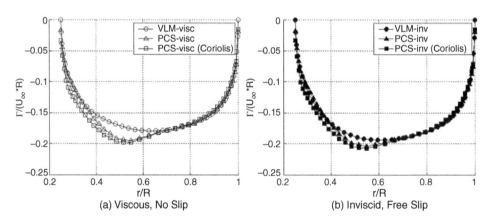

**Figure 7.3** Distribution of bound circulation (NREL Phase VI rotor S-Sequence, $V_0 = 7$ m s$^{-1}$) from Schmitz and Chattot (2006b). Source: Published with permission from ASME Journals.

will see in Chapter 8, $\Gamma(r)$ is also valuable in scaling aerodynamics from model- to full-scale conditions. For separated and stalled flow, it has been shown that a hybrid methodology captures 3-D flow phenomena (Schmitz and Chattot 2005a, 2006c) and can inform reduced-order models on how to correct 3-D airfoil data. The importance of capturing 5–10 full vortex wake revolutions in a CFD domain to obtain the correct rotor inflow is highlighted by a simple compelling study where the vortex wake in a hybrid methodology is advected for varying number of revolutions, see for example, Schmitz and Chattot (2005b).

## 7.2    Numerical Modeling of Wind Turbine Wakes

Blade-resolved CFD simulations as described in the previous section are computationally expensive and, in general, used to assess the performance of final design candidates or to validate against (scaled) experimental data. Note that a significant shift in scales occurs when going from a turbine blade boundary layer, that is, $O(10^{-6}m)$ near-wall grid spacing, to the length scale of a wind turbine array, that is, $O(10^3m)$. Solving (and modeling) turbulence across this range of disparate scales and computing interactions between individual turbines in an array is possible today but far beyond routine engineering practice. In this context, low-fidelity engineering-type methods and mid-fidelity actuator-type methods have been developed that reduce the required range of computed scales. Both sets of models are introduced next.

### 7.2.1    Engineering-Type Wake Models

Various engineering-type wind turbine wake codes are available today, which are to a large extent based on theory and standards developed in the 1980–1990s, see for example, Ainsle (1988), Quarton and Ainsle (1990), Larsen et al. (1999), Madsen (1996), Magnusson et al. (1996), and Schepers (1998). The so-called *kinematic* models are based on superposition of self-similar velocity-deficit solutions of co-flowing jets. Some authors used Gaussian profiles to include the effect of turbulence on wake growth based on earlier measurements (Vermeulen 1980), while others (e.g. Crespo and Hernandez 1996) find decay ratios for the velocity deficit and turbulence intensity assuming axi-symmetric flow. On the other hand, *field* models are rooted in the parabolized Navier–Stokes equations. Here the local ABL inflow is based on empirical methods, for example due to Veers (1988) and Mann (1994, 1998); the turbine is modeled as a distribution of momentum sinks, see for example, Ammara et al. (2002) and Masson et al. (2001). The resulting wake is nearly axi-symmetric with a Gaussian velocity profile, see for example, Jiminez et al. (2007), Lange et al. (2003), Schlez et al. (2007), Vermeer et al. (2003), and Zahle and Sørensen (2007). Wake models in use today are often a blend between the various model components mentioned before. They range from desktop applications of engineering-type wake models, which have computing times ranging from minutes to hours, to high-fidelity wind farm models with computing times on the order of days. The interested reader is referred to Göçmen et al. (2016) for a comprehensive review of various wake models developed at the Technical University of Denmark. Some other examples of engineering-type wake models include the Wind Atlas Analysis and Application Program *WAsP* from the Risø National Laboratory for Sustainable Energy in Denmark, the *WAKEFARM* code from the Dutch Energy Institute ECN (Energy Research Centre of the Netherlands), and the commercial software package *WindFarmer* from Garrad Hassan.

The *WAsP* program generates a local wind climate from a meteorological measurement station and data of the European Wind Atlas (Troen and Petersen 1989). Wake models are based on the work of Katic et al. (1986) and newer developments on Rathmann et al. (2007). *WAsP* is fast and robust; however, it exhibits limitations when applied over complex terrain, see Bowen and Mortensen (1996). The *WAKEFARM* code is rooted in the UPMWAKE code originally developed at the Universidad Politecnica de Madrid, see Crespo et al. (1988, 1999) and Crespo and Hernandez (1996). *WAKEFARM* is based on the parabolized Navier–Stokes equations, and accounts for turbulence through the $k$-$\epsilon$ turbulence model. The parabolic assumption postulates a dominant flow direction, and therefore requires modeling of the near wake by other means, for example by starting the simulation behind the rotor and setting Gaussian velocity-deficit profiles as boundary conditions. Lastly, the *WindFarmer* software package models the wake development by solving for the parabolized axi-symmetric Reynolds-Averaged Navier–Stokes equations and assuming thin shear layers. A standard eddy-viscosity turbulence closure due to Ainsle (1988) is used. A number of empirical expressions for surface roughness, forest canopies, deep-array effects, and so on make it an efficient and very versatile program.

### 7.2.2  Actuator Wake Models

The basic concept of all actuator models is to not resolve the individual turbine blades but to apply sectional rotor/blade forces as external body forces to the right-hand-side (i.e. source terms) of the momentum equation of the underlying flow solver, whether that be an URANS, Large-Eddy Simulation (LES), or even an inviscid Euler solver. This allows the smallest computational grid scale to be between that of the turbine blade chord, that is, $O(10^0 m)$, and a fraction of the rotor diameter, that is, $O(10^1 m)$. As a momentum source, an elemental rotor/blade force, $f_{N,m}$, is projected to a volumetric body force, $F_p$, following:

$$\text{Body Force: } F_p(x_p, y_p, z_p) = -\sum_N \sum_m f_{N,m}(x_{N,m}, y_{N,m}, z_{N,m})\eta_{N,m} \tag{7.14}$$

Here $N$ is the blade index, $m$ is an actuator (point) index, and $\eta_{N,m}$ is a projection function, most commonly represented as either a two-dimensional or three-dimensional Gaussian, see next. The actual form of the rotor/blade force, $f_{N,m}$, depends on the particular type of actuator method. In general, actuator-type models can be classified into actuator disk (AD), actuator line (AL), and actuator surface (AS) methods, see Figure 7.4.

For the simple AD, typically a single or distributed thrust force is used to represent the wind turbine rotor. This force distribution is injected into the computational domain based on an assumed thrust coefficient, $C_T$, obtained from for example, Eq. (2.15) and the average wind speed over the rotor disk, $\overline{V}_0$. Hence the general blade load/force, $f_{N,m}$, becomes:

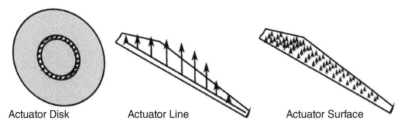

Actuator Disk                Actuator Line                Actuator Surface

**Figure 7.4** Concept of actuator-type methods.

$$\text{Actuator Disk: } \boldsymbol{f}_{N,m}(x_{N,m}, y_{N,m}, z_{N,m}) = \frac{1}{2}\rho_\infty \overline{V}_0^2 C_T A_{N,m} \boldsymbol{e}_T \tag{7.15}$$

$$\text{Gaussian Projection (2-D) : } \eta_{N,m} = \frac{1}{\pi \epsilon^2 \Delta} \exp[-(|r|/\varepsilon)^2] \tag{7.16}$$

In Eq. (7.15), $\boldsymbol{e}_T$ is the unit vector in the direction of rotor thrust (i.e. streamwise). Note that $N = 1$ for an (actuator) "disk" representation where, in general terms, the rotor disk area, $A$, is equal to $\sum_N \sum_m A_{N,m}$. In Eq. (7.16), $\varepsilon$ denotes a planar Gaussian projection radius, and the parameter $\Delta$ is a (spreading) length scale in the direction of $\boldsymbol{e}_T$, that is, normal to the AD. An AL representation, on the other hand, is rooted in blade element theory, see Section 3.2, such that $\boldsymbol{f}_{N,m}$ becomes the sectional blade force vector consisting of lift and drag forces:

$$\text{Actuator Line: } \boldsymbol{f}_{N,m}(x_{N,m}, y_{N,m}, z_{N,m}) = \frac{1}{2}\rho_\infty V_{rel}^2 \, c \, (c_l \boldsymbol{e}_l + c_d \boldsymbol{e}_d)\Delta_b \tag{7.17}$$

$$\text{Gaussian Projection (3-D) : } \eta_{N,m} = \frac{1}{\epsilon^3 \pi^{3/2}} \exp[-(|r|/\varepsilon)^2] \tag{7.18}$$

In Eq. (7.17), $c$ is the local/sectional chord length, $c_l$ and $c_d$ are the blade element section lift and drag coefficients, and $\boldsymbol{e}_l$ and $\boldsymbol{e}_d$ are respective unit vectors perpendicular to and in the direction of the sectional relative velocity, $V_{rel}$. Furthermore, the parameter $\Delta_b$ describes the width of a given actuator element (respectively, the distance between actuator points). In Eq. (7.18), the parameter $\epsilon$ denotes a spherical Gaussian projection radius. In addition, $|r| = |r_P - r_{N,m}|$ is the distance between a given grid cell/node, $r_P = (x_P, y_P, z_P)^T$, and an actuator point, $r_{N,m} = (x_{N,m}, y_{N,m}, z_{N,m})^T$. Note that a volume integral over the projection function $\eta_{N,m}$ [1/m$^3$] equals exactly one for either a two- or three-dimensional projection, thus ensuring that the total physical force(s), $\boldsymbol{f}_{N,m}$, are indeed embedded in the computational domain as a body force, $\boldsymbol{F}_P$, distributed over several grid cells. As far as AS methods are concerned, a chordwise distribution function is used in addition to a radial force projection along the blade. Some compelling examples can be found in the work of Shen et al. (2009) and Watters and Masson (2010). Nevertheless, any type of AS method requires considerably more computational grid nodes than AD or AL approaches, resulting in a computational expense more akin to blade-resolved CFD simulations.

In the following, an emphasis is given to actuator-line modeling as it has become the prominent actuator model for the wind energy science community, see also

Martinez-Tossas and Leonardi (2012) and Breton et al. (2017) for more detailed reviews on actuator-type methods.

### 7.2.2.1 ALM – Actuator-Line Model (Sørensen and Shen)

The concept of an ALM for the purpose of wind turbine wake simulation was first introduced by Sørensen and Shen (2002), see also Sørensen et al. (2015). This model has become the standard in the wind energy science community, see for example, works of Calaf et al. (2010), Lu and Porté-Agel (2011), Meyers and Meneveau (2012), and Chatelain et al. (2013). In practice, the standard ALM follows Eqs. (7.17) and (7.18), using the $F_1$ tip correction (see Section 3.4.4.1) in some implementations to account for tip losses, see also Shen et al. (2005). Sectional lift and drag coefficients, $c_l$ and $c_d$, along the AL are functions of the local angle of attack, $\alpha$, (angle between local $V_{rel}$ and chord line) and determined by a table lookup procedure similar to a standard Blade Element Momentum Theory (BEMT) method. The volumetric projection of a sectional blade force, $f_{N,m}$, using a spherical (3-D) Gaussian, $\eta_{N,m}$, according to Eq. (7.18) is illustrated in Figure 7.5. It can be seen that the standard ALM uses a constant Gaussian projection radius, $\epsilon$, along the actuator line, resulting in some measurable force spreading of $f_{N,m}$ beyond the blade tips.

As for the Gaussian projection radius, $\epsilon$, it is generally recommended, see for example, Troldborg (2008) and Troldborg et al. (2010, 2011a,b), that it be chosen based on the grid spacing, $\Delta_{grid}$, used in the respective computational method as:

$$\text{ALM: } \epsilon/\Delta_{grid} = 2 \tag{7.19}$$

Equation (7.19) is a simple compromise between numerical discretization and stability that requires that $\epsilon \geq \Delta_{grid}$ and that the projected body force, $F_p$, is injected as compact as possible into the computational domain with minimum repercussions on the accuracy of computed induced velocity fields. Note, however, that ALM computed blade loads and wakes are sensitive to grid refinements through Eq. (7.19). The ALM methodology is easy to implement in CFD solvers and is the standard in the wind energy science community.

### 7.2.2.2 ALM* – Variable-ε Actuator-Line Model

One fundamental limitation of the classical ALM is that the volumetric force projection, $F_p$, cannot account for the blade planform. In this context, an extension to the classical ALM was first proposed by Shives and Crawford (2013) who suggested that the Gaussian projection radius, $\epsilon$, be a function of the radial blade chord distribution, $c$, and hence of the blade planform. A generalization of this idea is that $\epsilon$ is related to the actual radial load distribution (Jha et al. 2014a). Figure 7.6 illustrates the ALM* concept of a radially variable Gaussian projection radius, $\epsilon(r)$. Note that ALM* was developed for use without any additional tip correction.

ALM $\epsilon(r)/\Delta_{grid}$ = const

**Figure 7.5** Volumetric body-force projection (ALM) – Jha and Schmitz (2018). Source: Published with permission from Cambridge University Press.

In ALM*, a radially variable Gaussian projection radius, $\epsilon(r)$, along the blade is chosen based on an equivalent elliptic planform, $c^*(r)$, that has the same aspect ratio, $AR$, as the actual blade. The following relation is recommended (Jha et al. 2014a):

$$\text{ALM}^* : \epsilon/c^* = C_1(\pi AR) = \epsilon_0/c_0 \tag{7.20}$$

In Eq. (7.20), the parameter $C_1 = 0.25(n_{max}\Delta_{grid}/R)$ and $\epsilon_0$, $c_0$ refer to respective mid-blade values of the equivalent elliptic planform, with $\epsilon_0 = n_{max}\Delta_{grid}$ and $n_{max}$ chosen such that $C_1 \in [0.025; 0.030]$. In order to limit force overlap, see Figure 7.6, $\Delta_b/\Delta_{grid} \geq 1.5$ is recommended where $\Delta_b$ is the distance between actuator points $m$ and $m+1$. In general, good results at affordable computational expense are obtained for actuator-type approaches when $\Delta_{grid}/R = 1/32$ or smaller, with the ALM* approach being less sensitive to grid refinement than the standard ALM. Figure 7.7 illustrates the wake structures computed by ALM (a) and ALM* (c) downstream of the NREL 5-MW turbine at a Region II wind speed of $V_0 = 8m/s$ (Jha et al. 2014a). It can be seen that the ALM* computed root-/tip vortices appear somewhat more compact than in the standard ALM case, which is attributed to a smaller Gaussian projection radius at blade root and tip.

The effect of the actual actuator-line modeling approach on wake characteristics and performance prediction of a wind turbine array is described briefly in Section 7.3.3. A somewhat different approach, though related to ALM*, is suggested in Churchfield et al. (2017) and Martinez-Tossas et al. (2017) where an optimized Gaussian projection radius, $\epsilon(r)$, becomes a function of the blade chord distribution, though requiring considerably finer grids and hence higher computational cost.

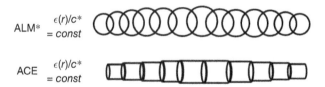

**Figure 7.6** Volumetric body-force projection (ALM*, ACE) – Jha and Schmitz (2018). Source: Published with permission from Cambridge University Press.

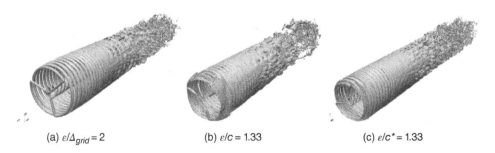

(a) $\varepsilon/\Delta_{grid} = 2$      (b) $\varepsilon/c = 1.33$      (c) $\varepsilon/c^* = 1.33$

**Figure 7.7** Wake structure and strength for the NREL 5-MW turbine ($V_0 = 8$ m s$^{-1}$) showing isosurface of vorticity magnitude 0.5 s$^{-1}$ (Jha et al. 2014b). Source: Published with permission from ASME Journals.

### 7.2.2.3 ACE – Actuator Curve Embedding (Jha and Schmitz)

The conceptual idea of ACE is rooted in a reverse approach compared to ALM and ALM*. Instead of computing a sphere of influence around each actuator point and searching for grid points within the individual spheres, the ACE method is constructed such that each grid point is interrogated only "once" and associated with only "one" actuator element (Jha and Schmitz 2018). This simple concept of "switching loops" eliminates both the force overlap between actuator points/elements and prohibits force spreading beyond the blade tips. The ACE concept for wind turbine wake modeling was inspired by the methodology of vorticity embedding due to Steinhoff and Ramachandran (1988), see also Schmitz et al. (2009) as an example of vorticity embedding implemented into an Eulerian-/Lagrangian free-wake method to compute rotorcraft airloads in hover. One key element of the ACE method is a 2-D Gaussian projection function for the body force, $F_P$, such that:

$$\text{Body Force: } F_P(x_P, y_P, z_P) = -\sum_N \frac{1}{\pi \epsilon^2 \Delta_b} f_{p_s} \exp[-(p_n/\epsilon)^2] \tag{7.21}$$

Note that this type of 2-D projection had already been proposed by Mikkelsen (2003) and used by Shives and Crawford (2013) for finite span wings. Here the blade force, $f_{p_s}$, is projected in a disk/cylinder along the respective actuator element, see Figure 7.6. In Eq. (7.21), $p_n$ is the (minimum) normal distance of grid point $r_P = (x_P, y_P, z_P)^T$ to a particular actuator element, and $\Delta_b$ is again the width of that respective actuator element. Note that $f_{p_s}$ is conserved for the general case of $\Delta_{grid} \neq \Delta_b$ by multiplying the body force, $F_P$, by a scaled cell volume $\Delta V^{2/3} \cdot \Delta_b$ when injecting the former into the momentum equation of a CFD solver. A variable Gaussian projection radius, $\epsilon(r)$, along the blade is determined according to:

$$\text{ACE: } \epsilon/c^* = C_1(\pi AR)[C_2 \cdot \lambda + C_3] \tag{7.22}$$

Here the parameter $C_1 \in [0.025; 0.030]$ is used as in ALM*. For parked blades ($\lambda = 0$), $C_2 = 0$ and $C_3 = 1$. For typical turbine operation with $\lambda \geq 4$, $C_2 = 0.1410$ and $C_3 = -0.3609$ (Jha and Schmitz 2018). In the ACE concept, the blade force, $f_{p_s}$, is an interpolated value between $f_{N,m}$ and $f_{N,m+1}$ at adjacent actuator points following:

$$\text{Actuator Curve: } f_{p_s} = f_{N,m} + p_s(f_{N,m+1} + f_{N,m}) \tag{7.23}$$

In Eq. (7.23), $p_s$ is the normalized tangential distance between actuator elements $m$ and $m+1$ for the minimum normal distance, $p_n$, of a given grid point $r_P = (x_P, y_P, z_P)^T$ to the respective actuator element. The primary difference between the ACE concept and ALM-type of approaches is that every grid point (vector) $r_P$ is associated "with exactly one" actuator element as opposed to having each grid point being possibly influenced by more than one actuator point in ALM-type approaches. This conceptual difference eliminates volumetric force spreading beyond the blade tips, which can affect the downwash and hence radial circulation distribution. The parameters $p_s$ and $p_n$ representing local tangential and normal coordinates can be determined from:

$$\text{Tangential Coordinate, } p_s : p_s = \frac{(r_P - r_{N,m}) \cdot (r_{N,m+1} - r_{N,m})}{\Delta_b^2} \tag{7.24}$$

$$\text{Normal Coordinate, } p_n : p_n = \frac{\|(r_P - r_{N,m}) \times (r_P - r_{N,m+1})\|}{\Delta_b} \tag{7.25}$$

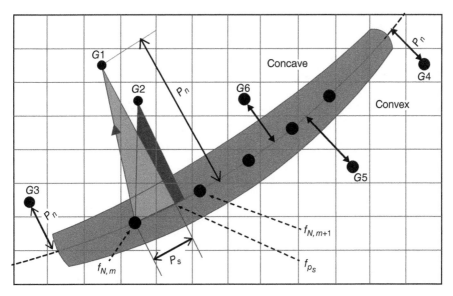

**Figure 7.8** Actuator curve embedding (ACE) computation of $p_n$ and $p_s$ at each $(x_p, y_p, z_p)$ – Jha and Schmitz (2018). Source: Published with permission from Cambridge University Press.

Note that $\Delta_b = \|r_{N,m+1} - r_{N,m}\|$ is the width of the respective actuator element (or distance between actuator points). In the ACE algorithm, a particular grid point, $r_p$, is associated with a particular actuator element between actuator points $m$ and $m+1$ for which $p_s \in [0, 1]$ at minimum normal distance, $p_n$. Figure 7.8 illustrates the embedding process.

The generality of the ACE concept avoids force overlap for straight blades but also for convex/concave curvature, see Figure 7.8. In the case of convex/concave curvature, the parameters $p_s$ and $p_n$ are chosen following:

Convex: $(p_{s_m} > 1) \wedge (p_{s_{m+1}} < 0)$

$$\rightarrow p_n = \|r_P - r_{N,m+1}\|, p_s = 1 \tag{7.26}$$

Concave: $p_{s_m}, p_{s_{m+1}} \in [0; 1]$

$$\rightarrow p_n = \min(p_{n_m}, p_{n_{m+1}}), p_s = p_{s_m} \vee p_{s_{m+1}} \tag{7.27}$$

The ACE concept has demonstrated its predictive capability by, for example, recovering exactly the constant downwash solution of an elliptically loaded wing (Jha and Schmitz 2018). A comparative application of various actuator-line approaches to the NREL 5-MW turbine (Jonkman et al. 2009) is shown in Figure 7.9 (Jha and Schmitz 2018), comparing sectional normal/tangential forces along the blade span (normal/tangential force definition according to Figure 3.3). Both *XTurb* results and those obtained by ALM/ALM*/ACE are compared against a high-fidelity CFD analysis by Chow and Van Dam (2012). The associated *XTurb* input file is *NREL-5MW-Ch7-2-2.inp*. The output file *XTurb_Output.dat* reads:

**XTurb Example 7.1 NREL 5-MW – XTurb_Output.dat (NREL-5MW-Ch7-2-2.inp)**

```
7.2.2 - NREL 5MW                    ***** XTurb V1.9    -   OUTPUT *****
   Blade Number        BN =  3

                     +       BLADE ELEMENT MOMENTUM THEORY (BEMT)      +

PREDICTION
  Blade Radius      BRADIUS    =    63. [m]
  Air Density       RHOAIR     =    1.225 [kg/m**3]
  Air Dyn. Visc.    MUAIR      =    0.000018 [kg/(m*s)]
  Number of Cases   NPRE       =    23

  Thrust          T  = 0.5*RHOAIR*VWIND**2.*(pi*BRADIUS**2.)*CT
  Power           P  = 0.5*RHOAIR*VWIND**3.*(pi*BRADIUS**2.)*CP
  Torque          TO = P / RPM
  Bending Moment  BE = 0.5*RHOAIR*VWIND**2.*(pi*BRADIUS**2.)*BRADIUS*CB
```

| Number | VWIND[m/s] | RPM[1/min] | TSR | PITCH[deg] | ... | T[N] | ... | TO[Nm] | BE[Nm] |
|---|---|---|---|---|---|---|---|---|---|
| 1 | 3.0000 | 6.9720 | 15.3322 | 0.0000 | ... | 76415.3327 | ... | -49177.7295 | -1196392.1646 |
| 2 | 4.0000 | 7.1830 | 11.8472 | 0.0000 | | 119952.4711 | | -241512.3019 | -1799747.1429 |
| 3 | 5.0000 | 7.5060 | 9.9039 | 0.0000 | | 172014.1600 | | -534937.1172 | -2520693.0154 |
| 4 | 6.0000 | 7.9420 | 8.7327 | 0.0000 | | 233118.1043 | | -932028.3362 | -3368779.8852 |
| 5 | 7.0000 | 8.4690 | 7.9818 | 0.0000 | | 302896.7243 | | -1422502.3144 | -4339753.4136 |
| 6 | 8.0000 | 9.1560 | 7.5507 | 0.0000 | | 383316.6441 | | -1971737.2309 | -5465138.2426 |
| 7 | 9.0000 | 10.2960 | 7.5474 | 0.0000 | | 485005.4442 | | -2496557.1260 | -6914710.0110 |
| 8 | 10.0000 | 11.4310 | 7.5414 | 0.0000 | | 598482.6980 | | -3084560.1329 | -8531985.9856 |
| 9 | 11.0000 | 11.8900 | 7.1311 | 0.0000 | | 698631.3297 | | -3927508.0236 | -9914895.4654 |
| 10 | 12.0000 | 12.1000 | 6.6523 | 3.8230 | | 589907.7926 | | -4190671.7292 | -8203774.4725 |
| 11 | 13.0000 | 12.1000 | 6.1406 | 6.6020 | | 504306.3465 | | -4150501.9255 | -6827349.9947 |
| ... | ... | ... | ... | ... | | ... | | ... | ... |

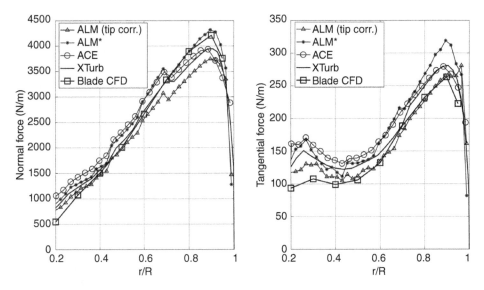

**Figure 7.9** Sectional forces on rotating NREL 5-MW rotor, $V_0 = 8$ m s$^{-1}$ and *RPM* = 9.16 (ALM, $\epsilon/\Delta_{grid} = 2$; ALM*, $\epsilon/c^* = 1.33$, $\epsilon_{min} = \Delta_{grid}$; ACE, $\epsilon/c^* = 1.2$, $\epsilon_{min} = \Delta_{grid}$) – Jha and Schmitz (2018). Source: Published with permission from Cambridge University Press.

In the absence of experimental data for true validation purposes, it appears from Figure 7.9 that some differences between the various models are present. However, there is no particular "winner" when it comes to ALMs, though ACE computed results appear somewhat smoother along the blade radius as a result of continuous force spreading.

### 7.2.3 Limitations of Actuator Methods

Actuator methods are a powerful computational modeling tool to represent rotating wind turbine blades within CFD solvers without actually resolving the blades. Nevertheless, this primary strength is also their main weakness, in the sense that parameters have always to be found empirically (at least to some extent). From a "modeler mind" perspective, one can argue that this step can never be eliminated/replaced due to the nature of the problem and method. In this respect, free model parameters should be kept at a minimum. Recent activities in the research community have also looked at a conceptual hybridization between AL and AS methods, using essentially a multi-dimensional Gaussian to represent both radial and chordwise load variations (Churchfield et al. 2017; Martinez-Tossas et al. 2017). However, it is also important to realize that accounting for chordwise loading naturally results in denser computational grids below the chord scale of $O(1m)$, making the respective approaches computationally expensive and difficult to apply to LES-type grid requirements, now commonly used for high-fidelity simulations of wind turbine wake interactions, see Section 7.3.

In the end, the author wishes to cite Leonardo da Vinci by saying "Simplicity is the ultimate sophistication." Hence a simple, though elegant and consistent, model is the most general. In this context, the ALM, ALM*, and ACE concepts are promising, though still probably not the final answer. In the same context, we are cursed in the wind energy community with the fact that sectional blade loads/forces are $\sim V_{rel}^2$ and turbine power $\sim V_0^3$ so that small uncertainties and inaccuracies can actually have a measurable effect on power prediction of wind turbine arrays.

## 7.3 Wake Modeling – Effect of Atmospheric Stability State

The impact of the atmospheric stability state on wind turbine array performance was recognized in the early 2000s largely due to the work of Barthelmie et al. (2003, 2004, 2006, 2007, 2009, 2010); Jensen et al. (2004); and Rathmann et al. (2007). For example, the Horns Rev Offshore wind farm in Denmark revealed that wind farm efficiency can range from 61% in a stable ABL to 71% in a neutral and 74% in an unstable ABL for that particular wind farm. In this context, efficiency is defined as the ratio of actual produced wind farm power to a theoretical value where all turbines operate at the power associated with the mean wind speed entering the wind farm. Data analyses showed that increased vertical (buoyant) turbulent mixing in an unstable ABL is in part responsible for enhancing the recovery process of the wake momentum deficit downstream of wind turbines. The observed variation in wind turbine power over a diurnal cycle or in sunny versus cloudy weather is important for operators of large wind farms as it can notably affect predicted power production and hence project revenue. As for the predictive capability of engineering-type methods (see Section 7.2.1), a representative example is shown in Figure 7.10 for the Horns Rev Offshore wind farm in Denmark. The power produced by individual turbines in an array is normalized by the power of the first turbine in that respective array. It can be seen that predictions obtained using engineering-type methods can vary notably from measured data. Though the stability state is one contributor to the observed discrepancy, one has to also account for uncertainties in wind speed and direction as well as other phenomena such as wake meandering. Of note in Figure 7.10 is a common observation that wind turbine power distribution in an array levels at approximately the third turbine row, for a turbine spacing of $>7D$.

The wind energy science community has since recognized that actuator methods (see Section 7.2.2) embedded in turbulence resolving LES, instead of standard URANS solvers, give rise to better predict the anisotropic turbulence associated with varying stability states and hence atmosphere-turbine interactions. Consequently, reduced-order engineering-type methods can be improved using analyses of highly resolved LES data. Some of the first numerical efforts using the AD concept within a LES solver are the works of Lu and Porté-Agel (2011), Ivanell et al. (2009), Meyers and Meneveau (2010), Singer et al. (2010), and Stovall et al. (2010). The following section introduces the atmospheric boundary layer LES solver in OpenFOAM (Churchfield et al. 2010) as an example of a wake modeling solver used in the wind energy community.

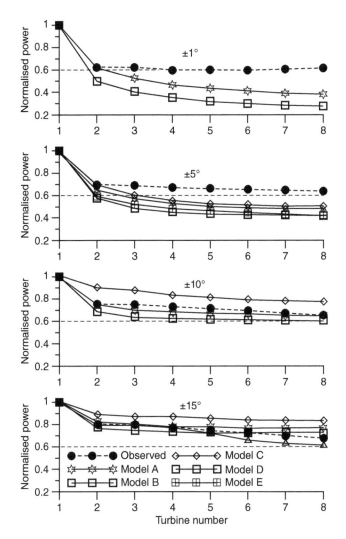

**Figure 7.10** Comparison of models and measurements for Horns Rev. (direction 270°). The initial wind speed calculated from the power output of the first turbine is $8 \pm 0.5$ m s$^{-1}$. From the top down, the widths of the wake sectors considered in the four panels are $\pm 1°$, $\pm 5°$, $\pm 10°$, and $\pm 15°$ (Barthelmie et al. 2009). Source: Reproduced with permission by R. Barthelmie.

### 7.3.1 Atmospheric Boundary Layer LES Solver in OpenFOAM

The ABL-LES solver in OpenFOAM was originally developed by Churchfield et al. (2010) and later applied to a number of wind farm applications, for example, Churchfield et al. (2012). In this solver, the LES momentum equation is written as:

$$\text{LES:} \quad \frac{\partial \bar{u}_i}{\partial t} + \frac{\partial (\bar{u}_j \bar{u}_i)}{\partial x_j}$$

$$= -2\varepsilon_{ijk}\Omega_j \bar{u}_k - \frac{1}{\rho}\frac{\partial \tilde{p}}{\partial x_i} - \frac{1}{\rho}\frac{\partial p}{\partial x_i} - \frac{\partial R_{ij}^D}{\partial x_j} + \left(\frac{\rho_b}{\rho} - 1\right)g_i + \frac{1}{\rho}(F_P)_i \qquad (7.28)$$

Here, the left-hand-side consists of the time-rate-of-change of the filtered velocity, $\partial \overline{u}_i / \partial t$, and the convection term, $\partial (\overline{u}_j \overline{u}_i)/\partial x_j$. On the right-hand-side, $2\varepsilon_{ijk}\Omega_j \overline{u}_k$ is the Coriolis term as a result of planetary rotation, followed by the pressure gradient terms of the modified pressure variable (Churchfield et al. 2010), $\widetilde{p}$, and that of the driving pressure (gradient), $p$. Furthermore, $\partial R_{ij}^D / \partial x_j$ represents the divergence of the deviatoric part of the stress tensor, while $(\rho_b/\rho - 1)g_i$ is used to model buoyancy effects using the Boussinesq approximation, with $g_i$ being gravitational acceleration. The last term on the right-hand-side is the volumetric body force, $\boldsymbol{F}_P$, that accounts for the momentum source associated with an actuator-type model, see Section 7.2.2. Note that the ABL-LES solver in OpenFOAM is capable of resolving anisotropic turbulence in a neutral and unstable atmosphere (Churchfield et al. 2012). Finite-volume discretization is used in OpenFOAM, with typically second-order-accurate temporal (Crank-Nicolson) and spatial discretization. In the original implementation (Churchfield et al. 2010), the Smagorinsky model is used for subgrid-scale effects; the classical pressure-implicit splitting operation (PISO) algorithm solves for the pressure.

A typical LES computed wind turbine wake is shown in Figure 7.11 for the case of the NREL 5-MW turbine in uniform inflow. Note that in Figure 7.11, ABL turbulent inflow is not considered for the purpose of introducing the three main regions of a general wind turbine wake. The first region is a near wake that typically extends between two and three rotor diameters downstream of the rotor disk and is governed by wake expansion with an associated pressure increase. The second region is an intermediate wake where pressure and centerline velocity remain approximately constant, and where a turbulent mixing

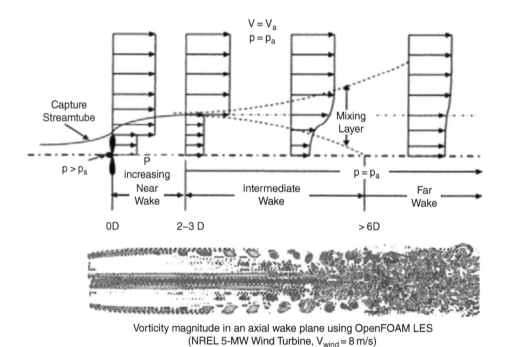

Vorticity magnitude in an axial wake plane using OpenFOAM LES
(NREL 5-MW Wind Turbine, $V_{wind} = 8$ m/s)

**Figure 7.11** Wind turbine wake computed with LES (Jha et al. 2015). Source: Published through MDPI open access.

layer increases the wake outer boundary and reaches the centerline at about seven rotor diameters. The third region is the so-called far-wake region, in which the wake is now nearly axi-symmetric and the centerline velocity recovery continues at approximately constant pressure.

### 7.3.2 Example of Turbine–Turbine Interaction for Neutral/Unstable Stability

An example of a turbine-turbine interaction problem computed with the ABL-LES solver in OpenFOAM is shown next. Figure 7.12 illustrates the wake structures downstream of two NREL 5-MW turbines by means of an iso-surface of vorticity magnitude (0.51/s) applied to an instantaneous volume field (Jha et al. 2015). It can be seen that in this particular case with neutral ABL inflow, the vortex structure downstream of Turbine 1 remains stable for approximately four rotor revolutions, while the breakup of tip vortices occurs after a shorter wake age downstream of Turbine 2. This is attributed to higher turbulent inflow to Turbine 2 than to Turbine 1. Indeed, for an unstable inflow with notable vertical (buoyant) turbulent structures, breakdown of tip vortices also occurs faster downstream of Turbine 1 due to the increased turbulent kinetic energy in the inflow.

In Figure 7.12, the NREL 5-MW turbines are spaced seven rotor diameters from one another and subject to a zero-yaw mean hub-height wind speed of $8\,\mathrm{m\,s^{-1}}$. The ABL domain is $3 \times 3 \times 1$ km and forced by neutral or unstable precursor ABL data. Note that these types of ABL-LES simulations have $>20 \times 10^6$ grid points and may take several weeks of computation clock time on parallel computers using on the order of 500 CPUs. In order to obtain converged statistics for turbine power and Reynolds stresses, simulations are run on the order of 2000 s of real (simulated) time, corresponding to about 300 full rotor revolutions. The mean and standard deviation of turbine power for both turbines is shown in Figure 7.13 for neutral boundary-layer (NBL) inflow and (unstable) moderately-convective boundary-layer (MCBL) inflow.

It can be seen that the atmospheric stability state has some effect on the power produced by Turbine 1. This is due to a "fuller" inflow velocity profile in MCBL as opposed

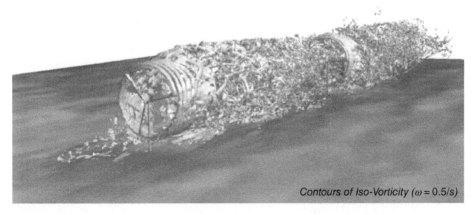

*Contours of Iso-Vorticity (ω = 0.5/s)*

**Figure 7.12** Turbine-turbine interaction (Jha et al. 2015) in neutral boundary-layer flow (NREL 5-MW turbines, $V_0 = 8\,\mathrm{m\,s^{-1}}$). Source: Published through MDPI open access.

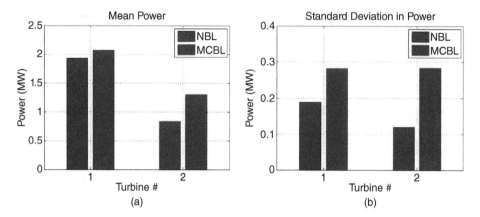

**Figure 7.13** Turbine-turbine interaction – Mean and standard deviation of turbine power (NREL 5-MW turbines, $V_0 = 8$ m s$^{-1}$). Left Bar: NBL and Right Bar: MCBL. Source: Published through MDPI open access.

to NBL inflow for the same mean hub-height wind speed, see also Figure 1.14. The primary observation, however, is that the power produced by Turbine 2 is close to 40% higher in MCBL when compared to NBL inflow. The reason for this is that an unstable (or convective) boundary layer has enhanced vertical mixing as opposed to a neutral boundary layer, simply because of surface heating and the associated stronger vertical turbulent structures. The enhanced "mixing" and subsequent faster recovery of the wake momentum deficit downstream of Turbine 1 can be quantified by computing Reynolds stresses along vertical lines at different downstream positions. Analyses of high-fidelity LES data, for example Jha et al. (2015), provide useful information toward improved low-order wake models. Though an unstable (MCBL) atmospheric inflow seems advantageous with respect to turbine array performance, Figure 7.13 also demonstrates that the associated standard deviation in turbine power (and hence loads) is increased significantly in MCBL when compared to NBL inflow, with negative repercussions on blade fatigue and hence turbine/blade lifetime.

### 7.3.3 Effect of ALM Approach on Wind Turbine Array Performance Prediction

The previous sections discussed the effect of the atmospheric stability state on wind turbine performance within a turbine–turbine interaction problem. From an analysis and prediction perspective, it is also important to understand how different actuator-line modeling approaches as introduced in Section 7.2.2 affect the performance prediction of an array of wind turbines. For example, one can ask the question of whether or not differences in actuator-line modeling approaches are significant compared to differences associated with the atmospheric stability state, and whether or not actuator-line associated "deltas" propagate through an array of wind turbines. As a model problem, Figure 7.14 shows the same simulation setup as used in Section 7.3.2, though comparing qualitative instantaneous flowfields, that is, at the exact same simulation time, computed with either the standard ALM ($\epsilon/\Delta_{grid} = const.$, Section 7.2.2.1) or the ALM* approach ($\epsilon/c^* = const.$, Section 7.2.2.2) based on a radially elliptic $\epsilon$ distribution, see also Jha et al. (2014b) and Jha and Schmitz (2016).

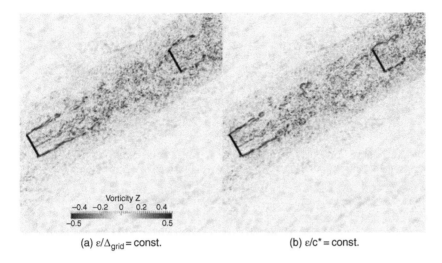

(a) $\varepsilon/\Delta_{grid}$ = const.                                    (b) $\varepsilon/c^*$ = const.

**Figure 7.14** Turbine-turbine interaction – ALM/ALM* flowfield in horizontal plane at hub height (NREL 5-MW turbines, $V_0 = 8$ m s$^{-1}$). Source: Published with permission from ASME Journals.

It is apparent form Figure 7.14 that instantaneous flowfields appear qualitatively very similar. In particular, the extent of the near-/intermediate-/far-wake regions is practically the same between both actuator-line approaches. This indicates that discrepancies in blade loads associated with different actuator-line approaches that are notable for blade aerodynamics predictions may not be as important for the wake development downstream of Turbine 1. Figure 7.15 shows a quantitative comparison of mean turbine power and (normalized) standard deviation between the two actuator-line approaches and in neutral (NBL) versus unstable (MCBL) atmospheric inflow.

It is important to understand from Figure 7.15 that "deltas" at Turbine 1 associated with the actuator-line approach are notably smaller than those due to a neutral versus moderately convective (unstable) atmosphere. While the mean power of Turbine 1 only differs by $2.6-2.7\%$ between both actuator-line approaches, it appears that 'deltas' accumulate through an array as the wake of Turbine 1 propagates the associated discrepancy in the momentum deficit to Turbine 2, which is then subject to the same "delta" associated with the actuator-line approach. Indeed, the mean power of Turbine 2 differs by $4.0-4.2\%$ between both ALM approaches. Assuming that the solution is subject to minimal artificial dissipation, the predicted total power of an array of more than two turbines may accumulate a total "delta" of >6% associated with the actuator-line modeling approach and then be of the same magnitude than the atmospheric stability state itself, see Figure 7.15. In conclusion, the wind energy science community has to be careful in identifying and quantifying the uncertainties of the particular actuator-line modeling approach in conjunction with the atmospheric stability state.

### 7.3.4 Bridging the Gap – Meso-Microscale Coupling

The interaction of atmospheric turbulence of varying stability states with wind turbines has been recognized in its effect on blade aerodynamics and subsequently on power production and blade fatigue. Actuator modeling using either AD or AL

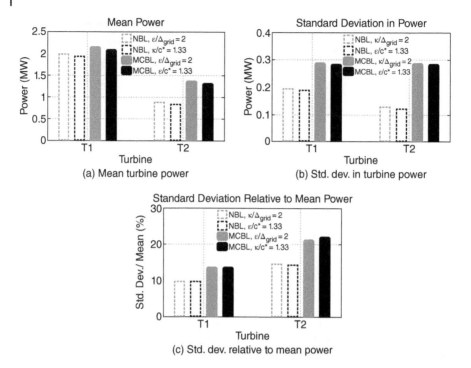

**Figure 7.15** Turbine-turbine interaction – ALM/ALM* mean and standard deviation of turbine power (NREL 5-MW Turbines, $V_0 = 8$ m s$^{-1}$). Source: Published with permission from ASME Journals.

methods embedded in LES solvers have increased the community's understanding of the sensitivities associated with atmospheric stability and have contributed to improved low-order models for wind siting. For further improvements and optimal wind turbine siting and forecasting, the assessment of a local "wind resource" has to include *mesoscale* processes (10 km− 1 km) such as frontal passages and local-scale circulations induced by surface-atmosphere interactions as well as *microscale* processes (1 km − 1 m) such as turbine wake interactions, wake meandering, and terrain effects.

Achieving a successful coupling between the meteorological *meso-* and *microscale* will not only result in optimal wind siting but also advance the ability for accurate power forecasting, with the end goal of reduced losses in the electrical grid and ultimately a lower cost of energy. *Mesoscale* modeling is typically the lower-scale end of weather prediction codes such as the Weather Research and Forecasting (WRF) model (Skamarock et al. 2005, 2008; Skamarock and Klemp 2008), containing adequate surface parameterizations for planetary boundary layers in an otherwise global *macroscale* circulation model. To bridge the modeling "gap," numerical weather prediction codes have to essentially enable nested mesoscale-LES capability during typical diurnal cycles at wind sites and site candidates. This implies that modeling methods have to be advanced to bridge the so-called *terra incognita* (Wyngaard 2004) within which the turbulent scales transition from the *mesoscale* to the *microscale* or, from the perspective of modeling tools, where a transition occurs between parameterized turbulence in *mesoscale* models to resolved turbulence in LES-type *microscale* models of wind farms.

Some examples of significant advances in this area are the works of Singer et al. (2010), Mirocha et al. (2010), Gopalan et al. (2014), and Rodrigo et al. (2016).

A strong area of research for the next generation in the wind energy science community is to establish a *Verification & Validation* framework across the range of scales (including the *terra incognita*) affecting wind power forecasting, wind farm turbulence, and wind turbine blade aerodynamics. Here, a case study (Lee and Lundquist 2017) of four diurnal cycles with significant power production is a compelling example that assessed the capability of a wind farm parameterization (WFP) distributed with the WRF model, documenting sensitivity and deriving guidelines. Another fundamental question is that of grid nesting to bridge mesoscales and microscales with grid-cell sizes in the *terra incognita* that can violate both the assumptions at mesoscale and microscale with respect to the subgrid-scale parametrization. While this can lead to under-resolved convective structures, the question is whether or not this is important for power forecasting (Mazzaro et al. 2017). Computing across all scales from the mesoscale to the blade boundary-layer scale is computationally prohibitive but can shed light into the primary dependencies from the atmosphere down to local blade aerodynamics (Kirby et al. 2017). As far as wind turbine blade design is concerned, it may well be that simple, though elegant, blade design modifications can have a profound impact on reducing blade fatigue in turbulent atmospheric inflow, and the author challenges the young generation with this task.

# References

Ainsle, J.J. (1988). Calculating the flow in the field of wind turbines. *Journal of Wind Engineering and Industrial Aerodynamics* 27: 213–224.

Ammara, I., Leclerc, C., and Masson, C. (2002). A viscous three-dimensional method for the aerodynamic analysis of wind farms. *Journal of Solar Energy Engineering* 124: 345–356.

Barthelmie, R.J., Larsen, G., Pryor, S. et al. (2004). ENDOW (efficient development of offshore wind farms): modeling wake and boundary layer interactions. *Wind Energy* 7: 225–245.

Barthelmie, R.J., Folkerts, L., Ormel, F. et al. (2003). Offshore wind turbine wakes measured by SODAR. *Journal of Atmospheric and Oceanic Technology* 30: 466–477.

Barthelmie, R.J., Folkerts, L., Rados, K. et al. (2006). Comparison of wake model simulations with offshore wind turbine wake profiles measured by sodar. *Journal of Atmospheric and Oceanic Technology* 33: 888–901.

Barthelmie, R.J., Frandsen, S.T., Nielsen, N.M. et al. (2007). Modelling and measurements of power losses and turbulence intensity in wind turbine wakes at Middelgrunden offshore wind farm. *Wind Energy* 10: 217–228.

Barthelmie, R.J., Hansen, K., Frandsen, S.T. et al. (2009). Modelling and measuring flow and wind turbine wakes in large offshore wind farms. *Wind Energy* 12: 431–444.

Barthelmie, R.J., Pryor, S.C., Frandsen, S.T. et al. (2010). Quantifying the impact of wind turbine wakes on power output at offshore wind farms. *Journal of Atmospheric and Oceanic Technology* 27: 1302–1317.

Benjanirat, S. and Sankar, L. (2003) Recent Improvements to a Combined Navier–Stokes/Full-Potential Methodology for Modeling Horizontal Axis Wind Turbines. AIAA-2004-0830.

Bowen, A. J. and Mortensen, N. G. (1996) Exploring the Limits of WAsP: The Wind Analysis and Application Program. *Proceedings of the European Union Wind Energy Conference*, Gothenburg, Sweden.

Breton, S.-P., Sørensen, J.N., Hansen, K. et al. (2017). A survey of modeling methods for high-fidelity wind farm simulations using large eddy simulation. *Philosophical Transactions of the Royal Society of London. Series A* 375: 20160097.

Calaf, M., Meneveau, C., and Meyers, J. (2010). Large eddy simulation study of fully developed wind-turbine array boundary layers. *Physics of Fluids* 22: 015110.

Chatelain, P., Backaert, S., Winckelmans, G., and Kern, S. (2013). Large eddy simulation of wind turbine wakes. *Physics of Fluids* 91: 587–605.

Chow, R. and Van Dam, C.P. (2012). Verification of computational simulations of the NREL 5-MW rotor with a focus on inboard flow separation. *Wind Energy* 18: 967–981.

Churchfield, M. J., Moriarty, P. J., Vijayakumar, G., and Brasseur, J. G. (2010) Wind energy related atmospheric boundary-layer large-eddy simulation using OpenFOAM. Technical Report NREL/CP-500-48905, National Renewable Energy Laboratory.

Churchfield, M.J., Lee, S., Michalakes, J., and Moriarty, P.J. (2012). A numerical study of the effects of atmospheric and wake turbulence on wind turbine dynamics. *Journal of Turbulence* https://doi.org/10.1080/14685248.2012.668191.

Churchfield, M. J., Schreck, S., Martinez-Tossas, L. A., Meneveau, C., and Spalart, P. R. (2017) An advanced actuator line method for wind energy applications and beyond. AIAA-2017-1998.

Crespo, A. and Hernandez, J. (1996). Turbulence characteristics in wind-turbine wakes. *Journal of Wind Engineering and Industrial Aerodynamics* 61: 71–85.

Crespo, A., Hernandez, J., Fraga, E., and Andreu, C. (1988). Experimental validation of the UPM computer code to calculate wind turbine wakes and comparison with other models. *Journal of Wind Engineering and Industrial Aerodynamics* 27: 77–88.

Crespo, A., Hernandez, J., and Frandsen, S. (1999). Survey of modeling methods for wind turbine wakes and wind farms. *Wind Energy* 2: 1–24.

Duque, E.P.N., Burklund, M.D., and Johnson, W. (2003). Navier–Stokes and comprehensive analysis performance predictions of the NREL Phase VI experiment. *ASME Journal of Solar Energy Engineering* 125 (4): 457–467.

Göçmen, T., van der Laan, P., Réthoré, P.E. et al. (2016). Wind turbine wake models developed at the Technical University of Denmark: a review. *Renewable and Sustainable Energy Reviews* 60: 752–769.

Gopalan, H., Gundling, C., Brown, K. et al. (2014). A coupled mesoscale–microscale framework for wind resource estimation and farm aerodynamics. *Journal of Wind Engineering and Industrial Aerodynamics* 132: 13–26. https://doi.org/10.1016/j.jweia .2014.06.001.

Hand, M. M., Simms, D. A., Fingersh, L. J., Jager, D. W., Cotrell, J. R., Schreck, S., and Larwood, S. M. (2001) Unsteady Aerodynamics Experiment Phase VI: Wind Tunnel Test Configurations and Available Data Campaigns. Technical Report NREL/TP-500-29955, National Renewable Energy Laboratory.

Ivanell, S., Mikkelsen, R., Sorensen, J., and Henningson, D. (2009). ACD modelling of wake interaction in the horns rev wind farm. In: *Extended Abstracts for Euromech Colloquim 508 on Wind Turbine Wakes*. Madrid, Spain: European Mechanics Society.

Jensen, L., Mørch, C., Sørensen, P., and Svendsen, K. H. (2004) Wake measurements from the Horns Rev wind farm. EWEC 2004, November 22–25, 2004, London, UK.

Jha, P.K. and Schmitz, S. (2016). Blade load unsteadiness and turbulence statistics in an actuator-line computed turbine–turbine interaction problem. *ASME Journal of Solar Energy Engineering* 138: 0311002.

Jha, P.K. and Schmitz, S. (2018). Actuator curve embedding – an advanced actuator line model. *Journal of Fluid Mechanics* 834: https://doi.org/10.1017/jfm.2017.793.

Jha, P.K., Churchfield, M.J., Moriarty, P.J., and Schmitz, S. (2014a). Guidelines for volume force distributions within actuator line modeling of wind turbines in large-eddy simulation-type grids. *ASME Journal of Solar Energy Engineering* 136: 0310014.

Jha, P. K., Churchfield, M. J., Moriarty, P. J., and Schmitz, S. (2014b) The Effect of Various Actuator-Line Modeling Approaches on Turbine-Turbine Interactions and Wake-Turbulence Statistics in Atmospheric Boundary-Layer Flow. *32nd ASME Wind Energy Symposium*, AIAA 2014-0710.

Jha, P.K., Duque, E.P.N., Bashioum, J., and Schmitz, S. (2015). Unraveling the mysteries of turbulence transport in a wind farm. *Energies* 8: 6468–6496. https://doi.org/10.3390/en8076468.

Jiminez, A., Crespo, A., Migoya, E., and Garcia, J. (2007). Advances in large-eddy simulation of a wind turbine wake. *The Science of Making Torque from Wind. Conference Series* 75: 012041. https://doi.org/10.1088/1742-6596/75/1/012041.

Jonkman, J., Butterfield, S., Musial, W., and Scott, G. (2009) Definition of a 5-MW reference wind turbine for offshore system development. Technical Report NREL/TP-500-38060, National Renewable Energy Laboratory.

Katic, I., Hoejstrup, J., and Jensen, N. O. (1986) A Simple Model for Cluster Efficiency. *Proceedings of the European Wind Energy Association*, Rome, Italy.

Kirby, A. C., Brazell, M. J., Yang, Z., Roy, R., Ahrabi, B. R., Mavriplis, D. J., and Stoellinger, M. K. (2017) Wind Farm Simulations Using an Overset hp-Adaptive Approach with Blade-Resolved Turbine Models. AIAA-2017-3958.

Lange, B., Waldl, H., Guerrero, A.G. et al. (2003). Modeling of offshore wind turbine wakes with the wind farm program FLaP. *Wind Energy* 6: 87–104.

Langtry, R.B. and Menter, F.R. (2009). Correlation-based transition modeling for unstructured parallelized computational fluid dynamics codes. *AIAA Journal* 47 (12): 2894–2906. https://doi.org/10.2514/1.42362.

Larsen, G. C., Carlen, I., and Schepers, G. J. (1999) European Wind Turbine Standards 2. Project Results. Rep. ECN-C-99-073 89.

Launder, B.E. and Spalding, D.B. (1974). The numerical computation of turbulent flows. *Computer Methods in Applied Mechanics and Engineering* 3 (2): 269–289.

Le Pape, A. and Gleize, V. (2006) Improved Navier–Stokes computations of a stall-regulated wind turbine using low Mach number preconditioning. AIAA-2006-1502.

Lee, J.C.Y. and Lundquist, J.K. (2017). Evaluation of the wind farm parameterization in the weather research and forecasting model (version 3.8.1) with meteorological and turbine power data. *Geoscientific Model Development* 10: 4229–4244. https://doi.org/10.5194/gmd-10-4229-2017.

Lu, H. and Porté-Agel, F. (2011). Large-eddy simulation of a very large wind farm in a stable atmospheric boundary layer. *Physics of Fluids* 23: 065101.

Madsen, H. Aa. (1996) A CFD analysis of the actuator disc flow compared with momentum theory results. *Proceedings of the IEA Joint Action of Tenth Symposium on Aerodynamics of Wind Turbines*.

Magnusson, H., Rados, K.G., and Voutsinas, S.G. (1996). A study of the flow downstream of a wind turbine using measurements and simulations. *Wind Engineering* 20 (6): 389–403.

Mann, J. (1994). The spatial structure of neutral atmospheric surface-layer turbulence. *Journal of Fluid Mechanics* 273: 141–168.

Mann, J. (1998). Wind field simulation. *Probabilistic Engineering Mechanics* 13: 269–282.

Martinez-Tossas, L. A. and Leonardi, S. (2012) Wind Turbine Modeling for Computational Fluid Dynamics. Technical Report NREL/SR-5000-55054, National Renewable Energy Laboratory.

Martinez-Tossas, L.A., Churchfield, M.J., and Meneveau, C. (2017). Optimal smoothing length scale for actuator line models of wind turbine blades based on Gaussian body force distribution. *Wind Energy* 20: 1083–1096.

Masson, C., Smaili, A., and Leclerc, C. (2001). Aerodynamic analysis of HAWTs operating in unsteady conditions. *Wind Energy* 4: 1–22.

Mazzaro, L.J., Munoz-Esparza, D., Lundquist, J.K., and Linn, R.R. (2017). Nested mesoscale-to-LES modeling of the atmospheric boundary layer in the presence of under-resolved convective structures. *Journal of Advances in Modeling Earth Systems*, AGU Publications, https://doi.org/10.1002/2017MS000912.

Menter, F. R. (1993) Zonal Two Equation $k$-$\omega$ Turbulence Models for Aerodynamic Flows. AIAA-1993-2906.

Menter, F.R. (1994). Two-equation eddy-viscosity turbulence models for engineering applications. *AIAA Journal* 32 (8): 1598–1605.

Meyers, J. and Meneveau, C. (2010) Large Eddy Simulations of Large Wind-Turbine Arrays in the Atmospheric Boundary Layer. AIAA-2010-0827.

Meyers, J. and Meneveau, C. (2012). Optimal turbine spacing in fully developed wind farm boundary layers. *Wind Energy* 15: 305–317.

Mikkelsen, R. (2003) *Actuator Disc Methods Applied to Wind Turbines*. Ph.D. thesis, Technical University of Denmark.

Mirocha, J., Kirkil, G., and Lundquist, J. (2010) Nested high-resolution mesoscale/large-eddy simulations in WRF: Challenges and opportunities. *Fifth International Symposium on Computational Wind Engineering*, Charlotte, NC.

Le Pape, A. and Lecanu, J. (2004). 3-D Navier–Stokes computations of a stall-regulated wind turbine. *Wind Energy* 7: 309–324.

Quarton, D. and Ainsle, J. (1990). Turbulence in wind turbine wakes. *Wind Engineering* 14: 15–23.

Rathmann, O., Frandsen, S. T., and Barthelmie, R. J. (2007) Wake modelling for intermediate and large wind farms. *Proceedings of the European Wind Energy Conference and Exhibition*, Milan, Italy.

Rodrigo, J.S., Chavez Arroyo, R.A., Moriarty, P. et al. (2016). Mesoscale to microscale wind farm flow modeling and evaluation. *WIREs Energy and Environment* 6: https://doi.org/10.1002/wene.214.

Schepers, J. G. (1998) Wakefarm, Nabij zog model en ongestood windsnelheisveld (in Dutch). Technical Report ECN-C—98-016, Energy Center of the Netherlands.

Schlez, W., Umana, A., Barthelmie, R., Larsen, G., Rados, K., Lange, B., Schepers, G., and Hegberg, T. (2007) ENDOW: improvement of wake models within offshore wind farms. *Proceedings of the European Wind Energy Conference EWEC 2007*.

Schlichting, H. (1979). *Boundary-Layer Theory*. New York: Mc-Graw Hill.

Schmitz, S. (2006) *Coupling of Navier–Stokes Solver with Helicoidal Vortex Model for the Computational Study of Horizontal Axis Wind Turbines.* Ph.D. Dissertation, University of California Davis.

Schmitz, S. (2012). *XTurb-PSU: A Wind Turbine Design and Analysis Tool.* The Pennsylvania State University.

Schmitz, S. and Chattot, J.J. (2005a). A parallelized coupled Navier–Stokes/vortex-panel solver. *ASME Journal of Solar Energy Engineering* 127: 475–487.

Schmitz, S. and Chattot, J.J. (2005b). Influence of the vortical wake behind wind turbines using a coupled Navier–Stokes/vortex-panel methodology. In: *Computational Fluid and Solid Mechanics*, 832–836. Abingdon/Oxford: Elsevier Ltd.

Schmitz, S. and Chattot, J.J. (2006a). Characterization of three-dimensional effects for the rotating and parked NREL phase VI wind turbine. *ASME Journal of Solar Energy Engineering* 128: 445–454.

Schmitz, S. and Chattot, J.J. (2006b). A coupled Navier–Stokes/vortex-panel solver for the numerical analysis of wind turbines. *Computers and Fluids* 35: 742–745.

Schmitz, S. and Chattot, J.J. (2007a). Flow physics and Stokes' theorem in wind turbine aerodynamics. *Computers and Fluids* 36: 1583–1587.

Schmitz, S. and Chattot, J.J. (2007b). Method for aerodynamic analysis of wind turbines at peak power. *AIAA Journal of Propulsion and Power* 23: 243–246.

Schmitz, S., Bhagwat, M., Moulton, M.A. et al. (2009). The prediction and validation of hover performance and detailed blade loads. *Journal of the American Helicopter Society* 54: 32004.

Shen, W.Z., Mikkelsen, R., Sørensen, J.N., and Bak, C. (2005). Tip loss corrections for wind turbine computations. *Wind Energy* 8: 457–475.

Shen, W.Z., Zhang, J.H., and Sørensen, J.N. (2009). The actuator surface model: a new Navier–Stokes based model for rotor computations. *ASME Journal of Solar Energy Engineering* 131: 011002.

Shives, M. and Crawford, C. (2013). Mesh and load distribution requirements for actuator line CFD simulations. *Wind Energy* 16: 1183–1196.

Simms, D., Schreck, S., Hand, M., and Fingersh, L. J. (2001) NREL Unsteady Aerodynamics Experiment in the NASA-Ames Wind Tunnel: A Comparison of Predictions to Measurements. Technical Report NREL/TP-500-29494, National Renewable Energy Laboratory.

Singer, M., Mirocha, J., Lundquist, J., and Cleve, J. (2010) Implementation and assessment of turbine wake models in the weather research and forecasting model for both mesoscale and large-eddy simulation. *International Symposium on Computational Wind Engineering*, Chapel Hill, NC, May 2010.

Skamarock, W.C. and Klemp, J.B. (2008). A time-split nonhydrostatic atmospheric model for weather research and forecasting applications. *Journal of Computational Physics* 227: 3465–3485. https://doi.org/10.1016/j.jcp.2007.01.037.

Skamarock, W. C., Klemp, J., Gill, D., Barker, D., Wang, W., and Powers, J. (2005) A description of the Advanced Research WRF Version 2. NCAR Technical Note NCAR/TN-468&STR, National Center for Atmospheric Research.

Skamarock, W. C., Klemp, J. B., Dudhia, J., Gill, J. B., Barker, D. M., Duda, M. G., Huang, X-Y., Wang, W., and Powers, J. G. (2008) A description of the advanced research WRF version 3. NCAR Technical Note NCAR/TN–475+STR, National Center for Atmospheric Research.

Sørensen, N.N. and Schreck, S. (2014). Transitional DDES computations of the NREL phase-VI rotor in axial flow conditions. *The Science of Making Torque from Wind Conference Series* 555: 012096.

Sørensen, J.N. and Shen, W.Z. (2002). Numerical modeling of wind turbine wakes. *Journal of Fluids Engineering* 124: 393–399.

Sørensen, N.N., Michelsen, J.A., and Schreck, S. (2002). Navier–Stokes predictions of the NREL phase VI rotor in the NASA Ames 80 ft × 120 ft wind tunnel. *Wind Energy* 5: 151–169.

Sørensen, J.N., Mikkelsen, R.F., Henningson, D.S. et al. (2015). Simulation of wind turbine wakes using the actuator line technique. *Philosophical Transactions of the Royal Society of London. Series A* 373: 20140071.

Sørensen, N.N., Zahle, F., Boorsma, K., and Schepers, G. (2016). CFD computations of the second round of MEXICO rotor measurements. *The Science of Making Torque from Wind (TORQUE 2016)*. IOP Publishing Journal of Physics: Conference Series 753: 022054. https://doi.org/10.1088/1742-6596/753/2/022054.

Steinhoff, J. S. and Ramachandran, K. (1988) Free Wake Analysis of Helicopter Rotor Blades in Hover using a Finite Volume Technique. Technical Report DAAG29-84-K-0019. U.S. Army Research Office, Research, Triangle Park, NC.

Stovall, T. D., Pawlas, G., and Moriarty, P. J. (2010) Wind farm wake simulations in OpenFOAM. AIAA-2010-0825.

Troen, I. and Petersen, E.L. (1989). *European Wind Atlas*. Roskilde, Denmark: Risoe National Laboratory for Sustainable Energy.

Troldborg, N. (2008) Actuator Line Modeling of Wind Turbine Wakes. Ph.D. Dissertation, Technical University of Denmark (DTU).

Troldborg, N., Sørensen, J.N., and Mikkelsen, R. (2010). Numerical simulation of wake characteristics of a wind turbine in uniform inflow. *Wind Energy* 13: 86–99.

Troldborg, N., Larsen, G.C., Madsen, H.A. et al. (2011a). Numerical simulation of wake interaction between two wind turbines at various inflow conditions. *Wind Energy* 14: 859–876.

Troldborg, N., Sørensen, J.N., and Mikkelsen, R. (2011b). Large-eddy simulation of wind-turbine wakes. *Boundary-Layer Meteorology* 138: 345–366.

Veers, P. S. (1988) Three-Dimensional Wind Simulation. *Technical Report SAND88-0152*, Sandia National Laboratories.

Vermeer, L.J., Soerensen, J.N., and Crespo, A. (2003). Wind turbine wake aerodynamics. *Aerospace Sciences* 39: 467–510.

Vermeulen, P.E.J. (1980). An experimental analysis of wind turbine wakes. In: *Proceedings of the 3rd Symposium on Wind Energy Systems*, 431–450. Lyngby, Denmark.

Watters, C.S. and Masson, C. (2010). Modelling of lifting-device aerodynamics using the actuator surface concept. *International Journal of Numerical Methods in Fluids* 62: 1264–1298.

Wilcox, D.C. (1988). Re-assessment of the scale-determining equation for advanced turbulence models. *AIAA Journal* 26 (11): 1299–1310.

Wilcox, D.C. (1998). *Turbulence Modeling for CFD*, 2e, 174. Anaheim: DCW Industries.

Wyngaard, J.C. (2004). Toward numerical modeling in the "Terra Incognita.". *Journal of the Atmospheric Sciences* 61: 1816–1826.

Xu, G. and Sankar, L. (2002) Application of a Viscous Flow Methodology to the NREL Phase VI Rotor. AIAA-2002–0030.

Zahle, F. and Sørensen, N.N. (2007). On the influence of far-wake resolution on wind turbine flow simulations. In: *The Science of Making Torque from Wind Conference Series*, IOP Publishing Journal of Physics: Conference Series, vol. 75, 012042. https://doi.org/10.1088/1742-6596/75/1/012042.

## Further Reading

Chattot, J.J. and Hafez, M.M. (2015). *Theoretical and Applied Aerodynamics – and Related Numerical Methods*. Dordrecht: Springer.
Schaffarczyk, A.P. (2014). *Introduction to Wind Turbine Aerodynamics*. Dordrecht: Springer.

# 8

# Design Principles, Scaled Design, and Optimization

*A designer knows he has achieved perfection not when there is nothing left to add, but when there is nothing left to take away.*

— Antoine de Saint-Exupéry

## 8.1  Design Principles for Horizontal-Axis Wind Turbines

In general, the first rotor design parameter to be determined is the rotor diameter, $D$. Basic momentum theory in Chapter 2 gave the following relation for the rotor power, $P$, as a function of fluid density, $\rho$, wind speed, $V_0$, rotor disk area, $A$, and the power coefficient, $C_P$, as:

$$\text{Wind Turbine Power}: \quad P = \frac{1}{2}\rho V_0^3 A C_P \tag{8.1}$$

For a given/desired power, $P$, the required rotor disk area, $A = (\pi/4)D^2$, and associated rotor diameter, $D$, can be determined by rearranging Eq. (8.1) such that:

$$\text{Rotor Diameter}: \quad D = \sqrt{\frac{8}{\pi}\frac{P}{\rho V_0^3 C_P}} \tag{8.2}$$

Note that in general, the fluid density is a function of temperature and elevation such that $\rho = \rho(T, z)$, see also Section 1.2.2.3, so that one has to be cautious with both surface temperature and elevation. On the other hand, the power coefficient, $C_P$, is primarily a function of the achievable airfoil lift-to-drag ratio, $c_l/c_d$, as outlined in Section 3.7. The lift-to-drag ratio, $c_l/c_d$, is airfoil dependent and surely depends on the Reynolds number, $Re$, see also Section 4.2, which in turn is a function of wind speed and the rotor solidity, $\sigma$, and therefore on the diameter-based Reynolds number, $Re_D$ (see Section 4.2.1, i.e. $C_P = C_P(Re_D)$).

Assuming standard atmospheric conditions at sea level with $\rho = 1.225$ kg/m$^3$, an approximated rated wind speed of $V_{Rated} = 11$ m/s and an average power coefficient across scales of $C_P = 0.4$, we can estimate the required rotor diameter, $D$, for a given/desired power, $P$, by the following relation:

$$\text{Rotor Diameter (Average) Estimate}: \quad D[m] \approx 0.062\sqrt{P[W]} \tag{8.3}$$

*Aerodynamics of Wind Turbines: A Physical Basis for Analysis and Design*, First Edition. Sven Schmitz.
© 2020 John Wiley & Sons Ltd. Published 2020 by John Wiley & Sons Ltd.
Companion website: www.wiley.com/go/schmitz/wind-turbines

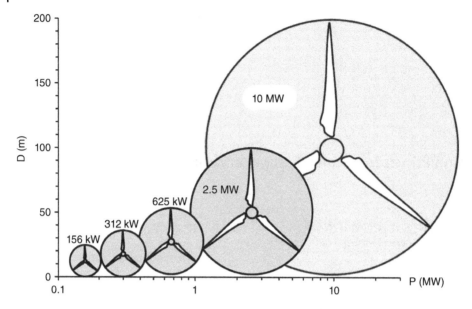

**Figure 8.1** Dependence of rotor diameter, $D$, with power, $P$ ($\rho = 1.225$ kg/m$^{-3}$, $V_0 = 11$ m s$^{-1}$, $C_p = 0.4$).

Equation (8.3) is a good first-order estimate of the required rotor diameter, $D$, for a given/desired power, $P$. The square-root dependence is captured well in Figure 8.1 where it can be seen that a 10 MW large-scale wind turbine requires twice the rotor diameter compared to a standard 2.5 MW wind turbine.

Figure 8.1 also provides some first guidance for model-scale (MS) experiments where the rotor diameter is typically constrained by blockage or boundary-layer considerations; hence Eq. (8.3) is helpful for initial sizing of the model generator requirements. A more refined estimate can be made taking into account a variable power coefficient, $C_p$, with rotor size and power according to:

$$\text{Rotor Diameter Estimate}: \quad D[m] \approx 0.040 \sqrt{\frac{P[W]}{C_p}} \tag{8.4}$$

As an example, modern utility-scale wind turbine designs with a rated power of $P \geq 2.5$ MW have now power coefficients of $C_p \cong 0.50$ and possibly higher in some cases, reducing the required associated rotor diameter, $D$, by a factor of $\sqrt{0.4/0.5} \cong 0.9$, which can be significant with respect to blade length and transportation constraints.

### 8.1.1 Wind Turbine Design Standards

The International Electrotechnical Commission (IEC) began work on an international wind turbine certification standard in the late 1980s, resulting in the *IEC 1400-1 Wind Turbine Generator Systems – Part 1 Safety Requirements* (Second Edition IEC, 1997). A significantly revised edition was published in 2005 under *IEC 61400-1* and is still in use today.

**Table 8.1** IEC standards for wind turbines.

| IEC number | Title |
| --- | --- |
| IEC-WT01 | IEC System for Conformity Testing and Certification of Wind Turbines Rules and Procedures |
| IEC 61400-1 | Wind Turbines – Part 1: Design Requirements (3rd edition) |
| IEC 61400-2 | Wind Turbines – Part 2: Design Requirements for Small Wind Turbines |
| IEC 61400-3 | Wind Turbines – Part 3: Design Requirements for Offshore Turbines |
| ISO/IEC 81400-4 | Wind Turbines – Part 4: Gearboxes for Turbines from 40 kW to 2 MW |
| IEC 61400-12 TS | Wind Turbines – Part 12: Wind Turbine Power Performance Testing |
| IEC 61400-11 TS | Wind Turbines – Part 11: Acoustic Noise Measurement Techniques |
| IEC 61400-13 TS | Wind Turbines – Part 13: Measurement of Mechanical Loads |
| IEC 61400-14 | Wind Turbines – Part 14: Declaration of Apparent Sounds Power Levels and Tonality of Wind Turbines |
| IEC 61400-21 | Wind Turbines – Part 21: Power Quality Measurements |
| IEC 61400-22 TS | Wind Turbines – Part 22: Conformity Testing and Certification of Wind Turbines |
| IEC 61400-23 TS | Wind Turbines – Part 23: Full-Scale Structural Testing of Rotor Blades |

### 8.1.1.1 IEC Standards for Wind Turbines

The most important wind turbine design standard in use today is the *IEC 61400-1* (IEC 2005), which includes (among other items) a range of design wind conditions, design load specifications under extreme wind/weather, and the verification of acceptable stresses due to turbine loading. A selection of subcomponents is listed in Table 8.1 where the designation TS stands for Technical Specifications.

Three classes of wind conditions (I, II, and III) are defined in the *IEC 61400-1* standard (IEC 2005), each of which includes a reference wind speed and respective turbulence intensity. Specific design wind conditions affecting turbine aerodynamics are:

- Normal wind profile (NWP)
- Normal turbulence profile (NTM)
- Extreme wind shear (EWS)
- Extreme operating gust (EOG)
- Extreme direction change (EDC)
- Extreme coherent gust (ECG)
- Extreme wind shear (EWS)
- Design load cases (start-up, normal/emergence shutdown, parked, operation, transport, maintenance)

As the most important wind turbine design standard, the purpose of the *IEC 61400-1* is to ensure that wind turbines are appropriately protected against load damage and hazards during the operating lifetime.

### 8.1.1.2 Wind Turbine Design Loads

In general, *ultimate loads* and *fatigue loads* affect the design of wind turbine components; here ultimate loads are those that refer to probable maximum loads and material stress (considering a safety factor) under design load cases, while fatigue loads refer to a component's ability to withstand a prescribed number of load cycles of varying amplitude. For this purpose, wind turbine loads are generally classified in the following categories:

- *Steady loads.* Mean Wind (Rotation and Parked)
- *Cyclic loads.* Wind Shear, Yaw Error, Gravity, Rotation
- *Stochastic loads.* Turbulence, Rotation
- *Transient loads.* Gusts, Start/Stop, Pitch Motion, Teeter
- Resonance-Induced Loads (Excitation of Structural Modes due to Cyclic/Stochastic/ Transient Loads)

Due to the large range of wind turbine load conditions, it is usually not possible to avoid all natural frequencies during all design load conditions, particularly during the transients associated with turbine start-up and shutdown. In this context, it is important to understand how wind turbine performance, blade loads/moments, and blade natural frequencies scale with the rotor diameter, respectively blade radius, $R$. Here we assume an optimum (and constant) tip speed ratio, $\lambda$, so that the rotor speed, $\Omega$, scales $\sim R^{-1}$. Table 8.2 lists the primary loads and resonances and their scaled relation with the blade radius, $R$.

From Chapter 2, we know that for a given tip speed ratio, $\lambda$, rotor power, $P$, and thrust, $T$, are proportional to the squared value of the rotor diameter (and hence blade radius, $\sim R^2$), while rotor torque, $Q$, scales as $\sim R^3$. The aerodynamic moments, $M_A$, in Table 8.2 include flap bending and torsion and are $\sim R^3$ as a result of blade section forces ($\sim R^2$) being multiplied by a lever arm ($\sim R$). Note that blade/rotor weight scales as $\sim R^3$ in contrast to power/thrust as $\sim R^2$, being a "square-cube-law" of wind turbine design that may eventually limit wind turbine size. Also, centrifugal forces, $F_C \sim m\Omega^2 R$, where $m$ refers to mass; hence $F_C \sim R^2$ for constant $\lambda$. As for blade natural frequencies, $\omega_N$, they scale as $\sim R^{-1}$ assuming constant material properties and structural layout. Hence the "per-rev" excitation with $\Omega \sim R^{-1}$ behaves the same such that a "scaled excitation," $\Omega/\omega_N$, becomes $\sim R^0$ and therefore independent of the blade scale.

**Table 8.2** Scaling wind turbine loads and resonances with blade radius, $R$, at constant tip speed ratio, $\lambda$.

| Blade loads and resonances | Symbol | Scaling ($\lambda = const.$) |
|---|---|---|
| Power | $P$ | $\sim R^2$ |
| Thrust | $T$ | $\sim R^2$ |
| Torque | $Q$ | $\sim R^3$ |
| Aerodynamic moments | $M_A$ | $\sim R^3$ |
| Centrifugal forces | $F_C$ | $\sim R^2$ |
| Weight | $F_g$ | $\sim R^3$ |
| Natural frequency | $\omega_N$ | $\sim R^{-1}$ |
| Rotor speed ($\lambda = const.$) | $\Omega$ | $\sim R^{-1}$ |
| Scaled excitation | $\Omega/\omega_N$ | $\sim R^0$ |

## 8.1.2 Rotor Design Procedure

In the following, an introduction is given to a simplified general rotor design procedure. Here we focus exclusively on pitch- and speed-controlled horizontal-axis wind turbines (HAWT). Some constraints with respect to structural integrity, transportation, and noise are given alongside a brief introduction to cost analysis and COE. In the early design process, blade-element momentum (BEM) theory and vortex wake methods (VWMs) and associated efficient computational tools are instrumental as a starting point for every iterative rotor design process.

### 8.1.2.1 General Rotor Design Process

An optimum rotor design solely based on aerodynamics may result in unrealistic blade planforms with respect to blade weight as well as structural, manufacturing, and lifetime considerations. In the following, we will draw from aerodynamics knowledge gained in previous chapters and develop a general rotor design process; however, it makes sense to first introduce some common design constraints that have to be considered.

**Common Design Constraints**

- *Maximum blade chord, $c_{max}$.* For transportation purposes of rotor blades on roads, bridges, and so on, the maximum allowable blade chord is typically limited. In addition, $c_{max}$ may also be limited by moment loads in parked conditions under extreme weather events. This constraint, however, may be relaxed for small-scale wind turbines where a larger chord may show airfoil performance benefits due to higher Reynolds number, *Re*.
- *Blade thickness distribution, t.* Large utility-scale wind turbines require thicker airfoils at inboard blade stations due to the scaled behavior of structural stresses of both aerodynamic and gravitational (weight) moments. This has implications on airfoil choice and associated airfoil performance characteristics.
- *Low cut-in wind speed, $V_{In}$.* Wind turbines are desired to cut-in at a minimum wind speed in order to maximize energy production. Consequently, the blade design has to accommodate for a large start-up torque in low winds to overcome rotor inertia. Note that an optimum "parked" design does not necessarily coincide with an optimum "rotating" design due to the induced inflow distribution along the blades.
- *Maximum tip speed, $V_{Tip}$.* For onshore applications near residential areas, the tip speed is typically limited to 65 ... 75 m s$^{-1}$ due to noise considerations. This constraint is of course relaxed for offshore applications. Hence large offshore wind turbines can be designed for higher tip speed ratio, $\lambda$, thus potentially making two-bladed rotors more attractive due to a broader $C_P$ versus $\lambda$ curve and $C_{P,max}$ shifted to a higher $\lambda$ (see Section 3.7).
- *Maximum rotor speed, $\Omega_{max}$.* If the maximum tip speed, $V_{Tip}$, is not limited, the rotor speed might be as blade loads increase as $\sim\Omega^2$. Indeed, blade loads (including rotor thrust) are highest at the rated wind speed, $V_{Rated}$, at the end of Region II (see Section 3.7). Note that on the other hand, for a given power, shaft torque decreases with increasing rotor speed. In the end, combined constraints of $V_{Tip}$ and $\Omega_{max}$ practically determine a design tip speed ratio, $\lambda$, at $V_{Rated}$.
- *Rotor Thrust, T.* Modern utility-scale wind turbines typically operate at high thrust coefficients, $C_T$, ranging from 0.8 ... 1.2. Rotor thrust typically peaks at $V_{Rated}$ and

one has to consider $C_{T,max}$ in the context of tower/foundation design. In addition, the rotor thrust coefficient, $C_T$, has to be constrained/fixed for a "fair" comparison between rotor/blade designs. As for small-scale wind turbines, the achievable $C_{P,max}$ is a function of airfoil $(c_l/c_d)_{max}$, see Section 3.7; hence a corresponding optimal $C_T$ may actually be less than 0.8 depending on the specific application.

Rotor design is an iterative process considering aerodynamics, system dynamics, materials and structures, transportation, and more. The following steps of a general rotor design process do have an emphasis on rotor aerodynamics, which can be thought of as an "Iteration 1" (or "napkin design") in an all-encompassing design process that may not be aimed at generating maximum power, but with the overall goal of manufacturing and operating a rotor design with the lowest COE (see Section 8.1.2.2). As a start, we consider Eqs. (3.25) and (3.26) for an ideal rotor (see Section 3.3.1) and assume that the tip speed ratio, $\lambda$, is sufficiently large such that $1 \ll (3/2)^2 \lambda_r^2$ at the outer part of the blade where most of the rotor torque (and hence power) is generated. Then the following approximate interdependent relation can be found:

$$\text{Ideal } (B, \lambda, c_l, c) \text{ Relation}: \quad B\lambda c_l \cdot c(r/R) = \left(\frac{4}{3}\right)^2 \pi \cdot \frac{1}{\lambda_r} \tag{8.5}$$

In Eq. (8.5), $B$ is the blade number, $\lambda = (\Omega R)/V_0$ is the tip speed ratio, $c_l$ is the blade section lift coefficient, $c$ is the local blade section chord, and $\lambda_r = \lambda \cdot r/R$ is the local speed ratio. It becomes apparent that the ideal radial blade chord distribution in the outer portion of the blade is inversely proportional to the radial location, $r/R$. Note, however, that Eq. (8.5) assumes ideal loading conditions (see Section 3.3.1) and an associated high thrust coefficient of $C_T = 8/9$. Therefore, one has to be cautious in applying Eq. (8.5) for an optimum blade chord design. Nevertheless, Eq. (8.5) is introduced here for the purpose of illustrating that the basic rotor design parameters $B$, $\lambda$, $c_l$ are interdependent with respect to an optimal (aerodynamic) setting and have to be chosen subject to constraints and such that realistic blade planforms $c(r/R)$ are obtained. Next, the rotor design process is layed out as follows:

**Rotor Design Process**

1) *Rotor/generator size, D.* The projected rated power of a pitch- and speed-controlled wind turbine is to a large extent governed by the available wind resource (see Section 1.2.1). Given a chosen rated generator power, $P$, the rotor diameter can be estimated from Eq. (8.2) for a particular wind site $(\rho, V_0)$ where $V_0 = V_{Rated}$ in this context. Note that the fluid density, $\rho$, is a function of both elevation and air temperature, resulting in some notable effect of desert versus cold climates and sea level versus mountain/ridge wind sites. As for the anticipated power coefficient, $C_P$, it is practically a function of rotor size itself as the achievable $C_{P,max}$ is closely related to airfoil $(c_l/c_d)_{max}$, which is a function of blade section Reynolds number and hence rotor size (see Sections 3.7 and 4.2).
2) *Design tip speed ratio, $\lambda$.* For pitch- and speed-controlled wind turbines, a design tip speed ratio, $\lambda$, is held approximately constant throughout Region II and is hence an important design parameter. Note that from basic momentum theory (see Section 2.2.4), $\lambda$ should be as large as possible. In reality, a design tip speed ratio ranging between 5 … 10 gets fairly close to the theoretical limit. Here, constraints

in the tip speed, $V_{Tip}$, and load scaling properties (Table 8.2) become important. Furthermore, the achievable airfoil $(c_l/c_d)_{max}$ influences the choice of $\lambda$ with respect to the maximum possible power coefficient, $C_{P, max}$ (see Figure 3.14 in Section 3.7).

3) *Number of blades, B.* From an aerodynamics perspective, tip losses are reduced with increasing number of blades (see Section 3.4). This stands in contrast to transportation and manufacturing costs where it is obvious that fewer blades are desired. Some further considerations come into effect when considering how the local blade chord, $c$, and the design lift coefficient, $c_l$, are influenced by the tip speed ratio, $\lambda$, and the number of blades, $B$, in Eq. (8.5). For example, increasing the number of blades, $B$, for the same design lift coefficient, $c_l$, reduces the chord, $c$, for the same rotor solidity. Then, for structural considerations, the blade thickness is expected to increase, with tentative negative repercussions on airfoil performance. Some discussion on two-bladed versus three-bladed rotors has already been given in Section 3.7.5, showing that a two-bladed rotor has a higher optimum tip speed ratio, $\lambda$, thus requiring higher tip speeds, $V_{Tip}$, that may be subject to noise constraints. Also, rotor centrifugal loads scale as $\sim\Omega^2$ that may also in this case increase rotor weight/structure to withstand the loads, while on the other hand, an associated increased chord, $c$, and higher $\lambda$ both increase the blade section Reynolds number and hence airfoil performance.

4) *Design lift coefficient and airfoils, $c_l$ and $c_{l, max}$, $c_l/c_d$.* As far as aerodynamics is concerned, the design lift coefficient, $c_l$, for the outer blade halve should be chosen close to the respective $(c_l/c_d)_{max}$ of suitable airfoils at expected Reynolds numbers based on $Re_D$ (see Section 4.2). Note, however, that this cannot be done without considering relations such as Eq. (8.5) and in the context of $B$, $\lambda$ so that "realistic" blade chord distributions, $c$, are obtained (see Section 3.3). For the inner blade halve, this typically results in higher design $c_l$ criteria in order to keep the local blade chord, $c$, within realistic bounds, see dependence in Eq. (8.5). For structural reasons, thick inboard airfoil sections are used with higher maximum lift coefficients, $c_{l, max}$, as outlined in Section 4.2.

5) *Preliminary "One-Point" blade design, $c(r)$ and $\beta(r)$.* For a preliminary "napkin design," the rotor diameter-based Reynolds number, $Re_D$, defines blade section Reynolds numbers and hence expected airfoil $(c_l/c_d)_{max}$ of suitable airfoils (see Section 4.2). This in turn defines the approximate maximum power coefficient, $C_{P, max}$, that can be expected for the rotor (see Figure 3.14 in Section 3.7). While values for $C_{P, max}$ are fairly high for utility-scale wind turbines, a notably reduced $C_{P, max}$ due to the Reynolds scale can be an important consideration for small-scale and MS wind turbines ($D \ll 1$ m) as it affects the associated (average) thrust coefficient, $C_T$, according to momentum theory in the windmill state (see Figure 2.3 in Section 3.1). This in turn affects the (optimal) average axial induction factor, $a$, across the rotor disk area. Preliminary rotor design as described in Section 3.3 for ideal rotors with/without wake rotation then assumes (ideal) radial distributions of axial- and angular induction factors, $a$ and $a'$, that are functions of $\lambda$, $B$, $c_l$. Note that those can be (as a first approximation) scaled to other average axial induction factors if needed. As an alternative, assumed (ideal) radial distributions for $a$ and $a'$ can be obtained from computational methods described in Section 8.1.2.3.

6) *Evaluation of blade design, AEP.* As a next step, the performance of the "napkin design" from Step 5 before has to be analyzed with a suitable computational method

and over an expected wind speed probability (see Section 1.2). These analyses should include root and tip loss (see Section 3.4), rotational augmentation effects (see Chapter 5), and reduced-order aeroelastic analyses (see Section 8.1.2.3). Here, the *Annual Energy Production (AEP)* is a primary assessment criterion as it affects the COE and hence profitability of a planned wind site (see Section 8.1.2.2). Furthermore, blade loads/moments have to be assessed considering load responses to unsteady gusts and associated fatigue load amplitudes.

7) *Iterative refined and constrained blade design.* Based on the overall assessment in Step 6, design iterations are initiated considering ease of manufacturing, blade/hub assembly, transportation, and so on. If the wind site has higher probability of low winds, a refined blade design with lower cut-in wind speed can help gain some cost relevant *AEP* increment; this can be realized using for example, a larger rotor diameter or higher-solidity blades (both of which may, however, adversely affect rotor weight and cost).

In summary, it is important to understand that an optimal aerodynamic design conflicts with a number of multi-disciplinary design constraints. This section gave a brief introduction to a general rotor design process with some emphasis on aerodynamics, but also discussing some other aspects, constraints, and interdependencies that have to be considered. For the aerodynamicist, the art is to travel through the multi-disciplinary design space with an understanding of the dependencies and sensitivities that can result in an overall design for lowest (levelized) COE.

### 8.1.2.2 COE versus Levelized Cost of Energy (LCOE)

The *AEP* of a wind project is primarily a function of the capacity factor, *CF*, at a given operational wind site and the rated installed power (see Section 1.2.4). Based on a variable rate [$/kWh] at which energy can be sold to the electricity market over a project lifetime, the total income of a wind project can be estimated/calculated. However, the actual profitability of a wind project is governed by a balance between *AEP* and the costs associated with the original capital investment, operation and maintenance, and others. Here, the challenge has been to estimate all costs correctly prior to a typical project lifecycle of 20 years, particularly as the energy market, but also wind resource, is subject to uncertain fluctuations. There are many considerations that "go beyond aerodynamics" but affect the aerodynamic design of new wind turbines through constraints, some of which are listed in the previous section. Again, aerodynamics is an art that travels the optimum design space by keeping the rotor/blade design as optimal as possible given a set of multi-disciplinary constraints. Broadly defined, the *COE* of a wind energy project is the ratio of total costs to the energy produced and sold to the electricity market:

$$Cost\ of\ Energy = (Total\ Costs)/(Energy\ Produced) \tag{8.6}$$

In a simplified manner, the *COE* can be defined according to:

$$COE : COE = \frac{(C_{Capital} \times FCR) + C_{O\&M}}{AEP}\ [\$/kWh] \tag{8.7}$$

Here $C_{Capital}$ is the original project capital cost, *FCR* is a fixed (annual) charge rate, $C_{O\&M}$ are annual operation and maintenance costs, and *AEP* is the annual energy production. The original project capital cost, $C_{Capital}$, includes the investment costs associated with all turbines, transportation to the wind site, and installation at the wind site up to the point at which the project becomes operational. In this context, the fixed (annual) charge rate, *FCR*, is an annual fraction of $C_{Capital}$ accounting for costs associated with taxes, debt, and equity costs to the operator. Operation and maintenance costs, $C_{O\&M}$, include the annual periodic maintenance of mechanical and electrical equipment, periodic testing, blade cleaning, and uncertain unscheduled maintenance costs. Consider a simple approximate example of a single wind turbine with $P_{Rated} = 1.5$ MW, assuming $C_{Capital} = \$1,500,000$, $FCR = 12\%$, and $C_{O\&M}$ being 3.0% of $C_{Capital}$. Furthermore, consider a capacity factor, *CF*, of 0.30 at the wind site. Then using Eqs. (1.40) and (8.7), the *COE* becomes $COE = (0.15 \cdot \$1.5 \times 10^6)/(8760 \text{ h} \cdot 0.30 \cdot 1.5 \times 10^3 \text{ kW}) = 0.057\$/\text{kWh}$. It is apparent that the *COE* is a very sensitive balance subject to uncertainties associated with (unexpected) operation and maintenance and the associated reduction in turbine availability, capacity factor, and ultimately energy production when turbines are less operational than anticipated. Therefore, both numerator and denominator are affected by unexpected operations and maintenance costs. Aerodynamic and structural design can reduce operation and maintenance costs, $C_{O\&M}$, by carefully executed blade designs that minimize material fatigue based on knowledge gained from high-fidelity simulations of wind turbine response in the atmospheric boundary layer. Another issue with the simplified *COE* relation in Eq. (8.7) is that it does not account for the time value of money. Here the method of *levelizing*, see for example Rabl (1985), is of interest as it expresses both cost and revenue within a *Life Cycle Costing (LCC)* method to result in equivalent equal payments. A *LCOE* is broadly defined as the ratio of all levelized costs to the energy produced:

$$\textit{Levelized Cost of Energy} = (\textit{Total Levelized Costs})/(\textit{Energy Produced}) \qquad (8.8)$$

A simplified definition of the LCOE becomes:

$$\textit{Levelized Cost of Energy}: \quad LCOE = \frac{C_{NPV} \times CRF}{AEP} \quad [\$/\text{kWh}] \qquad (8.9)$$

where $C_{NPV}$ is the "net present value," essentially an economic value based on projected (annual) gross savings/costs and used to compare investment options in a *LCC* analysis, and *CRF* is a "capitol recovery factor" governed by the loan interest rate for the particular wind project. The inclusion of levelized costs, time value of money, and so on in the *LCOE* make it a better cost model compared to the classical *COE*, though not necessarily easy to grasp with an aerodynamicist's mind. Nevertheless, the bottom line is that operation and maintenance costs, $C_{O\&M}$, and their effect on the "net present value," $C_{NPV}$, have to be reduced to a minimum as they affect both numerator and denominator of either cost-of-energy equation. The interested reader is referred to Burton et al. (2001), Manwell et al. (2009), and Cory and Schwabe (2009) for more in-depth discussion on cost modeling and COE.

### 8.1.2.3 Computational Tools for Rotor Analysis and Design

Efficient computational modeling tools play an important role in the wind turbine design process and are used for analysis/design of specific subcomponents or provide supplementary data such as wind flow fields or structural properties required by other analysis methods. In general, computational tools used in the wind energy community can be classified as: (i) wind simulators generating realistic atmospheric turbulent inflow, (ii) aerodynamic/aeroelastic performance codes to compute blade sectional forces and

**Table 8.3** Examples of wind turbine analysis and design codes.

| Code | Organization | Class |
|---|---|---|
| TurbSim | NREL | Wind simulator |
| IECWind | NREL | Wind simulator |
| PROPID | UIUC | Rotor performance and design |
| XTurb | Penn State | Rotor performance and design |
| AeroDyn | NREL | Rotor performance (coupled to FAST) |
| Cp-Max | TUM | Rotor performance, structure (dynamics) |
| PHATAS | ECN | Rotor performance and structure |
| HAWC2 | Risø | Rotor performance and structure |
| FAST | NREL | Rotor performance and structure |
| PreComp | NREL | Structure (properties) |
| BModes | NREL | Structure (modes) |
| NuMad | Sandia | Structure (finite-element preprocessor) |
| FLEX | DTU | Rotor performance, structure (dynamics) |

integrated thrust/torque/moment data, and (iii) structure tools providing multi-body rotor/turbine dynamic analyses, efficient modal decomposition, material properties, or finite-element codes for stress analyses. Table 8.3 lists some examples of wind turbine analysis and design codes, including the primary developing organization and broad code classification.

As most tools listed in Table 8.3 are still under continued development, no specific web links or references are given, and the interested reader is encouraged to contact the respective organization for more information.

## 8.2   Scaled Design of Wind Turbine Blades

Model-scale (MS) testing of scaled designs in controlled environments is an integral part of the engineering research and development process, for example, wind-tunnel testing of airfoils, wings, aircraft, cars, and so on. At times, simplified canonical (component) experiments are conducted, serving specific objectives in the *Verification & Validation (V&V)* process of computational tools, while properly scaled aerodynamic and/or dynamic model experiments are also used for the purpose of V&V studies or to test actual full-scale (FS) representative prototypes in an instrumented and controlled environment. The following section is an introduction to MS testing of wind turbines. At first, all relevant aerodynamic and structural (dynamic) dimensionless groupings are given, followed by a discussion on the physical limitations associated with scaling FS conditions to a MS environment. Next, some of today's most referenced MS (aerodynamic) experiments ($1m \leq D \leq 10m$) are presented along with a selected number of *XTurb* examples for further analysis.

## 8.2.1 Limitations of Scaled Blade Aerodynamics and Dynamics

Using the classical Buckingham $\Pi$ Theorem (Buckingham 1914), dimensionless group-ings relevant to aerodynamics and structural dynamics can be derived. They are listed in Table 8.4 and subdivided into three primary categories with respect to rotor aero-dynamics, blade/wake aerodynamics, and rotor dynamics. In Table 8.4, $r' = r/R$ is the normalized radial blade station.

Matching the tip speed ratio, $\lambda$, between FS and MS turbines is mandatory to achiev-ing scaled rotor operation as well as scaled rotor/blade aerodynamics (see Section 3.7). The Mach number, $Ma$, on the other hand is the ratio of blade tip speed to the local speed of sound (with $a_0$ denoting the local speed of sound) and is an aerodynamic parameter describing the effects of compressibility on airfoil aerodynamics. As a rule-of-thumb, however, airfoil aerodynamics is considered "incompressible" as long as $Ma < 0.3$ (for most airfoils). Hence, even assuming a blade tip speed of $V_{Tip} = \lambda V_0 = 90$ m s$^{-1}$ and an approximate speed of sound of $a_0 = 330$ m s$^{-1}$ at sea-level conditions, blade sectional Mach numbers of $Ma < 0.27$ are obtained. In this respect, wind turbine blades of all scales practically operate in incompressible flow, and therefore exact $Ma$ matching is not mandatory in practice from the perspective of airfoil aerodynamics. The situation is different, however, for blade section Reynolds numbers, $Re_c(r') \sim c(r')$, and thus proportional to the geometric scaling ratio itself (at least to first order). It becomes apparent that $Re_c(r')$ cannot be matched in practice between FS ($D \cong 100$ m) and MS ($1$ m $\leq D \leq 10$ m) conditions as testing MS turbines at notably higher-than-FS wind speeds to match Reynolds number might result in undesirable compressibility effects. As the Reynolds number is generally defined as the ratio of inertial and viscous forces, a notably reduced Reynolds scale does nonetheless affect airfoil aerodynamics (see Chapter 4) and particularly the achievable airfoil $(c_l/c_d)_{max}$, with implications on the maximum turbine power coefficient, $C_{P,max}$, at a given $\lambda$ (see Section 3.7). Fortunately, this dilemma can be compensated to some extent with suitable blade planform designs and airfoil choice, see the *XTurb* example in Section 8.2.2. The normalized radial blade circulation, $\Gamma(r')/(V_0 R)$, in Table 8.4, is a key parameter to be matched between FS and MS in order to achieve correctly scaled blade/wake aerodynamics, see also Kelley et al. (2016) and Hassanzadeh et al. (2016). Following the vortex theory presented in Chapter 6, the radial circulation distribution, $\Gamma(r)$, determines both the radial loading

**Table 8.4** Dimensionless groupings relevant to wind turbine aerodynamics and structural dynamics ($r' = r/R$).

| Dimensionless grouping | Definition | Relevance |
|---|---|---|
| Tip speed ratio, $\lambda$ | $\lambda = \Omega R / V_0$ | Rotor aerodynamics |
| Tip Mach number, $Ma$ | $Ma = \lambda V_0 / a_0$ | Rotor aerodynamics |
| Reynolds number, $Re_c(r')$ | $Re_c(r') = \rho V_{rel} c(r') / \mu$ | Blade aerodynamics |
| Blade circulation, $\Gamma(r')/(V_0 R)$ | $\dfrac{\Gamma(r')}{V_0 R} = \dfrac{1}{2} c_l(r') \dfrac{V_{rel}(r')}{V_0} \dfrac{c(r')}{R}$ | Blade/wake aerodynamics |
| Lock number, $Lo$ | $Lo = c_{l,a} \rho c R^4 / I$ | Rotor dynamics |
| Norm. Natural Frequency, $\widetilde{\omega}$ | $\widetilde{\omega} = \omega / \Omega$ | Rotor dynamics |
| Froude number, $Fr$ | $Fr = V_0 / \sqrt{gR}$ | Rotor dynamics |

distribution and the trailing vorticity, $\delta\Gamma(r)$, in the wake. Therefore, one can make a strong argument that a dimensionless scaled radial blade circulation, $\Gamma(r')/(V_0 R)$, is indeed a primary scaling parameter between FS and MS that should be matched to the extent possible. Here, we are fortunate that this is indeed possible, see definition in Table 8.4, and without too much difficulty for most of the blade as $V_{rel}(r')/V_0$ scales with $\lambda_r$ to first order, and the product $c_l \cdot c$ can be easily kept constant using a suitable blade planform design and airfoil choice (see Section 8.2.2).

If blade aeroelastic scaling is of interest in addition to aerodynamic scaling, some relevant dimensionless groupings are listed in Table 8.4, that is, the Lock number, the normalized natural frequency, and the Froude number. For full aeroelastic scaling between FS and MS, all these groupings have to be matched (in theory) to the extent possible. The Lock number, $Lo = c_{l,\alpha} \rho c R^4 / I$, is defined as the ratio of the blade aerodynamic and mass inertia forces where $c_{l,\alpha}$ is the airfoil lift-curve slope and $I = \int_0^R mr^2 dr$ is the mass moment of inertia about the flapping axis. In general, matching the Lock number, $Lo$, is important for the prediction of rotor loads and aeroelastic stability (Singleton and Yeager 1998; Yeager and Kvaternik 2001). The normalized natural frequency, $\widetilde{\omega} = \omega/\Omega$, where $\omega$ is the blade natural frequency about the flapping axis, ensures that the "per revolution" blade excitation is consistent between FS and MS, practically meaning that FS and MS blades have the same Campbell diagram. Last, the rotorcraft community typically refers to the Froude number squared, $Fr^2 = V_0^2/gR$, where $g$ is the gravitational acceleration, as the ratio of rotor aerodynamic forces to rotor weight (or gravitational forces); see, for example, Hunt (1973). In this context, $Fr^2$ is relevant when scaling for the effect of gravitational forces about the rotor axis; that is, lead/lag motion.

For full similarity according to Buckingham's Π Theorem, all dimensionless groupings in Table 4.1 have to be matched; however, it is not surprising that this is impossible in practice. Nevertheless, appropriately scaled rotors can be designed for specific dynamic scaling objectives, an active area of research in the wind energy science community that has experienced growing interest and relevance (Bottasso et al. 2014; Nanos et al. 2016; Canet et al. 2018).

### 8.2.2 Example of Scaled Aerodynamics from Utility-Scale to MS Turbine

In the following, we refer to FS conditions to represent the utility scale with $D \cong 100$ m and to MS conditions for the test turbine with $1\,\text{m} \leq D \leq 10\,\text{m}$. For aerodynamic similarity between FS and MS turbines, the following conditions have to be satisfied:

- *Tip speed ratio.* $\lambda_{FS} = \lambda_{MS}$
- *Tip Mach number.* $Ma_{FS}, Ma_{MS} < 0.3$
- *Reynolds Re-scale.* $Re_{c,FS} \approx Re_{c,MS}$ (to the extent possible)
- *Blade circulation.* $\left(\frac{\Gamma(r)}{V_0 R}\right)_{FS} = \left(\frac{\Gamma(r)}{V_0 R}\right)_{MS}$
- *Airfoil choice/characteristics.* $\left(\frac{c_l}{c_d}\right)_{max,FS} \approx \left(\frac{c_l}{c_d}\right)_{max,MS}$ (as much as *Re*-scale allows)
- *Thrust/bending moment coefficients.* $C_{T,FS} = C_{T,MS}$, $C_{B,FS} = C_{B,MS}$
- *Power coefficient.* $C_{P,FS} \approx C_{P,MS}$ (as much as *Re*-scale allows)

Note that the nature of testing in "air" is such that the Reynolds *Re*-scale cannot be matched in practice given (possible) geometric scaling ratios >20 between FS and MS.

As the chord-based Reynolds number is defined as $Re_c = \rho V_{rel} c / \mu$, changing the testing fluid to "water" actually assists in $Re$-scaling as water has $1/15^{th}$ the kinematic viscosity, $\nu = \mu/\rho$, compared to air (Reich et al. 2014). Another possibility is testing in a pressurized tunnel (i.e. increasing $\rho$); such facilities exist, for example, at Princeton University (Miller et al. 2017) or in Goettingen (Germany). From a blade aerodynamics perspective, however, matching the radial circulation, $\Gamma(r')/(V_0 R)$, is the primary aerodynamic parameter that accounts for both the load distribution and the trailing vorticity emanating from the blade and into the wake:

$$\text{Blade Circulation}: \quad \frac{\Gamma(r')}{V_0 R} = \frac{1}{2} c_l(r') \frac{V_{rel}(r')}{V_0} \frac{c(r')}{R} \tag{8.10}$$

In this respect, one actually has the opportunity to increase $Re_c$ and hence airfoil $(c_l/c_d)_{max}$ by increasing the blade solidity of the MS turbine via $c(r')$ and lowering $c_l(r')$ accordingly, noting that this may change the operating airfoil $c_l/c_d$. Looking at Eq. (8.10), one has to also be aware that $V_{rel}(r')$ is a function of axial/angular induction factors, $a(r')$ and $a'(r')$, see Section 3.1.2, and thus on the local blade flow angle, $\phi(r')$:

$$\text{Relative Velocity}: \quad V_{rel}(r') = V_0 \sqrt{(1 - a(r'))^2 + (1 + a'(r'))^2 \, \lambda_r^2} \tag{8.11}$$

$$\text{Velocity Triangle}: \quad \tan \phi(r') = \frac{1}{\lambda_r} \frac{1 - a(r')}{1 + a'(r')} \tag{8.12}$$

As root- and tip-loss factors, $F_R$ and $F_T$, are also in general a function of $\phi$ (see Section 3.4), one can make a strong argument that keeping $\phi(r')$ constant between FS and MS is helpful in accounting consistently for root/tip losses (at least to first order). Furthermore, this actually assists in determining the blade pitch distribution of the MS blade as:

$$\text{Blade Pitch/Twist}: \quad \beta(r') = \phi(r') - \alpha(r') \tag{8.13}$$

Note that in Eq. (8.13), the local angle of attack, $\alpha(r')$, is determined from the lift coefficient, $c_l(r')$, of the respective MS airfoils where different design lift coefficients may be chosen for the MS blade. As for a rule-of-thumb with respect to scaled airfoil characteristics, the maximum lift-to-drag ratio is $(c_l/c_d)_{max} \approx 150$ for the FS ($D \cong 100$ m) blade and reduces to $(c_l/c_d)_{max} \approx 50 \ldots 100$ for the MS ($1 \text{ m} \leq D \leq 10$ m) test blade. The question then arises as to how well $C_{T,FS} = C_{T,MS}$ and $C_{P,FS} = C_{P,MS}$ can be matched under the constraint of significant geometric scaling ratios and associated $Re$-scale effects on airfoil $(c_l/c_d)_{max}$. Here, we refer to Section 6.2.5 (Relationship between Vortex Theory and Blade-Element Theory) and rewrite Eqs. (6.46) and (6.47) for the incremental thrust and torque, $dT$ and $dQ$, using Eqs. (6.37) and (6.38) to obtain:

$$\text{Incremental Thrust}: \quad dT = B \frac{1}{2} \rho V_0^2 R^2 \frac{2\Gamma}{V_0 R} \left[ \lambda_r (1 + a') + \frac{c_d}{c_l} (1 - a) \right] dr' \tag{8.14}$$

$$\text{Incremental Torque}: \quad dQ = B \frac{1}{2} \rho V_0^2 R^3 \frac{2\Gamma}{V_0 R} \left[ (1 - a) - \frac{c_d}{c_l} \lambda_r (1 + a') \right] dr' \tag{8.15}$$

**Table 8.5** Example – scaled NREL 5-MW turbine ($V_0 = 8$ m s$^{-1}$, $\lambda = 7.55$).

| Turbine | $\lambda$ | $V_0$ [m s$^{-1}$] | $Ma$ | $R$ [m] | rpm | $Re_{MS}/Re_{FS}$ |
|---|---|---|---|---|---|---|
| NREL 5-MW (FS) | 7.55 | 8.0 | 0.18 | 63.0 | 9.16 | 1 |
| Scaled NREL 5-MW (MS) | 7.55 | 13.0 | 0.30 | 0.63 | 1488.5 | 1/62 |

In Eqs. (8.14) and (8.15), the dependence of $\Gamma$, $\lambda_r$, $a$, $a'$, $c_d/c_l$ on $r' = r/R$ was omitted for ease of presentation. Assuming perfectly scaled $\Gamma$, $a$, $a'$, it becomes clear that $C_{T,FS} = C_{T,MS}$ can be satisfied fairly well as $\lambda_r(1 + a') \gg \frac{c_d}{c_l}(1 - a)$ in Eq. (8.14) for most of the blade and typical values of $a$ and $a'$. The power coefficient, however, is more sensitive to $Re$-scale effects on airfoil $c_l/c_d$ as the corresponding terms in Eq. (8.15) may become of the same order for the outer blade portion, that is, $\lambda_r \to \lambda$, and reduced airfoil performance, $c_l/c_d$. As a result, $C_{P,FS} > C_{P,MS}$, particularly when the scaled FS condition is close to the FS turbine $C_{P,max}$ and simply cannot be achieved at the MS condition due to reduced airfoil performance.

In the following, a simplified scaling example is presented to illustrate the concepts discussed previously and to foster our understanding of scaled aerodynamics from utility scale to model test scale. Here the National Renewable Energy Laboratory (NREL) 5-MW turbine (Jonkman et al. 2009) is considered at a wind speed of $V_0 = 8$ m s$^{-1}$ at $\lambda = 7.55$, and the task is to design a scaled wind turbine for wind-tunnel testing. In this simplified example, we choose a $R_{MS}/R_{FS} = 1/100$ geometric scale in order to be able to conduct this test in a number of existing medium-size wind tunnels. Table 8.5 summarizes some of the basic parameters of both the NREL 5-MW turbine and its MS counterpart. Note that by increasing tunnel (wind) speed under the constraint of tip $Ma \leq 0.3$, a large geometric scaling ratio can be softened to an actual Reynolds scale of $Re_{MS}/Re_{FS} = 1/62$. From a data acquisition standpoint, one has to be aware of the respective increase in MS rotor speed.

The next question that arises is that of choosing appropriate airfoils for the MS turbine, with reasonably good performance at the reduced Reynolds scale. For simplicity, let us design the MS turbine using only one airfoil along the blade span (i.e. National Advisory Committee for Aeronautics, NACA 4412), which may not be the optimum choice but serves the purpose of demonstrating the scaled design process. Both the (benchmark) NACA 64618 tip airfoil of the NREL 5-MW turbine and the chosen MS turbine airfoil NACA 4412 are shown in Figure 8.2. As a rule-of-thumb, choosing a thinner airfoil for the MS turbine is helpful as these typically (not true all the time) have a more modest decrease in $c_{l,max}$ with decreasing Reynolds number when compared to thicker airfoils due to boundary-layer displacement and adverse pressure-gradient effects (see also Chapter 4). In some cases, additional airfoil camber might also help due to cambered airfoils having $c_l$ versus $c_d$ polars shifted to higher $c_l$ values, resulting (in some cases) in higher $(c_l/c_d)_{max}$.

Figures 8.3 and 8.4 show lift curves ($c_l$ vs. $\alpha$) and airfoil polars ($c_l$ vs. $c_d$) at different Reynolds numbers for the NACA 64618 and NACA 4412 airfoils. All airfoil data were generated using the Xfoil code (Drela 1989) where a critical amplification factor of $N_{crit} = 9$ was used at FS Reynolds number and $N_{crit} = 7$ for all lower Reynolds numbers.

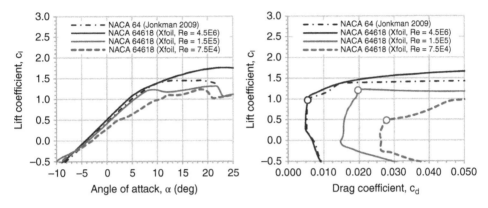

**Figure 8.2** Scaling example – NACA 64618 (FS tip airfoil) and NACA 4412 (MS airfoil).

**Figure 8.3** *Re*-dependence of airfoil characteristics (NACA 64618) – circle denotes point of $(c_l/c_d)_{max}$.

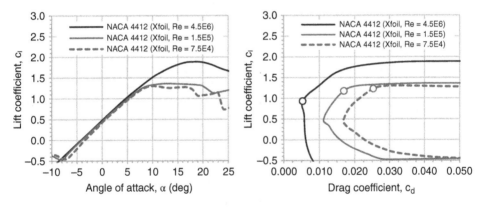

**Figure 8.4** *Re*-dependence of airfoil characteristics (NACA 4412) – circle denotes point of $(c_l/c_d)_{max}$.

Here, FS Reynolds number as well as $1/30^{th}$ and $1/60^{th}$ of FS Reynolds number are considered. It becomes apparent that the original NACA 64618 airfoil shows a significant reduction in $(c_l/c_d)_{max}$ with Reynolds number, therefore not being an appropriate airfoil for any part of the MS turbine blade. Furthermore, being 18% thick, the NACA 64618 airfoil experiences a notable reduction in lift-curve slope, $dc_l/d\alpha$, at the lowest Reynolds number, which is attributed to a large "decambering effect" (see also Section 4.1.2.3) of a thick airfoil at decreasing Reynolds number.

Fortunately, the situation is a bit different for the NACA 4412 airfoil in Figure 8.4. Here the lift-curve slope, $dc_l/d\alpha$, is practically insensitive to significant reductions in FS Reynolds number. Consequently, the maximum lift-to-drag ratio, $(c_l/c_d)_{max}$, does not reduce as much as for the NACA 64618 airfoil in Figure 8.3. Furthermore, it can be seen in Figure 8.4 for the NACA 4412 airfoil that the shape of the $c_l$ versus $c_d$ polar at the lower-than-FS Reynolds numbers is such that $(c_l/c_d) \approx (c_l/c_d)_{max}$ over a range of $c_l$ from $0.5 - 1.2$, which is also helpful in designing a well-performing MS turbine blade.

Before actually designing the MS turbine blade, the baseline FS case of the NREL 5-MW turbine is introduced using the *XTurb* code. The associated *XTurb* input file is *NREL-5MW-Ch8-2-2.inp*. The output file *XTurb_Output.dat* reads:

---

**XTurb Example 8.1  NREL 5-MW – XTurb_Output.dat (NREL-5MW-Ch8-2-2.inp)**

```
8.2.2 - NREL 5MW                        ***** XTurb V1.9    -   OUTPUT *****
   Blade Number        BN =  3


                                    +     BLADE ELEMENT MOMENTUM THEORY (BEMT)    +
PREDICTION
   Blade Radius       BRADIUS   =   63. [m]
   Air Density        RHOAIR    =   1.225 [kg/m**3]
   Air Dyn. Visc.     MUAIR     =   0.000018 [kg/(m*s)]
   Number of Cases    NPRE      =   1

   Thrust             T = 0.5*RHOAIR*VWIND**2.*(pi*BRADIUS**2.)*CT
   Power              P = 0.5*RHOAIR*VWIND**3.*(pi*BRADIUS**2.)*CP
   Torque             TO = P / RPM
   Bending Moment     BE = 0.5*RHOAIR*VWIND**2.*(pi*BRADIUS**2.)*BRADIUS*CB

   Number  VWIND[m/s] RPM[1/min] TSR     PITCH[deg]   CT       CP       CB       ...
     1     8.0000     9.1560     7.5507  0.0000       0.7842  -0.4835  -0.1775   ...
```

---

Here, coefficients of thrust ($C_T$), power ($C_P$), and root-flap bending moment ($C_B$) are integrated performance parameters that will be compared to those of MS turbine blade designs presented in the next subsections. In this context, it is important to understand that "The design space is infinite." This means in particular that a given non-dimensional radial circulation distribution, $\Gamma(r')/(V_0 R)$, see *XTurb_Output_Circulation.dat*, can be realized with an infinite number of blade designs. For a circulation-scaled rotor of the same axial and angular induction properties, Eq. (8.10) suggests that $c_l c = const.$ between FS and MS blades and that either the blade pitch, $\beta(r')$, or angle of attack, $\alpha(r')$, distribution can be determined from the FS blade flow angle, $\phi(r')$, and the respective other known parameter using Eq. (8.13). The following subsections provide some examples of circulation-scaled blade designs.

### 8.2.2.1  Scaled Design with Given $c_l$ (Lift Coefficient) Distribution (Scaled NREL 5-MW)

A circulation-scaled MS blade of the NREL 5-MW turbine is designed starting from a given (or desired) radial distribution of the lift coefficient, $c_{l, MS}(r')$. For simplicity and with reference to Figure 8.4 for the NACA 4412 airfoil, a constant $c_{l, MS}$ of 0.78 is chosen along the entire MS blade, resulting in a lift-to-drag ratio of $(c_{l, MS}/c_{d, MS}) \approx 60$ at MS Reynolds number. The MS blade design for chord, $c_{MS}(r')$, and pitch, $\beta_{MS}(r')$, is realized following:

**Find: Rotor Blade $c_{MS}(r')$ and $\beta_{MS}(r')$**

- For given lift coefficient $c_{l, MS}(r')$, at each $r'$ ...
- Compute $c_{MS}$ from $c_{l, MS} c_{MS} = c_{l, FS} c_{FS}$

- Find $\alpha_{MS}$ from lift curve $c_l(\alpha)$ of local MS airfoil
- Compute $\beta_{MS}$ from $\phi_{MS} = \alpha_{MS} + \beta_{MS}$ (from known $\phi_{MS} = \phi_{FS}$)

With all conditions satisfied, a circulation-scaled MS blade of the NREL 5-MW turbine is obtained. An example using the *XTurb* input file *Scaled-NREL-5MW-Ch8-2-2.inp* can be run with the following output (*XTurb_Output.dat*):

---

**XTurb Example 8.2 Scaled NREL 5-MW – XTurb_Output.dat**
**(Scaled-NREL-5MW-Ch8-2-2.inp)**

```
8.2.2 - Scaled NREL 5MW            ***** XTurb V1.9    -  OUTPUT *****
     Blade Number       BN =   3


                            +     BLADE ELEMENT MOMENTUM THEORY (BEMT)    +
PREDICTION
     Blade Radius     BRADIUS   =   0.63 [m]
     Air Density      RHOAIR    =   1.225 [kg/m**3]
     Air Dyn. Visc.   MUAIR     =   0.000018 [kg/(m*s)]
     Number of Cases  NPRE      =   1

     Thrust           T  = 0.5*RHOAIR*VWIND**2.*(pi*BRADIUS**2.)*CT
     Power            P  = 0.5*RHOAIR*VWIND**3.*(pi*BRADIUS**2.)*CP
     Torque           TO = P / RPM
     Bending Moment   BE = 0.5*RHOAIR*VWIND**2.*(pi*BRADIUS**2.)*BRADIUS*CB

     Number  VWIND[m/s] RPM[1/min]  TSR     PITCH[deg]   CT      CP       CB      ...
       1    13.0000    1488.5000   7.5540   0.3410      0.7848  -0.4450  -0.1776  ...
...
```

---

Comparison against the baseline NREL 5-MW turbine reveals that both thrust and root-flap bending moment coefficients, $C_T$ and $C_B$, are matched very well, while the reduced airfoil $(c_l/c_d)_{max}$ results in a reduced power coefficient, $C_P$, of about 0.04 counts (absolute).

### 8.2.2.2 Scaled Design with Given c (Chord) Distribution (PScaled NREL 5-MW)

Another approach of designing a circulation-scaled MS blade of the NREL 5-MW turbine consists of specifying a radial chord distribution, $c_{MS}(r')$. A simple example is that of keeping it the same as the radial chord distribution of the original FS blade, $c_{FS}(r')$. The radial pitch distribution, $\beta_{MS}(r')$, of the MS blade can then be found following:

**Find: Rotor Blade $\beta_{MS}(r')$**

- For given blade chord $c_{MS}(r')$, at each $r'$ ...
- Set $c_{l,MS} = c_{l,FS}$ (from $c_{l,MS}c_{MS} = c_{l,FS}c_{FS}$)
- Find $\alpha_{MS}$ from lift curve $c_l(\alpha)$ of local MS airfoil
- Compute $\beta_{MS}$ from $\phi_{MS} = \alpha_{MS} + \beta_{MS}$ (from known $\phi_{MS} = \phi_{FS}$)

Here it is important to understand that for a circulation-scaled rotor with the same chord distribution between FS and MS blades, the lift coefficient distribution, $c_{l,MS}(r')$ also remains unchanged; however, the corresponding angle-of-attack distribution, $\alpha_{MS}(r')$, is likely different as the MS airfoil(s) may have different camber and are further subject to "viscous decambering" as a function of Reynolds number and angle of attack itself. Consequently, a circulation-scaled MS blade of the same FS planform shape has to be twisted differently. An associated example using the *XTurb* input file is *PScaled-NREL-5MW-Ch8-2-2.inp* and can be run with the following output (*XTurb_Output.dat*):

---

**XTurb Example 8.3  PScaled NREL 5-MW – XTurb_Output.dat**
**(PScaled-NREL-5MW-Ch8-2-2.inp)**

```
8.2.2 - PScaled NREL 5MW              ***** XTurb V1.9    -  OUTPUT *****
   Blade Number       BN =  3

                            +    BLADE ELEMENT MOMENTUM THEORY (BEMT)    +
PREDICTION
Blade Radius       BRADIUS   =   0.63 [m]
Air Density        RHOAIR    =   1.225 [kg/m**3]
Air Dyn. Visc.     MUAIR     =   0.000018 [kg/(m*s)]
Number of Cases    NPRE      =   1

Thrust             T = 0.5*RHOAIR*VWIND**2.*(pi*BRADIUS**2.)*CT
Power              P = 0.5*RHOAIR*VWIND**3.*(pi*BRADIUS**2.)*CP
Torque            TO = P / RPM
Bending Moment    BE = 0.5*RHOAIR*VWIND**2.*(pi*BRADIUS**2.)*BRADIUS*CB

Number  VWIND[m/s] RPM[1/min] TSR     PITCH[deg]   CT      CP       CB      ...
   1   13.0000    1488.5000   7.5540  -0.0605     0.7837  -0.4497  -0.1774  ...
...
```

---

Comparison against the baseline NREL 5-MW rotor thrust and root-flap bending moment coefficients, $C_T$ and $C_B$, demonstrates that both can be satisfied very well in practice. As for the rotor power coefficient, $C_P$, an associated absolute error on the order of 0.04 is a result of significantly lower MS Reynolds numbers along the blades and associated reduction in airfoil $(c_l/c_d)_{max}$.

### 8.2.2.3  Scaled Design with Given $\beta$ (Pitch/Twist) Distribution (TScaled NREL 5-MW)

A final example of designing a circulation-scaled MS blade of the NREL 5-MW turbine starts with a given blade pitch distribution, $\beta_{MS}(r')$. The most straightforward example here is setting $\beta_{MS}(r')$ equal to the pitch distribution of the FS blade, $\beta_{FS}(r')$. The associated radial chord distribution of the MS blade, $c_{MS}(r')$, for a circulation-scaled rotor can be found following:

**Find: Rotor Blade $c_{MS}(r')$**

- For given blade pitch/twist $\beta_{MS}(r')$, at each $r'$ ...
- Compute $\alpha_{MS}$ from $\phi_{MS} = \alpha_{MS} + \beta_{MS}$ (from known $\phi_{MS} = \phi_{FS}$)
- Find $c_{l,MS}$ from lift curve $c_l(\alpha)$ of local MS airfoil
- Compute $c_{MS}$ from $c_{l,MS}c_{MS} = c_{l,FS}c_{FS}$

Note that in this case, both the radial distribution of lift coefficient, $c_{l,MS}(r')$, and blade chord, $c_{MS}(r')$, are likely to deviate from the original FS blade due to different airfoil(s) being chosen in the design and the effects of scaled Reynolds numbers. The associated example using the *XTurb* input file *TScaled-NREL-5MW-Ch8-2-2.inp* can be run with the following output (*XTurb_Output.dat*):

```
┌─────────────────────────────────────────────────────────────────────────┐
│ XTurb Example 8.4  TScaled NREL 5-MW – XTurb_Output.dat                   │
│ (TScaled-NREL-5MW-Ch8-2-2.inp)                                            │
├─────────────────────────────────────────────────────────────────────────┤
│                                                                           │
│   8.2.2 - TScaled NREL 5MW            ***** XTurb V1.9    -  OUTPUT *****  │
│     Blade Number        BN =  3                                           │
│                                                                           │
│                                +    BLADE ELEMENT MOMENTUM THEORY (BEMT)  +│
│   PREDICTION                                                              │
│     Blade Radius       BRADIUS   =   0.63 [m]                             │
│     Air Density        RHOAIR    =   1.225 [kg/m**3]                      │
│     Air Dyn. Visc.     MUAIR     =   0.000018 [kg/(m*s)]                  │
│     Number of Cases    NPRE      =   1                                    │
│                                                                           │
│     Thrust             T = 0.5*RHOAIR*VWIND**2.*(pi*BRADIUS**2.)*CT       │
│     Power              P = 0.5*RHOAIR*VWIND**3.*(pi*BRADIUS**2.)*CP       │
│     Torque            TO = P / RPM                                        │
│     Bending Moment    BE = 0.5*RHOAIR*VWIND**2.*(pi*BRADIUS**2.)*BRADIUS*CB│
│                                                                           │
│     Number  VWIND[m/s] RPM[1/min]  TSR    PITCH[deg]   CT      CP     CB  ...│
│        1  13.0000   1488.5000   7.5540   0.0072    0.7845  -0.4487 -0.1775 ...│
│   ...                                                                      │
│                                                                           │
└─────────────────────────────────────────────────────────────────────────┘
```

Once more, it can be seen that rotor thrust and root-flap bending moment coefficients, $C_T$ and $C_B$, of the MS blade match very closely to those of the baseline NREL 5-MW turbine. As for the power coefficient, $C_P$, the same absolute discrepancy/error of about 0.04 is observed similar to the other examples of circulation-scaled MS blades and is attributed to a significantly reduced $(c_l/c_d)_{max}$.

### 8.2.2.4 Differences in Scaled Designs w.r.t. Airfoil Aerodynamics and Blade Loads

The scaled design examples of the preceding sections have been chosen for the sole purpose of demonstrating that "The design space is infinite" indeed. In this context, it is important to realize that one is fortunate enough to be able to match blade circulation as well as rotor thrust and root-flap bending moment coefficients in practice. While matching the power coefficient is desirable, it is not always possible due to notably reduced airfoil performance, particularly $(c_l/c_d)_{max}$, at a significantly reduced (Reynolds) $Re$-scale. In the end, though, some reduction in scaled power coefficient is probably not that important as one might think initially. This is (see also Eqs. (8.14) and (8.15)) because all radial blade loads, trailing vorticity, and rotor thrust and root-flap bending moment coefficients are the primary governing quantities for both blade aerodynamics and wake development and recovery. In this respect, exact matching of the power coefficient becomes secondary and more of a scaled generator/drive-train objective. With respect to scaled wake aerodynamics, it has been demonstrated experimentally (Chamorro et al. 2012) that wake development becomes $Re$-independent for rotor diameter-based Reynolds numbers of approximately $Re_D \geq 5 \times 10^5$ as a guideline (which is actually satisfied for all preceding scaling examples).

Figure 8.5 shows radial distributions of lift coefficient, $c_l$, and non-dimensional blade circulation, $\Gamma/V_0R$, for all scaled design examples and the baseline NREL 5-MW turbine. It can be seen that all designs analyzed with the *XTurb* code do obtain the same scaled circulation distribution despite the disparities in radial lift coefficient and the fact that only one design airfoil was chosen at a 1/62 $Re$ scale.

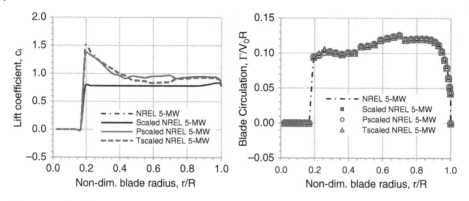

**Figure 8.5** Scaled NREL 5-MW turbines – radial distribution of lift coefficient and blade circulation.

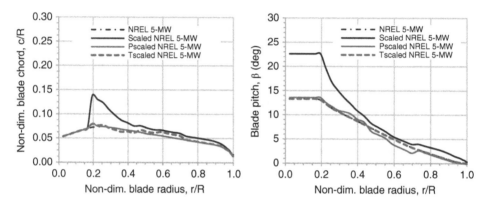

**Figure 8.6** Scaled NREL 5-MW turbines – blade planforms and pitch.

The corresponding scaled rotor blade planforms and pitch distributions are shown in Figure 8.6. Here, the free design starting from a constant $c_l$ distribution (see Section 8.2.2.1) stands out as it results in a higher-solidity blade design in the root region with correspondingly higher blade pitch. This effect could be reduced by prescribing a more gradual increase in design $c_l$ from the blade tip to the blade root, thus leading to lower chord (and solidity) following $c_l c = const.$ and, correspondingly, a higher angle of attack and therefore reduced blade pitch satisfying a prescribed $\phi = \alpha + \beta$.

Resulting normal-/tangential force parameters, $c_n \cdot c/R$ and $c_t \cdot c/R$, are shown in Figure 8.7. Note that these (aerodynamic) force parameters can be of interest with respect to dynamic scaling and are obtained by rotating the force triangle of lift-/drag force parameters, $c_l \cdot c/R$ and $c_d \cdot c/R$, by the local angle of attack at the respective blade section (see also Figure 3.3). In Figure 8.7, it becomes apparent that matching $c_n \cdot c/R$ is a mere result of the circulation-scaled methodology, while matching $c_t \cdot c/R$ can be only obtained in practice, if either the blade planform (chord) or blade pitch distribution is the same between MS and FS blades. In any case, tangential forces are an order of magnitude smaller than their normal counterparts and may therefore not be as important in scaled aerodynamic/dynamic turbine blades.

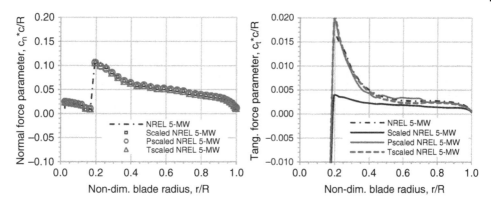

**Figure 8.7** Scaled NREL 5-MW turbines – radial distribution of normal and tangential force parameter, $c_n c/R$ and $c_t c/R$.

The scaled design examples presented in this section can be thought of as a starting point (or Iteration 1) in designing a MS test turbine of a large utility-scale baseline turbine. In fact, blade planform and pitch distributions in Figure 8.6 appear a bit "bumpy" and would have to be smoothened in order to be fully practical. Nevertheless, the interested reader is encouraged to explore further the aerodynamic design space of MS turbines. In addition, one has to be aware that exact sectional (Reynolds) *Re*-effects have not been taken into account. In fact, only one airfoil polar file was used along the entire blade. In this respect, the free design starting from a constant $c_l$ distribution (see Section 8.2.2.1) may be actually advantageous due to increased Reynolds numbers in the root region. Furthermore, the effects of scaled rotational augmentation effects (Section 5.1.4) have not been considered, neither in the design of the circulation-scaled blades nor in the performance analyses.

### 8.2.3 Model-Scale Wind Turbine Aerodynamics Experiments

Model-scale (MS) aerodynamic testing of wind turbine rotors in controlled test conditions is highly valuable with respect to answering basic research questions in aerodynamics, verifying the performance of (scaled) blade designs, dissecting the flow physics of rotational augmentation and stall delay, and (among others) providing unique validation data for computational methods at all levels of fidelity. It is correct to say that MS testing of wind turbine rotors has resulted in improved physical models of wind turbine aerodynamics, airfoil design, and rotor performance. Nevertheless, it is also fair to say that not all "mysteries" have been solved; examples include the persistent issues in predicting blade tip loads and rotational effects on the blade boundary layer in the root region. Therefore, MS testing continues to be a basic research need, with the end goal of advancing wind energy as a competitive and renewable source of energy.

In this context, one of the first MS horizontal-axis wind turbine rotor tests was conducted between 1986 and 1992 as part of a collaboration between Sweden and China (Dexin and Thor 1993; Ronsten 1992, 1994; Björck et al. 1995). The blades of the rotor ($D = 5.3$ m) were equipped with NACA 44## airfoils, ranging between 14 and 22% relative thickness. It is worth noting that one of the primary findings of this test campaign was the confirmation of rotational augmentation and stall delay effects on (scaled)

NREL Phase VI (D = 10.1 m)     MEXICO (D = 4.5 m)     Krogstad (D = 0.9 m)

(a)                            (b)                    (c)

**Figure 8.8** Examples of MS wind turbines. (a) NREL Phase VI, Source: Courtesy of Lee-Jay Fingersh, NREL 55064; (b) Mexico, Picture provided by the consortium that carried out the EU FP5 project Mexico: Model rotor EXperiments In COntrolled conditions; and (c) Krogstad, Photo by Per-Åge Krogstad. Published with permission.

rotating wind turbine blades. Since then, other MS tests have been designed and conducted in the wind energy science community. Today's most prominent MS wind turbine rotor tests used for validation studies of computational methods at all levels of fidelity are shown in Figure 8.8.

In this context, it is important to understand that MS testing is performed at similar wind (tunnel) speeds, $V_0$, when compared to FS operation; this is merely a result as a consequence of $Ma$ scaling (see e.g. Table 8.5). The only possibility to overcome this fundamental $Re$-scale issue for MS blades is to lower (somehow) the kinematic viscosity, $\nu_{MS}$, of the test fluid, so that higher sectional Reynolds numbers, $Re_c$, can be obtained and the total $Re$ scale and associated reduced airfoil performance be made less severe than the actual geometric scale. One approach is using a high-pressure test facility (Miller et al. 2017) as available at Princeton University or Göttingen (Germany); another possibility is testing in water rather than air (Reich et al. 2014) as water has about $1/15^{th}$ the kinematic viscosity of air.

The following subsections provide some additional information on the respective MS wind-tunnel tests of the NREL Phase VI, MEXICO, and Krogstad experiments. Examples using the *XTurb* code (Schmitz 2012) supplement the presentation and provide the interested reader with resources to conduct independent studies.

### 8.2.3.1 NREL Phase VI Rotor

The NREL Phase VI rotor ($D = 10$ m) experiment was conducted by the NREL in the NASA Ames $80 \times 120$ ft Wind Tunnel. The primary reference report concerning the Unsteady Aerodynamics Experiment (UAE) VI (Hand et al. 2001) describes details on the experimental design, setup, and data measurement campaigns. In total, five radial stations ($r/R = 0.30, 0.47, 0.63, 0.80, 0.95$) were equipped with chordwise pressure transducers, measuring respective normal and tangential forces at these radial stations (see also Figure 3.3). In addition, five-hole probes were installed in the leading-edge region at the same radial stations that allowed measuring the local relative velocity magnitude, $V_{rel}$, and blade flow angle, $\phi$. The two-bladed NREL Phase VI rotor was designed as a stall-controlled turbine exclusively equipped with the S809 airfoil. This airfoil has ever since challenged both experimentalists and computational modelers to predict its performance characteristics in steady/unsteady flow as well as dynamic stall conditions. Some examples are wind-tunnel tests conducted at Delft University of Technology (Somers 1997), the Ohio State University (Ramsay et al. 1995), and the University of Glasgow (Sheng et al. 2006a,b,c). Initial "Blind Comparison" efforts comparing measured data against performance codes (blade-element and vortex methods), aeroelastic models, and Computational Fluid Dynamics (CFD) simulations revealed a number of shortcomings in the predictive capability of computational modeling tools (Simms et al. 2001), even in the case of fully attached flow along the blade. Since then, the NREL Phase VI rotor has served as a validation database for many computational studies involving performance and aeroelastic codes as well as CFD methods.

An example analysis of the NREL Phase VI using the *XTurb* code is shown next for the input file *NREL_VI_J-Seq-Ch8-2-3.inp*. Radial distributions of normal and tangential force coefficients, $c_n$ and $c_t$, are documented in the corresponding *XTurb_Output2.dat* file:

---

**XTurb Example 8.5  NREL_VI (J-Seq) – XTurb_Output2.dat (NREL_VI_J-Seq-Ch8-2-3.inp)**

```
8.2.3 - NREL-PhaseVI-J              ***** XTurb V1.9    -  OUTPUT 2 *****
   Blade Number       BN =  2

   Solidity
   0.0519
                              +    BLADE ELEMENT MOMENTUM THEORY (BEMT)    +

   PREDICTION
          VWIND [m/s] RPM [1/min]   PITCH [deg]
          7.0000       72.0000      6.0000

   Number    TSR       PITCH [deg]      CT       CP       CPV      CB       CBV
      2    5.4168      6.0000        0.3561   -0.2635  0.0319  -0.1198  -0.0012

   "Thrust"   >   0    =>     Downwind Direction
   "Torque"   <   0    =>     Energy Extraction (Wind Turbine)   =>  Power = Torque * TSR
   "Bending"  <   0    =>     Flap Bending towards Downwind

   r/R     Chord/R Twist[deg] AOA[deg]  CL       CD     ...     ...      CNormal  CTangen
   ─────────────────────────────────────────────────────────────────────────────────────
   0.2500  0.1465  28.0230   0.0620   0.1502   0.0135 ...     ...      0.1502  -0.0133
   0.2512  0.1464  27.8893   0.0642   0.1504   0.0135 ...     ...      0.1504  -0.0133
   0.2546  0.1460  27.4891   0.8825   0.2455   0.0135 ...     ...      0.2457  -0.0097
    ...     ...     ...                ...      ...                      ...
 ...
```

---

Figure 8.9 shows a quantitative comparison between measured and *XTurb* computed normal and tangential force coefficients. Here the *XTurb* code was run in BEMT mode, using either the standard Prandtl/Glauert tip-loss factor ($TLOSS = 1$) or the tip-loss factor derived from free-wake computations, see Section 3.4.4.4 ($TLOSS = 2$).

The NREL Phase VI rotor is a good example of a common observation concerning the predictive capability of blade tip loads by BEM-type methods. It can be seen in Figure 8.9 that the standard Prandtl/Glauert tip-loss factor ($TLOSS = 1$) results in an overprediction of the normal force coefficient, $c_n$, in the tip region, while an adjusted tip-loss factor ($TLOSS = 2$) shows improved results. As far as the radial distribution of tangential force coefficients, $c_t$, in Figure 8.9 is concerned, the distribution itself is captured well; however, an absolute error seems to exist. This is also not an uncommon

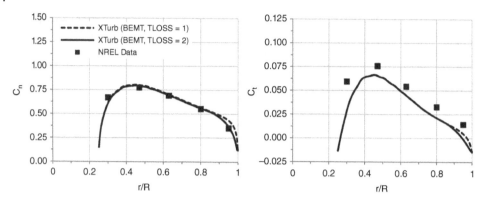

**Figure 8.9** NREL phase VI rotor – normal and tangential force coefficients, $c_n$ and $c_t$ (J-Sequence, $V_0 = 7$ m/s, rpm $= 72$, $\beta_{Tip} = 6°$).

observation, and it should be noted in this context that tangential force coefficients are typically an order of magnitude smaller than normal force coefficients. Note that normal and tangential force coefficients at a given radial station are integrated results from pressure-transducer data; now the chordwise distribution of (20–40) pressure transducers is typically optimized to integrate for normal force coefficients, while that same distribution may be suboptimal for respective tangential force coefficients that are very sensitive to sparse integration data. This issue is known in the rotorcraft community and can explain some of the persistent difficulties in predicting tangential force coefficients (Schmitz et al. 2009).

### 8.2.3.2 MEXICO Rotor

The original MEXICO experiment was conducted in 2006 in the open test section of the DNW-LLF $9.5 \times 9.5$ m Wind Tunnel (Schepers and Snel 2007). The MEXICO rotor ($D = 4.5$ m) was designed as a three-bladed rotor equipped with three different airfoils from root to tip. that is, the DU91-W2-250 airfoil, the Risø A1-21 airfoil, and the NACA64-418 airfoil. The MEXICO rotor had a total of five radial stations ($r/R = 0.25$, 0.35, 0.60, 0.82, 0.92) equipped with chordwise pressure transducers. Additional data campaigns are the MexNext experiment (Schepers et al. 2012, 2014) and the NewMEXICO campaign (Boorsma and Schepers 2014), both of which provided additional unique validation datasets for computational methods, including particle-image velocimetry (PIV) data.

An example *XTurb* analysis using input file *MEXICO-Ch8-2-3.inp* is shown next. Radial distributions of normal and tangential forces (per unit length) [N m$^{-1}$], $F_n$ and $F_t$, are documented in the *XTurb_Output_PREDICTION.dat* file:

---

**XTurb Example 8.6 MEXICO – XTurb_Output_PREDICTION.dat (MEXICO-Ch8-2-3.inp)**

```
2.3 - MEXICO               ***** XTurb V1.9    -   PREDICTION *****
    Blade Number        BN =   3

    Solidity
```

```
0.0569
                       +    BLADE ELEMENT MOMENTUM THEORY (BEMT)    +

PREDICTION
    VWIND [m/s] RPM [1/min]  PITCH [deg]
    10.0000       424.5000     -2.3000

Blade Radius      BRADIUS   =   2.25 [m]
Air Density       RHOAIR    =   1.225 [kg/m**3]
Air Dyn. Visc.    MUAIR     =   0.000018 [kg/(m*s)]

Number  VWIND[m/s] RPM[1/min] TSR     PITCH[deg] ...    T[N]        P[W]       TO[Nm]    BE[Nm]
    1   10.0000    424.5000   10.0020 -2.3000    ...  1005.9405 -2896.9759  -65.1686  -522.0769

   "Thrust"   >  0      =>     Downwind Direction
   "Torque"   <  0      =>     Energy Extraction (Wind Turbine)   =>   Power = Torque * TSR
   "Bending"  <  0      =>     Flap Bending towards Downwind

   Thrust            T  = 0.5*RHOAIR*VWIND**2.*(pi*BRADIUS**2.)*CT
   Power             P  = 0.5*RHOAIR*VWIND**3.*(pi*BRADIUS**2.)*CP
   Torque            TO = P / RPM
   Bending Moment    BE = 0.5*RHOAIR*VWIND**2.*(pi*BRADIUS**2.)*BRADIUS*CB

   Reynolds number Re = RHOAIR*VREL*Chord/MUAIR
   Thrust/m          Fth = 0.5*RHOAIR*VREL**2.*Chord*CThrust
   Torque/m          Fto = 0.5*RHOAIR*VREL**2.*Chord*CTorque*BRADIUS
   Fnormal/m         Fno = 0.5*RHOAIR*VREL**2.*Chord*CNormal
   Ftangen/m         Fta = 0.5*RHOAIR*VREL**2.*Chord*CTangen
   Bending/m         Fbe = 0.5*RHOAIR*VREL**2.*Chord*CBending*BRADIUS

   r[m]   Chord[m] Twist[deg] AOA[deg]  ...   Fno[N/m]    Fta[N/m]     Fbe[Nm/m]

   0.2092  0.1950  -2.3000    24.3888   ...    8.3155     -18.3404     -1.5846
   0.2124  0.1950  -2.3000    24.8968   ...    8.7257     -18.8001     -1.6916
   0.2218  0.1950  -2.3000    47.7021   ...    8.3272      -7.5763     -1.7782
   ...     ...                                   ...                      ...
```

Figure 8.10 shows quantitative comparisons between measured (New MEXICO) and *XTurb* computed results. In general, good comparisons are obtained using *XTurb* in BEMT mode (and Du/Selig stall delay) that are not too different from for example, full CFD analyses as presented in Sørensen et al. (2016) using the EllipSys3D code. Here the $r/R = 0.82$ station in particular appears to be difficult to predict in terms of the sectional normal force, $F_n$. As far as the sectional tangential forces, $F_t$, in Figure 8.10 are concerned, it is apparent that rotational augmentation effects are not well captured at $V_0 = 24$ m s$^{-1}$ ($\lambda = 4.2$) for the inboard part of the blade.

The interested reader is encouraged to conduct her/his own investigations of the MEXICO rotor. The *XTurb* example presented in this section is just a first step to answer some of the remaining questions about the MEXICO rotor. Note that some airfoil data of the MEXICO rotor are proprietary; hence a tentative user needs to obtain an appropriate agreement with the respective agencies.

### 8.2.3.3 Krogstad Turbine

The Krogtad turbine ($D = 0.9$ m) is a three-bladed MS wind turbine tested in the Norwegian University of Science and Technology (NTNU) $2.7 \times 2.0$ m Wind Tunnel (Krogstad and Lund 2012). The Krogstad turbine is exclusively equipped with the S826 airfoil (Somers 2005). Primary measurements included rotor thrust and power

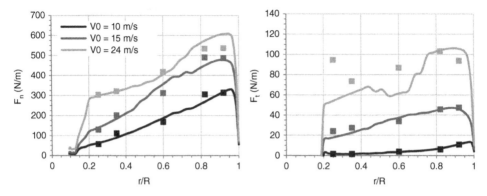

**Figure 8.10** New MEXICO rotor – normal and tangential force, $F_n$ and $F_t$ ($RPM = 424.5$, $\beta_{Tip} = -2.3°$).

coefficients, $C_T$ and $C_P$. The small rotor size did not allow for the installation of pressure transducers; however, a variety of wake flow diagnostics were performed, and the interactional effects of an upstream turbine wake on the performance of a downstream turbine was quantified (Adaramola and Krogstad 2010). Measured data on the Krogstad turbine(s) provide another unique dataset for ongoing computational model validation efforts.

For MS turbines with rotor diameters $D = O(1\ \text{m})$ such as the Krogstad turbine, it is instrumental that computational performance codes use airfoil tables that take into account the radial variation of $Re$-numbers and its effect on airfoil characteristics, specifically $(c_l/c_d)_{max}$. This is particularly important when part of the blade is operating at chord-based Reynolds numbers of $Re_c \le 2 \times 10^5$. Let us illustrate this by means of a suitable *XTurb* example where a previous example (see Section 6.2.6.1) is rerun using only one airfoil polar along the blade span whose Reynolds number is actually too small for the high-$\lambda$ cases. An excerpt from the corresponding *XTurb* input file *Krogstad_Re1x10E5-Ch8-2-3.inp* is shown next:

---

**XTurb Example 8.7  Krogstad – (Krogstad_Re1x10E5–Ch8-2-3.inp)**

```
&BLADE
    Name       = '8.2.3 - Krogstad',

    BN         = 3,

    ROOT       = 0.15000,
      ...              ...

      NAIRF    = 1,

      RAIRF    = 0.15000,

      AIRFDATA = './S826_NTNU_Re1x10E5.polar',

      ...              ...
```

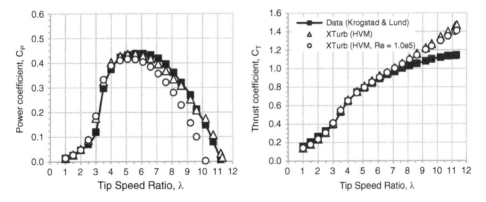

**Figure 8.11** Krogstad turbine – rotor power and thrust coefficients, $C_P$ and $C_T$ ($V_0 = 10$ m s$^{-1}$, $\beta_{Tip} = -1.62°$).

Results for rotor power and thrust coefficients, $C_P$ and $C_T$, obtained by *XTurb* in Helicoidal Vortex Model (HVM) mode are shown in Figure 8.11, with the baseline case being the same as presented in Figure 6.12 (see Section 6.2.6.1). It can be seen that compared to the baseline HVM case (triangle symbols), the $C_P$ versus $\lambda$ curve resulting from using exclusively one airfoil polar at $Re = 1 \times 10^5$ deviates notably for $\lambda > 5$. Indeed, this is approximately the condition at which parts of the Krogstad turbine blade start operating at $Re_c > 1 \times 10^5$. As airfoil $(c_l/c_d)_{max}$ typically decreases significantly below Reynolds numbers of approximately $2 \times 10^5$ (unless the airfoil was specifically designed for that $Re$ regime), one can also expect a notable reduction in power coefficient, $C_P$. Here, Figure 8.11 shows exactly this behavior.

As far as the thrust coefficient, $C_T$, in Figure 8.11 is concerned, there is only a small effect, which is attributed to the lift-curve slope staying approximately constant. The primary objective of using the *XTurb* example previously is to demonstrate that small-scale wind turbine testing has its own challenges associated with airfoil data, resulting in uncertainties in performance predictions that behave differently compared to larger MS turbines. The interested reader may explore the uncertainty associated with using *Xfoil* or CFD generated airfoil polars versus airfoil polars of the S826 airfoil measured in different facilities. In this context, an early summary of blind comparison runs (Krogstad and Eriksen 2013) applied to the Krogstad turbine is an excellent reference for further investigations.

## 8.3 Aerodynamic Optimization of Wind Turbine Blades

The design and optimization of wind turbines involves a number of disciplines, including aerodynamics, dynamics, structural integrity, manufacturability, transportation, operations and maintenance, and (of course) ultimately COE. Here, we acknowledge that a pure aerodynamic optimization is not (though close in most cases) the absolute optimum within a multi-disciplinary design space. However, we will address this issue to some extent by introducing a rotor thrust constraint on sample rotor designs, with the intent to compare rotor designs in a consistent manner and to include some awareness of the highest integrated rotor load, that is, combined rotor thrust, and its effects on both rotor performance and rotor solidity (and hence weight). The following sections are organized to reflect theory and methodology in both BEM methods as introduced in Chapter 3 and VWMs, see Chapter 6. The beauty of BEM theory is reflected in being able to sequentially add the effects of wake rotation, profile drag, and root-/tip losses to an otherwise (axial) momentum-theory optimized (constant-circulation) rotor design. On the other hand, a VWM allows the computation of an equilibrium wake of minimum induced losses by means of an optimized radial circulation distribution, which by itself offers an infinite design space for chord, twist, and lift coefficient.

### 8.3.1 Principles of Blade Element Momentum (BEM) Aerodynamic Design

The task of inverse blade design in BEM theory is typically started from an assumed optimum radial distribution of axial-/angular induction factors, $a(r')$ and $a'(r')$, along the blade span. A general inverse-design procedure for blade chord, $c(r')$, and blade pitch/twist, $\beta(r')$, can be described as:

**Find: Rotor Blade $c(r')$ and $\beta(r')$**

- Given: Blade number $B$, Tip speed ratio $\lambda$, Design $c_l(r')$
- Determine optimum $a(r')$ and $a'(r')$ from BEM theory
- Compute $\phi$ from $\tan\phi = (1-a)/[\lambda_r(1+a')]$
- Find $\alpha$ from lift curve $c_l(\alpha)$ of local airfoil
- Compute $\beta$ from $\phi = \alpha + \beta$
- Compute $c$ from "blade geometry parameter" $\Xi = f(a, a')$

This methodology can be used as a general blade design procedure where the physical accuracy and performance of the obtained blade design is directly related to the assumed optimum radial distribution of axial-/angular induction factors, $a$ and $a'$. For design purposes, a "blade geometry parameter," $\Xi$, is defined according to:

$$\text{Blade Geometry Parameter}: \quad \Xi = \lambda\sigma'r'c_l \tag{8.16}$$

In Eq. (8.16), $\sigma' = Bc/2\pi r$ is the local solidity according to Eq. (3.18) including the blade chord, $c$, and $r' = r/R$ is the non-dimensional blade radius (or radial coordinate); hence $\sigma'r' = Bc/2\pi R$ is essentially a local solidity weighted by the rotor circumference. Indeed, the blade geometry parameter, $\Xi$, is nothing else but $\lambda_r$ times the "optimum blade parameter" introduced in Section 3.3; however, $\Xi$ is more practical as it does not become singular at the blade root. The following subsections present a step-by-step refinement of both physical accuracy and optimality of the respective $a$ and $a'$ distributions, adding sequentially the effects of rotation, profile drag, and tip losses. As $\Xi \sim c \cdot c_l$ through $\sigma'$, the design methodology can be also modified for (i) given blade chord, $c$, and (ii) given blade pitch, $\beta$, following the methodologies for scaled designs in Section 8.2.2.

#### 8.3.1.1 Betz Optimum Rotor (Ideal Rotor Without Wake Rotation)

Let us begin by recalling the "Betz Optimum Rotor" (or Ideal Rotor without Wake Rotation) from Section 3.3.1 with the corresponding radial distribution of axial-/angular induction factors:

$$\text{Betz Limit}: \quad a = \frac{1}{3} \; ; \; a' = 0 \tag{8.17}$$

Equating the thrust relations in momentum theory and blade-element theory, Eqs. (3.10) and (3.14), under the assumption of zero drag ($c_d = 0$) results in the following relation for the blade geometry parameter:

$$\text{Blade Geometry Parameter (Thrust Eq.)}: \quad \Xi = 2\lambda_r \sin(\phi)\tan(\phi) \tag{8.18}$$

In Eq. (8.18), note that for given axial-/angular induction factors, $a$ and $a'$, the corresponding $\sin(\phi)$ and $\tan(\phi)$ can be evaluated from the local velocity triangle (see Figure 3.3). Furthermore, the local normalized relative velocity, $V_{rel}/V_0$, can be computed using for example, Eq. (8.11) or (3.6). It is a good exercise to show that the blade circulation according to Eq. (8.10) for the Betz Optimum Rotor becomes:

$$\text{Blade Circulation (Betz)}: \quad \frac{\Gamma(r')}{V_0 R} = \frac{1}{2}c_l(r')\frac{V_{rel}(r')}{V_0}\frac{c(r')}{R} = \frac{8}{9}\frac{\pi}{\lambda B} \tag{8.19}$$

In Eq. (8.19), $\lambda$ is the tip speed ratio and $B$ is the blade number. Hence, we have just shown that the Betz Optimum Rotor without wake rotation is indeed a "Constant-Circulation Rotor," that is, there is no wake rotation ($a' = 0$) and $d\Gamma/dr = 0$. Consequently, there is no trailing vorticity into the wake and associated induced losses. In the absence of rotation, profile drag, and tip losses, this means that all Betz Optimum Rotor designs without wake rotation have ideal power and thrust coefficients of $C_P = 16/27$ and $C_T = 8/9$, respectively. The interested reader can easily verify this using the relations in Section 6.2.5.2.

### 8.3.1.2 Effect of Rotation on BEM Optimum Blade Design
For an ideal rotor without profile drag ($c_d = 0$) and root/tip losses ($F = 1$), basic rotor disk theory in Section 2.2.4 resulted in the following optimum axial-/angular induction factors:

$$\text{Optimum (Wake Rotation)}: \quad a(\lambda_r)\text{ from Eq. (2.55)}; \; a' = \frac{1-3a}{4a-1} \tag{8.20}$$

Note that the axial induction factor, $a(\lambda_r)$ in Eq. (2.55) has to be determined using an iterative root-finding method (e.g. Newton iteration), which is not difficult and a good exercise for the reader. Alternatively, a useful analytic approximation is obtained by assuming that $a = 1/3$ along the entire blade radius and revisiting for $a'$ the 'first relation between $a$ and $a''$ from Section 2.2.2 with $a(1-a) = \lambda_r^2 a'(1+a')$. Assuming that $1+a' \approx 1$ for most of the blade, an approximation for optimum induction factors, $a$ and $a'$, reads:

$$\text{Approximate Optimum (Wake Rotation)}: \quad a = \frac{1}{3}; \; a' = \frac{2}{9}\frac{1}{\lambda_r^2} \tag{8.21}$$

Here, it is important to understand that this approximation loses validity for smaller tip speed ratios, $\lambda$, as can be inferred from Figure 2.10. Deriving the "blade geometry parameter," $\Xi$, from the momentum and blade-element thrust equations results in the same relation as Eq. (8.18); however, the addition of wake rotation and subsequent generation of torque calls for $\Xi$ to be derived from the respective torque Eqs. (3.11) and (3.15), still under the assumptions of zero profile drag and no root/tip losses, respectively. The

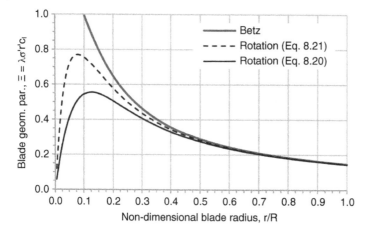

**Figure 8.12** Blade geometry parameter, $\Xi$ – Wake Rotation ($\lambda = 6$).

blade geometry parameter, $\Xi$, then becomes:

Blade Geometry Parameter (Torque Eq.), $\Xi = \lambda \sigma' r' c_l$ :

$$\Xi = \frac{4a'}{1-a} \lambda_r^2 \sin(\varphi) = \frac{4a' \lambda_r^2}{[(1-a)^2 + \lambda_r^2(1+a')^2]^{1/2}} \tag{8.22}$$

Figure 8.12 shows the radial distribution of the optimum "blade geometry parameter," $\Xi$, for the Betz Optimum Rotor and for two designs that include wake rotation, with axial-/angular induction factors from either Eqs. (8.20) or (8.21).

It can be seen that all distributions for $\Xi$ practically agree for the outer half ($r'$, $r/R > 0.50$) of the blade radius. For the inner half ($r'$, $r/R \leq 0.50$), however, the singular behavior of the Betz Optimum Rotor (no wake rotation) is apparent, while both designs including wake rotation result in finite (and hence realistic) solutions. Here, $\Xi$ determined from Eq. (8.20) is considered to be more correct from the standpoint of rotor disk theory. Note that the ratio $\Xi_{max}/\Xi_{min}$ would correspond to the geometric blade taper ratio, if the design lift coefficient were constant along the blade span. In practice, however, higher design lift coefficients are typically chosen at inboard blade stations, see also Chapter 4, resulting in smaller blade chord, $c$, and hence more slender blades, a design tradeoff between efficiency, blade weight, and required blade thickness for structural integrity.

### 8.3.1.3 Effect of Profile Drag on BEM Optimum Blade Design

As a next step, let us remove the assumption of zero drag but still consider no root/tip losses. Equating once more the momentum and blade-element theory torque Eqs. (3.11) and (3.15), results in the following relation for the "blade geometry parameter," $\Xi$:

Blade Geometry Parameter (Torque Eq.), $\Xi = \lambda \sigma' r' c_l$ :

$$\Xi = \frac{4a'}{1-a} \lambda_r^2 \sin(\varphi) \frac{1-a}{(1-a) + (c_d/c_l) \cdot \lambda_r(1+a')} \tag{8.23}$$

Note that Eq. (8.22) is recovered for zero drag ($c_d = 0$) and that Eq. (8.23) is written intentionally in a form where $(1-a)$ could be simply canceled. This is done to illustrate

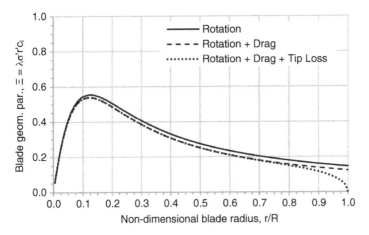

**Figure 8.13** Blade geometry parameter, $\Xi$ – Wake Rotation ($\lambda = 6$) + Profile Drag ($c_d/c_l = 0.02$) + Tip Loss ($B = 3$) – 1.

that the effect of profile drag on $\Xi$ is indeed small for typical airfoil $c_d/c_l$. Figure 8.13 reveals that the effect of profile drag on an optimum blade geometry parameter, $\Xi$, is quite small. Here a constant airfoil lift-to-drag ratio of 50 (or $c_d/c_l = 0.02$) was assumed along the blade span as a representative average across blades of different scales.

Nevertheless, it is important to realize that the assumed axial-/angular induction factors according to Eq. (8.20) were derived under the assumption of zero drag and that consequently, these may not be (and in fact are not, though close to) optimal in the present case of added profile drag (Burton et al. 2001, p. 75).

### 8.3.1.4 Effect of Root-/Tip Loss on BEM Optimum Blade Design

As a final step, the tip-loss factor, $F_T$, is retained in the respective balance of momentum and blade-element theory torque Eqs. (3.11) and (3.15). The resulting expression for the "blade geometry parameter," $\Xi$, is practically equal to Eq. (8.23), except for the $F_T$ factor:

Blade Geometry Parameter (Torque Eq.), $\Xi = \lambda \sigma' r' c_l$ :

$$\Xi = \frac{4 F_T a'}{1 - a} \lambda_r^2 \sin(\phi) \frac{1 - a}{(1 - a) + (c_d/c_l) \cdot \lambda_r (1 + a')} \tag{8.24}$$

Note that including the root loss factor, $F_R$, is not deemed necessary as $\Xi \to 0$ in the root region. Using Eq. (3.38) from Section 3.4.1 and approximating $V_{rel} \approx \Omega r$ in the tip region, lets us write $F_T$ as a direct function of $a$, an expression used in the context of BEM blade design, see for example, Chaviaropoulos and Sieros (2014):

$$\text{Glauert (Tip Loss)} : \quad F_T = \frac{2}{\pi} \cos^{-1} \left[ \exp \left( -\frac{B}{2} \frac{1 - r'}{r'} \frac{1}{\sin \phi(r)} \right) \right]$$

$$\approx \frac{2}{\pi} \cos^{-1} \left[ \exp \left( -\lambda \frac{B}{2} \frac{1 - r'}{1 - a} \right) \right] \tag{8.25}$$

The effect of tip loss on the blade geometry parameter, $\Xi$, is shown in Figure 8.13. It becomes clear that the resulting $\Xi$ is altered starting at the 80% radial station, the result being a rounded tip shape as seen on most turbine and propeller blades. A variation

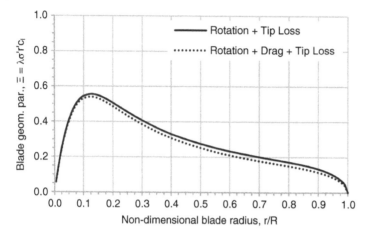

**Figure 8.14** Blade geometry parameter, $\Xi$ – Wake Rotation ($\lambda = 6$) + Profile Drag ($c_d/c_l = 0.02$) + Tip Loss ($B = 3$) – 2.

of the tip effect for an inviscid (i.e. $c_d/c_l = 0$) blade geometry parameter, $\Xi$, is shown in Figure 8.14 along with the viscous ($c_d/c_l = 0.02$) $\Xi$ from Figure 8.13. Here it is interesting to note that both $\Xi$ distributions are very similar in the immediate tip region in the outermost 5% of the blade span. Overall this gives rise to the idea that inviscid blade design may result in a close-to-optimal blade.

Unfortunately, such excitement may disappear instantaneously when remembering that assumed optimal axial-/angular induction factors from Eq. (8.20) stem from rotor disk theory that does not consider individual blades. In fact, induction factors in rotor disk theory (see Chapter 2) are "averages" over the rotor disk area. Tip effects, however, typically lead to locally higher-than-average induction; see for example, Figure 6.13. A first-order approximation would be to consider axial induction factors in the tip region to be $\sim a/F_T$, and the interested reader is referred to section 3.8.5 in Burton et al. (2001) for a formal analysis of this aspect.

### 8.3.1.5 Limitations of BEM Aerodynamic Optimization

One has to be always aware of the assumptions underlying a given exact theory. The preceding sections on the optimum blade geometry parameter, $\Xi$, obtained from BEM theory are a good starting point, though have to be taken with caution when used within inverse blade design methods. This is because assumed optimal "average" axial-/angular induction factors are typically rooted in rotor disk theory, assuming zero drag and an infinite number of blades. The danger here is that multi-disciplinary design methods may actually exploit a weakness of the analysis model/method itself when not all physics are included. The result can be less-than-optimal solutions that are not easily identifiable as such.

At least the following example shows that using the inverse-design methodology outlined earlier in this section results in the expected/assumed distribution of axial-/angular induction factors. Here we choose an ideal rotor with wake rotation and tip loss, following the respective $\Xi$ distribution in Figure 8.14. The corresponding *XTurb* input file is *IdealRotorWithRotation3.inp* (with $RLOSS = 0$, $TLOSS = 1$) with a 10% root cutout. Resulting *XTurb* integrated rotor performance parameters become:

**XTurb Example 8.8 Ideal Rotor with Rotation and Tip Loss (IdealRotorWithRotation3.inp)**

```
8.3.1 - Ideal Rot. w/ Rotat. 3     ***** XTurb V1.9     -  OUTPUT *****
   Blade Number       BN =  3

                              +     BLADE ELEMENT MOMENTUM THEORY (BEMT)     +

   Number   TSR    PITCH [deg]      CT       CP      CPV      CB       CBV
      1    6.0000    0.0080       0.7912  -0.5162   0.0000  -0.1702   0.0000
```

It is interesting to note that the power coefficient, $C_P$, obtained by *XTurb* is very similar compared to the example presented in Section 3.4.2 where the blade planform was designed based on the respective thrust equations in momentum and blade-element theories. In the present case for the design being based on the torque equations, however, both thrust and root-flap bending moment coefficients, $C_T$, and $C_B$, are a bit lower. This is attributed to the rounded blade tip shape. The interested reader can indeed verify in the corresponding *XTurb_Output_Method.dat* that computed axial-/angular induction factors are in fact quite close to the assumed (optimal) values.

Another question that may arise is that of designing rotors for thrust coefficients, $C_T$, lower than 0.8, with the objective of reducing root-flap bending moment coefficients, $C_B$. In this context, the question of optimal axial-/angular induction factors, $a$ and $a'$, under a thrust and/or bending moment constraint is of fundamental interest to future concepts such as the "Low Induction Rotor" (LIR), see Chaviaropoulos et al. (2013). A LIR assumes lower average axial induction (and hence rotor thrust and power), though with a tentatively reduced LCOE. A preliminary design example is given in Section 8.3.2.4. More fundamental research is needed in this area, and the interested reader is once more encouraged to take action. In this context, an additional and well-established resource is the PROPID code (Selig and Tangler 1995) as another example of BEM-based inverse blade design.

### 8.3.2 Principles of VWM Aerodynamic Design

The task of inverse blade design in VWM theory is typically approached from an optimum normalized radial circulation distribution, $\Gamma/(V_0 R)$, obtained either from assumptions of optimum rotor loading or computed by a fixed-/free- wake methodology. A general inverse-design procedure for blade chord, $c(r')$, and blade pitch, $\beta(r')$, can be described as:

**Find: Rotor Blade $c(r')$ and $\beta(r')$**

- Given: Blade number $B$, Tip speed ratio $\lambda$, Design $c_l(r')$
- Determine VWM optimum circulation $\Gamma(r')/(V_0 R)$
- Compute $a = -u_i/V_0$ and $a' = +w_i/(V_0 \lambda_r)$
- Compute $V_{rel}/V_0 = f(a, a', \lambda_r)$
- Find $c$ from $c_l\, c/R = 2\Gamma/V_{rel}$
- Compute $\phi$ from $\tan\phi = (1-a)/[\lambda_r(1+a')]$
- Find $\beta$ from $\phi = \alpha + \beta$
- Compute "blade geometry parameter" $\Xi = \lambda\sigma' r' c_l$

This methodology can be used as a general blade design procedure starting from an optimum $\Gamma/(V_0 R)$. In this context, it is again important to understand that the design space is infinite. If radial distributions of either $c_l$, $c$, or $\beta$ are given, the remaining two variables can be determined by adjusting this methodology, similar to what was presented for the circulation-scaled rotor in Section 8.2.2. The following section introduces a vortex wake methodology developed by Chattot (2003, 2006) that computes an optimum circulation distribution (and hence derived axial-/angular induction factors) under a thrust constraint, a method that is applied to rotor design examples in subsequent sections. In this context, the works of Lee (2015) and Rosenberg and Sharma (2016) are also of interest to the reader.

### 8.3.2.1 Optimum Circulation Distribution Under Thrust Constraint

Let us begin by recalling the relationship between vortex theory and blade-element theory (Section 6.2.5), particularly the relations for thrust/power coefficients, $C_T$ and $C_P$:

$$\text{Thrust Coefficient}: \quad C_T = \frac{T}{\frac{1}{2}\rho V_0^2 A}$$

$$= \frac{B}{\pi} \int_{r_{Root}/R}^{1} \frac{2\Gamma}{V_0 R} \left[ \left( \lambda_r + \frac{w_i}{V_0} \right) + \frac{c_d}{c_l} \left( 1 + \frac{u_i}{V_0} \right) \right] d\frac{r}{R} \tag{8.26}$$

$$\text{Power Coefficient}: \quad C_P = \frac{P}{\frac{1}{2}\rho V_0^3 A}$$

$$= \frac{B}{\pi} \lambda \int_{r_{Root}/R}^{1} \frac{2\Gamma}{V_0 R} \left[ \left( 1 + \frac{u_i}{V_0} \right) - \frac{c_d}{c_l} \left( \lambda_r + \frac{w_i}{V_0} \right) \right] \frac{r}{R} d\frac{r}{R} \tag{8.27}$$

Note that $u_i$ and $w_i$ are axial-/angular induced velocities at the lifting line whose relation to BEM axial-/angular induction factors, $a$ and $a'$, is also given in Section 6.2.5. In Eqs. (8.26) and (8.27), $\Gamma$ is the radial circulation distribution as defined in Eqs. (6.43) and (8.10). Next, an optimization problem is defined where an optimum circulation distribution, $\Gamma$, is sought that accounts for maximum power under a thrust constraint. The objective function becomes:

$$\text{Objective Function (Thrust Constraint)}: \quad F(\Gamma) = C_P(\Gamma) + \Lambda\, C_T(\Gamma) \tag{8.28}$$

In Eq. (8.28), $\Lambda$ is a Lagrange multiplier (here used as a capital letter, $\Lambda$, instead of $\lambda$, in order to not be confused with the tip speed ratio). The objective function, $F(\Gamma)$, is to be maximized under a thrust constraint, $C_{T,0}$. Limiting ourselves to the inviscid ($c_d = 0$) problem, the Frechet derivative becomes:

$$\text{Frechet Derivative}: \quad \frac{\partial F}{\partial \Gamma}(\delta\Gamma) = \frac{2B}{\pi V_0 R} \int_{r_{Root}/R}^{1} \left[ \delta\Gamma \left( 1 + \frac{u_i}{V_0} \right) \frac{r}{R} + \Gamma \frac{\partial u_i}{\partial \Gamma}(\delta\Gamma) \right] d\frac{r}{R}$$

$$+ \Lambda \frac{2B\lambda}{\pi V_0 R} \int_{r_{Root}/R}^{1} \left[ \delta\Gamma \left( \lambda_r + \frac{w_i}{V_0} \right) + \Gamma \frac{\partial w_i}{\partial \Gamma}(\delta\Gamma) \right] d\frac{r}{R} = 0, \quad \forall \delta\Gamma \tag{8.29}$$

Note that axial-/angular induced velocities, $u_i$ and $w_i$, are linear functions of $\Gamma$ according to the Biot–Savart Law (see Section 6.2.3). A discrete form of Eq. (8.29) is described in Chattot (2003, 2006) and constitutes a linear non-homogeneous system for the discrete radial circulation distribution, $\Gamma_j$, with $\Gamma_1 = \Gamma_{jx} = 0$ ($j = 1 \ldots jx$) as boundary conditions that can be solved iteratively using an over-relaxation method, for example. The next task is then to find a new value for the Lagrange multiplier, $\Lambda$, that results in the desired thrust constraint, $C_{T,0}$. This is done by decomposing discrete forms of the inviscid ($c_d = 0$) thrust/power coefficients in Eqs. (8.26) and (8.27) into their respective linear and bilinear forms:

Discrete Thrust Coefficient (Linear + Bilinear Form) :

$$C_T = C_{T,1} + C_{T,2} = \frac{2B}{\pi}\sum_{j=2}^{jx-1}\frac{\Gamma_j}{V_0 R}\lambda\, r'_j\, \Delta r'_j + \frac{2B}{\pi}\sum_{j=2}^{jx-1}\frac{\Gamma_j}{V_0 R}\frac{w_{i,j}}{V_0}\Delta r'_j \tag{8.30}$$

Discrete Power Coefficient (Linear + Bilinear Form) :

$$C_P = C_{P,1} + C_{P,2} = \frac{2B}{\pi}\sum_{j=2}^{jx-1}\frac{\Gamma_j}{V_0 R}r'_j\, \Delta r'_j + \frac{2B}{\pi}\sum_{j=2}^{jx-1}\frac{\Gamma_j}{V_0 R}\frac{u_{i,j}}{V_0}r'_j\,\Delta r'_j \tag{8.31}$$

Here, $r'_j = r_j/R$ and $\Delta r'_j = r'_{j+1} - r'_j$, while $u_{i,j}$ and $w_{i,j}$ are themselves linear functions of both the discrete bound circulation, $\overline{\Gamma}_j$, and trailing vorticity, $\Delta\Gamma_j = \Gamma_{j+1} - \Gamma_j$ (see also Section 6.2.4). Assuming that the current radial distribution of discrete $\Gamma_j$ varies approximately by a factor $\kappa$ from the actual optimum satisfying the thrust constraint, $C_{T,0}$, we write:

Thrust Constraint Parameter, $\kappa$ :   $C_{T,0} = \kappa C_{T,1} + \kappa^2 C_{T,2}$ \tag{8.32}

Solving Eq. (8.32) for $\kappa$ and following the analysis of Chattot (2003, 2006), an updated value for the Lagrange multiplier, $\Lambda$, can be found as:

Lagrange Multiplier, $\Lambda$ :   $\Lambda = -\dfrac{C_{P,1} + 2\kappa C_{P,2}}{C_{T,1} + 2\kappa C_{T,2}}$ \tag{8.33}

The updated $\kappa$-scaled radial circulation distribution and Lagrange multiplier, $\Lambda$, are then used in the discrete equivalent to Eq. (8.29) as an updated solution to the optimization problem. This methodology is repeated until a convergence criterion for $\Gamma_j$ is reached. The interested reader is referred to Chattot (2003, 2006) for a more detailed analysis of the methodology, including a viscous correction (i.e. $c_d > 0$) to the mathematics. The optimization methodology is available in *XTurb* as part of the HVM solution method (METHOD = 2), with input parameters defined in the OPTIM input list, see next. A total of three design optimization options are available:

---

**XTurb Example 8.9  VWM Optimization Under Thrust Constraint (OPTIM Input List)**

_____OPTIM = 1_____

Find c(r') & β(r')  (Given: α(r') respectively cl(r'))

```
&OPTI
   OPTIM   = 1,

   CTO = #.####,

   NALPHD = 2,

   RALPHD = #.##,
```

*(Continued)*

```
XTurb Example 8.9  (Continued)

                    #.##,

        DALPHD = #.#,
                 #.#,

       LAGMULT = #.#,
    &END
                                    _____OPTIM = 2_____

                        Find β(r') & cl(r')  (Given: c(r') from BLADE input list)
    &OPTI
       OPTIM    = 2,

       CT0 = #.####,

        LAGMULT = #.#,
    &END
                                    _____OPTIM = 3_____

                        Find c(r') & cl(r')  (Given: β(r') from BLADE input list)
    &OPTI
       OPTIM    = 3,

       CT0 = #.####,

        LAGMULT = #.#,
    &END
```

The following sections provide some optimization examples using the *XTurb* code in OPTIM mode. Note that solver convergence is not guaranteed when using multiple viscous airfoil polars along the blade span that are highly dissimilar from each other and/or for design points in stalled flow. Furthermore, results (and convergence) may depend on entries in the BLADE input list that serve as initial conditions. The reader/user is therefore cautioned when applying the method and is advised to start with simple problem setups as shown next.

### 8.3.2.2  Betz Minimum Energy Condition

Betz (1926) argued that under optimum conditions (i.e. minimum energy loss due to wake rotation), an elementary thrust, $\delta C_T$, produces a constant elementary torque, $\delta C_Q = \delta C_P/\lambda$. Neglecting profile drag (i.e. $c_d = 0$), it follows from Eqs. (8.26) and (8.27) that the Betz minimum energy condition reads ($r' = r/R$):

$$\text{Minimum Energy Condition}: \frac{1 + u_i/V_0}{\lambda_r + w_i/V_0} r' = \tan \phi(r') \cdot r' = const. \qquad (8.34)$$

Equation (8.34) is written in terms of the local blade flow angle, $\phi$, after using a trigonometric relation on the velocity triangle (see Figure 3.3). Note that a similar argument was made by Munk (1921) in the context of moving elemental trailing vortices within a planar wing (i.e. non-rotating) vortex sheet such that the downwash at the wing lifting line becomes constant for minimum induced drag. It is important to note, however, that this is only exactly true for lightly loaded, that is, low $C_T$, conditions with vanishing circulation and induced velocities.

Let us investigate this by performing inverse rotor designs for the example case of the ideal rotor as specified at the beginning of Section 3.3 ($B = 3$, $\lambda = 6$, $r_{Root}/R = 0.10$, $\alpha = 6.3°$), using an inviscid airfoil polar for a thin symmetric airfoil (*ThinSymmAirfInv.polar*). The corresponding *XTurb* input file is *IdealRotorWithRotation4.inp* with a 10% root cutout. The input parameters in the OPTIM input list are defined as:

---

**XTurb Example 8.10  Ideal Rotor with Rotation and Tip Loss (IdealRotorWithRotation4.inp)**

```
&OPTI
   OPTIM    = 1,

   CT0 = 0.7912,

   NALPHD = 2,

   RALPHD = 0.10,
            0.75,

   DALPHD = 6.3,
            6.3,

   LAGMULT = 0.0,
&END
```

---

Next, we consider a low-to-high range of thrust-coefficient constraints, $C_{T,0}$, with the objective to see how the Betz minimum energy condition holds for the respective cases. Results obtained for both the resultant optimum circulation distribution, $\Gamma/(V_0 R)$, (*XTurb_Output_Method.dat*) and the Betz minimum energy condition, $\tan\phi(r') \cdot r' = const.$, are shown in Figure 8.15.

It is not surprising to see that the optimum radial circulation distribution, $\Gamma/(V_0 R)$, increases with thrust coefficient, $C_T$. As for the Betz minimum energy condition,

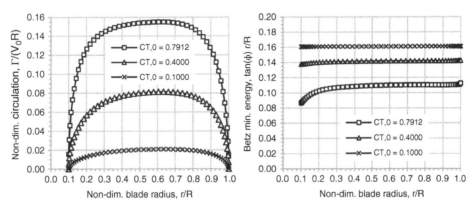

**Figure 8.15** Ideal rotor with wake rotation – optimum circulation distribution, $\Gamma/(V_0 R)$, and Betz minimum energy condition.

$\tan\phi(r') \cdot r' = const.$, it is indeed interesting to note that a nearly constant value for $\tan\phi(r') \cdot r'$ is obtained for $C_T = C_{T,0} = 0.10, 0.40$, while the respective Betz minimum energy condition does not hold for $C_T = C_{T,0} = 0.7912$. The interested reader is encouraged to perform individual studies on designing optimum rotors for different blade number, $B$, tip speed ratio, $\lambda$, and design angle of attack, $\alpha$, (i.e. design lift coefficient, $c_l$). The $C_T = 0.7912$ case is investigated a bit further as it relates to a corresponding BEM optimized rotor in Section 8.3.1.5. The corresponding *XTurb* output file *XTurb_Output_Method.dat* is shown next. Note that the chord and twist distributions shown next are actual results of the inverse-design method.

---

**XTurb Example 8.11 Ideal Rotor with Rotation and Tip Loss 4**
**(XTurb_Output_Method.dat) $C_{T0} = 0.7912$**

```
8.3.2 - Ideal Rot. w/ Rotat. 4      ***** XTurb V1.9    -   OUTPUT   -   METHOD *****
      Blade Number          BN = 3


      Solidity
      0.1171
                                 + Prescribed Wake Method [Chattot, Schmitz] +

      Number   TSR      PITCH [deg]      CT       CP       CPV      CB       CBV
           1   6.0000   0.1118        0.7912  -0.5188   0.0000  -0.1709  -0.0000

      "Thrust"   >  0    =>     Downwind Direction
      "Torque"   <  0    =>     Energy Extraction (Wind Turbine)   =>   Power = Torque * TSR
      "Bending"  <  0    =>     Flap Bending towards Downwind

      "Axial   Induction Factor"   a       = -ui
      "Angular Induction Factor"   a_prime = +wi/(r/R*TSR)

      r/R      Chord/R Twist [deg] AOA [deg]  PHI [deg] CIRC       ...        a        a_prime

      0.1000   0.0000  34.5781    6.3000    40.8781  -0.0000    ...      0.2489     0.4473
      0.1009   0.0394  34.4113    6.3000    40.7113   0.0157    ...      0.2498     0.4414
      0.1035   0.0775  33.9158    6.3000    40.2158   0.0310    ...      0.2525     0.4241
       ...      ...     ...                  ...                ...       ...
      0.9684   0.0400   0.2028    6.3000     6.5028   0.0814    ...      0.3337     0.0070
      0.9780   0.0337   0.1468    6.3000     6.4468   0.0693    ...      0.3330     0.0068
      0.9859   0.0272   0.1054    6.3000     6.4054   0.0564    ...      0.3320     0.0067
      0.9920   0.0206   0.0801    6.3000     6.3801   0.0429    ...      0.3306     0.0066
      0.9965   0.0138   0.0769    6.3000     6.3769   0.0289    ...      0.3280     0.0065
      0.9991   0.0071   0.1031    6.3000     6.4031   0.0148    ...      0.3232     0.0064
      1.0000   0.0000   0.1118    6.3000     6.4118  -0.0000    ...      0.3216     0.0063
```

---

Furthermore, note that the present VWM optimized blade by virtue includes both root- and tip losses, as opposed to the BEM optimized blade (Section 8.3.1.5) that did not include root losses but considers the same 10% root cutout. Here the thrust coefficient, $C_T$, is equal to the desired constraint, $C_{T,0}$, and it is interesting to note that both the power coefficient, $C_P$, and root-flap bending moment coefficient, $C_B$, obtained by *XTurb* are very similar to those in Section 8.3.1.5 for the BEM optimized blade. Also shown in *XTurb_Output_Method.dat* previously are radial distributions of axial-/angular induction factors, $a$ and $a'$, that agree in fact very well with ideal conditions derived for rotor disk theory in Section 2.2.2, though appear here for a three-bladed rotor (i.e. finite number of blades) and at a reduced thrust coefficient as a result of root/tip losses.

### 8.3.2.3 Effect of Profile Drag on VWM Optimum Blade Design (DTU 10-MW RWT)

The Technical University of Denmark (DTU) 10-MW Reference Wind Turbine (RWT) was designed as a notional public-domain offshore utility-scale wind turbine. Detailed information on the aerodynamics, structural layout, aeroelastic behavior, and control can be found in Bak et al. (2013). The idea of designing a relatively light-weight rotor required consideration of airfoils with high thickness ratios. In this context, the FFA-W3-xxx airfoil family with relative thickness ranging between 21.1 and 36.0% are suitable, while stiffness constraints in the blade root region required additional 48 and 60% thick airfoils as a transition to the cylindrical blade root. Airfoil data at representative Reynolds numbers were obtained from CFD computations and are also available to researchers (Bak et al. 2013). Here, we begin by performing some baseline *XTurb* computations of the DTU 10-MW RWT. The corresponding *XTurb* input file is *DTU-10MW-HVM-Ch8-3-2.inp*. Resulting *XTurb* integrated rotor performance parameters become:

```
XTurb Example 8.12  DTU 10-MW RWT (DTU-10MW-HVM-Ch8-3-2.inp)

8.3.2 HVM - DTU 10MW              ***** XTurb V1.9    -  OUTPUT *****
   Blade Number       BN =  3

                                  + Prescribed Wake Method [Chattot, Schmitz] +
   PREDICTION
   Blade Radius       BRADIUS   =   89.166 [m]
   Air Density        RHOAIR    =   1.225 [kg/m**3]
   Air Dyn. Visc.     MUAIR     =   0.000018 [kg/(m*s)]
   Number of Cases    NPRE      =   22

   Thrust            T = 0.5*RHOAIR*VWIND**2.*(pi*BRADIUS**2.)*CT
   Power             P = 0.5*RHOAIR*VWIND**3.*(pi*BRADIUS**2.)*CP
   Torque            TO = P / RPM
   Bending Moment    BE = 0.5*RHOAIR*VWIND**2.*(pi*BRADIUS**2.)*BRADIUS*CB

   Number  VWIND[m/s] RPM[1/min] TSR      PITCH[deg]   CT       CP        CB      ...
    ...                          ...                            ...
     6    9.0000      7.2290    7.5000   -3.4280      0.8563  -0.4989  -0.1895  ...
    ...                          ...                            ...
```

Computed rotor power, *P*, and thrust, *T*, for the DTU 10-MW RWT are shown in Figure 8.16, comparing results obtained by the EllipSys3D CFD code for the baseline case and fully turbulent (airfoil) flow as documented in Bak et al. (2013) to results obtained by *XTurb* in both BEMT (METHOD = 1) and HVM (METHOD = 2) modes. The associated *XTurb* input files used the blade pitch and rotor speed settings as specified in Bak et al. (2013). It can be seen that *XTurb* results show overall very good agreement when compared to blade-resolved EllipSys3D results, including the peak thrust force at the rated wind-speed condition. Note that $P > 10$ MW for the given blade tip pitch settings to account for transmission/generator losses and that some discrepancies in rotor power between *XTurb* and EllipSys 3D evolve at higher wind speeds in Region III ($V_0 > 20$ m s$^{-1}$); the interested reader can verify here that the rotor power is very sensitive to even small changes/inaccuracies in the blade tip pitch angle in this region.

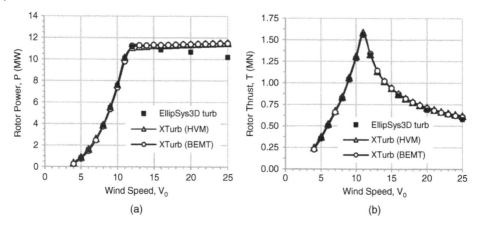

**Figure 8.16** (a) Rotor power, *P*, and (b) thrust, *T*, versus wind speed, $V_0$ (DTU 10-MW RWT).

An informative way of comparing engineering tools such as *XTurb* with high-fidelity blade-resolved CFD approaches such as EllipSys 3D is looking at radial distributions of local power and thrust coefficients, $dC_P$ and $dC_T$, respectively. Here, we consider a Region II wind speed of $V_0 = 9$ m s$^{-1}$. For *XTurb* results, for example, the corresponding coefficients can be found in the associated *XTurb_Output2.dat* file, see next, following the definition and transformation according to Figure 3.3.

---

**XTurb Example 8.13  DTU 10-MW RWT (XTurb_Output2.dat)**

```
8.3.2 HVM - DTU 10MW              ***** XTurb V1.9    -  OUTPUT 2 *****
    Blade Number        BN =  3

    Solidity
    0.0478
                        + Prescribed Wake Method [Chattot, Schmitz] +

    PREDICTION
          VWIND [m/s] RPM [1/min]   PITCH [deg]
          9.0000         7.2290      -3.4280

    Number   TSR       PITCH [deg]     CT       CP       CPV       CB       CBV
      6     7.5000     -3.4280        0.8563  -0.4989   0.0466  -0.1895  -0.0012

    "Thrust"  >  0    =>      Downwind Direction
    "Torque"  <  0    =>      Energy Extraction (Wind Turbine)   =>   Power = Torque * TSR
    "Bending" <  0    =>      Flap Bending towards Downwind

    r/R      Chord/R Twist [deg] AOA [deg]  ...   CThrust  CTorque  CNormal  CTangen   ...

    0.0314   0.0603   14.5000    62.2316    ...   0.5840   0.0043   0.5309  -0.2795   ...
    0.0329   0.0603   14.5000    61.6769    ...   0.5826   0.0047   0.5282  -0.2846   ...
    0.0374   0.0603   14.5000    60.0365    ...   0.5783   0.0060   0.5198  -0.2996   ...
    0.0448   0.0603   14.5000    57.4861    ...   0.5706   0.0083   0.5060  -0.3225   ...

    ...                                     ...
    0.9866   0.0174   -3.3046     6.5892    ...   1.1311  -0.0528   1.1261   0.1187   ...
    0.9940   0.0154   -3.3757     5.3194    ...   0.9855  -0.0227   0.9825   0.0809   ...
    0.9985   0.0138   -3.4149     2.7507    ...   0.6779   0.0174   0.6777   0.0230   ...
    1.0000   0.0133   -3.4280     1.9034    ...   0.5743   0.0247   0.5747   0.0097   ...
```

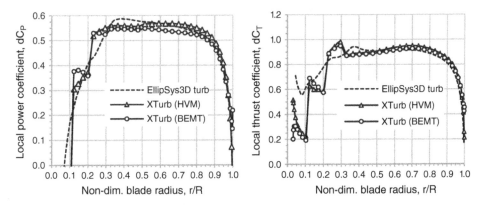

**Figure 8.17** Radial distribution of local power/thrust coefficient, $dC_P$ and $dC_T$, as computed by EllipSys 3D and XTurb (DTU 10-MW RWT, $V_0 = 9$ m s$^{-1}$).

In the general case, however, CFD results are normalized by the freestream dynamic pressure, $\frac{1}{2}\rho V_0^2$, instead of the actual local dynamic pressure, $\frac{1}{2}\rho V_{rel}^2$, as $V_{rel}$ is not easily determined in a CFD solution. Hence the comparisons here follow Eqs. (3.19) and (3.20), with an additional applied factor of $\pi R^2/(2\pi r dr)$, that is, normalizing by local rotor annulus area rather than rotor disk area, to compare to EllipSys 3D results as reported in Bak et al. (2013). Quantitative comparisons in Figure 8.17 show indeed good agreement between results obtained by *XTurb* and EllipSys 3D, particularly along the outer blade half for *XTurb* run in HVM mode (METHOD = 2). Note that inboard of $r/R = 0.30$, *XTurb* computed results appear discontinuous, which is a result of rapid transition between airfoil sections with dissimilar characteristics. This could be alleviated by adding appropriate blended airfoil sections to the analysis.

Next, let us consider an inverse design example for a Region II design point at $V_0 = 9$ m s$^{-1}$ where the baseline DTU 10-MW RWT performs at its maximum power coefficient. A simple and robust design example lies in considering only one airfoil, that is, the outermost airfoil (FFA-W3-241), in order to obtain smooth design solutions. Furthermore, let us investigate again the difference between designs obtained from inviscid ($c_d = 0$) versus viscous ($c_d > 0$) airfoil data, simply setting $c_d = 0$ in the associated airfoil data file, while retaining (viscous) lift coefficient data as is. The corresponding *XTurb* input file is *DTU-10MW-OPTIM-Ch8-3-2.inp*. Design parameters in the OPTIM input list are shown next:

---

**XTurb Example 8.14  Optimization of 10-MW Turbine – OPTIM = 1
(DTU-10MW-OPTIM-Ch8-3-2-Inv.inp)**

```
&OPTI
   OPTIM   = 1,

   CT0 = 0.8563,

   NALPHD = 3,
```

---

*(Continued)*

---

**XTurb Example 8.14  (Continued)**

```
      RALPHD  =  0.10,
               0.60,
               0.85,

      DALPHD  =  10.0,
                8.0,
                8.0,

   LAGMULT  =  0.0,
   &END
```

---

Note that $C_{T,0} = 0.8563$ is chosen for the thrust-coefficient constraint to be consistent with the baseline DTU 10-MW RWT. Furthermore, the angle-of-attack distribution is chosen such that the blade operates in the vicinity of $(c_l/c_d)_{max}$ of the actual airfoil cross sections. Figure 8.18 shows *XTurb* OPTIM results and those for the baseline DTU 10-MW RWT for both the blade geometry parameter, $\Xi$, and the associated radial circulation distribution, $\Gamma/(V_0 R)$. It is interesting to see again that inviscid and viscous designs are very similar. Furthermore, the baseline DTU 10-MW RWT parameters are also quite close to *XTurb* OPTIM results, attesting that the DTU 10-MW RWT is indeed a power-optimal design. Some differences exist at the inner blade portion that are attributed to planform constraints not considered in the optimization.

Computed blade chord and pitch distributions, $c/R$ and $\beta$, are shown in Figure 8.19 where the small difference between inviscid and viscous blade designs is once more visible and solely a consequence of high $c_l/c_d$ values for large utility-scale wind turbines. As for the blade chord, $c/R$, *XTurb* optimized planforms are similar to the baseline DTU 10-MW RWT. Note that the solidity in the root region can be reduced by elevating the design lift coefficient (respectively, the angle of attack) in this region as $c_i c = const.$ for an optimal blade geometry parameter, $\Xi$. Once again, the "design space is infinite." The total blade pitch angle, $\beta$, on the other hand, exhibits some differences between the *XTurb* optimized designs and the baseline DTU 10-MW RWT. Here, manufacturing constraints/costs can contribute to actual limitations on built-in blade twist.

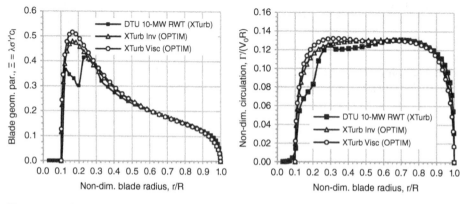

**Figure 8.18** Blade geometry parameter, $\Xi$, and radial circulation distribution, $\Gamma/(V_0 R)$ (DTU 10-MW RWT, $V_0 = 9$ m s$^{-1}$).

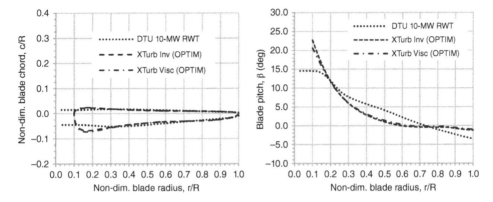

**Figure 8.19** Radial distribution of non-dimensional blade chord, *c/R*, and blade pitch, *β* (DTU 10-MW RWT).

Integrated *XTurb* performance coefficients (*DTU-10MW-OPTIM-Ch8-3-2.inp*) become:

---

**XTurb Example 8.15  Optimization of 10-MW Turbine – OPTIM = 1 (DTU-10MW-OPTIM-Ch8-3-2-Inv.inp)**

```
8.3.2 OPTIM - DTU 10MW - Inv      ***** XTurb V1.9    -  OUTPUT *****
   Blade Number        BN =  3


                                  + Prescribed Wake Method [Chattot, Schmitz] +
   PREDICTION ...

   Number  VWIND[m/s] RPM[1/min] TSR       PITCH[deg]    CT       CP       CB      ...
     1      9.0000     7.2290    7.5000    -1.1811      0.8563  -0.5317  -0.1861  ...



8.3.2 OPTIM - DTU 10MW            ***** XTurb V1.9    -  OUTPUT *****
   Blade Number        BN =  3


                                  + Prescribed Wake Method [Chattot, Schmitz] +
   PREDICTION ...

   Number  VWIND[m/s] RPM[1/min] TSR       PITCH[deg]    CT       CP       CB      ...
     1      9.0000     7.2290    7.5000    -0.9765      0.8543  -0.5055  -0.1840  ...
```

---

While an effective power coefficient of $C_P = 0.5317$ can be achieved for the inviscid blade design, this reduces to $C_P = 0.5055$ for the viscous design. The latter case is by itself very close to the baseline DTU 10-MW RWT performance, albeit with a few percent lower root-flap bending moment coefficient, $C_B$. This all attests once more to the performance quality of the baseline DTU 10-MW RWT design. The interested reader is encouraged to perform individual inverse blade designs with different airfoil sections and/or OPTIM design modes. Here another example of a large-scale reference rotor is the 100 m blade (13 MW) turbine designed by Sandia National Laboratories (Griffith and Ashwill 2011; Griffith and Richards 2014).

### 8.3.2.4 Design of Large-Scale Offshore "Low Induction Rotor" (LIR)

The concept of the LIR has been developed as a feasible design option for the next generation of large multi-MW offshore wind turbines (Chaviaropoulos and Sieros 2014). In general, offshore wind energy allows for higher tip speeds that are not noise constrained as well as longer blades due to relaxed transportation limits. Here the idea of developing larger rotors for the same rated power is intriguing to increase AEP by increasing rotor power in Region II. The culprit here, however, is that while rotor power $P \sim C_P R^2$, root-flap bending moment increases stronger with $Be \sim C_{Be} R^3$, which practically means that a longer blade has to be structurally stronger (i.e. thicker) compared to a shorter baseline blade, in order to withstand proportionally higher root-flap bending moments and associated fatigue loads. This means in particular that extra blade weight/cost is added to the larger rotor, if it were designed to perform at the same maximum power coefficient, $C_{P, max}$, occurring in Region II. This, however, is likely to increase the overall LCOE of the wind project. Fortunately, this dilemma can be overcome by allowing the larger rotor to perform at a less-than-optimal power coefficient, $C_P$, which in turn lowers the associated root-flap bending moment coefficient, $C_{Be}$, at a higher rate. A suitable design problem can then be formulated as:

$$\text{Design Problem}: \quad \frac{C_P(\lambda, a) \cdot R^2}{C_{P,0}(\lambda_0, a_0) \cdot R_0^2} \rightarrow max \text{ subject to } \frac{C_{Be}(\lambda, a) \cdot R^3}{C_{Be,0}(\lambda_0, a_0) \cdot R_0^3} \cong 1$$

$$(8.35)$$

In words, Eq. (8.35) states to maximize rotor power $P \sim C_P R^2$ compared to the baseline rotor (index "0"), while maintaining the same dimensional root-flap bending moment $Be \sim C_{Be} R^3$. The design problem can be recast to a more compact form by eliminating the dependence on the rotor blade radius, $R$, such that the following objective function is obtained, see Chaviaropoulos et al. (2013) and Chaviaropoulos and Sieros (2014):

$$\text{Objective Function}: \quad \frac{C_P(\lambda, a)}{C_{Be,0}(\lambda, a)^{2/3}} \rightarrow max \qquad (8.36)$$

The objective function in Eq. (8.36) can be simplified further assuming a constant $C_{Be}/C_T$ for a given rotor size under optimal conditions. Moreover, assuming a high design tip speed ratio, $\lambda$, it is reasonable to use basic momentum theory from Chapter 2 such that a simplified objective function becomes:

$$\text{Simplified Objective Function}: \quad \frac{C_P(a)}{C_T(a)^{2/3}} \rightarrow max \qquad (8.37)$$

Figure 8.20 shows Eq. (8.37) versus the rotor thrust coefficient, $C_T$, along with the associated mean axial induction factor, $a$, across the rotor disk according to basic momentum theory. It can be seen that the $C_P/C_T^{2/3}$ curve is quite flat around its maximum. Indeed, the momentum-theory optimum is at $C_T \cong 0.64$; however, taking into account wake rotation, profile drag, and root-/tip losses, a useful rule-of-thumb says that this results in a $\Delta C_T \cong -0.05$. Hence a design rotor thrust coefficient of $C_T = 0.59$ is chosen with a corresponding mean momentum-theory axial induction factor of $a = 0.18$.

Now let us consider the DTU 10-MW RWT from Section 8.3.2.3 as the baseline rotor (index "0"), with a reference Region II thrust coefficient of $C_{T,0} = 0.8563$. Then the size of the "larger" LIR rotor, which has the same peak root-flap bending moment than the baseline rotor, can be estimated from $(R/R_0)^3 \cong C_{T,0}/C_T = 0.8563/0.5900 = 1.451$,

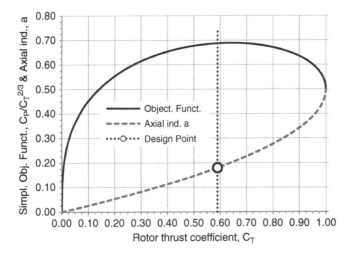

**Figure 8.20** Simplified objective function, $C_P/C_T^{2/3}$, and mean axial induction factor, $a$ (LIR Concept).

assuming $C_{Be} \sim C_T$ under optimal conditions. Hence, $R/R_0 \cong 1.13$, or a "larger" LIR blade radius of $R \cong 101$ m for a baseline blade length of $R_0 = 89.2$ m (DTU 10-MW RWT). Next, an inverse LIR blade design is considered in *XTurb* using the input file *LIR-10MW-OPTIM-Ch8-3-2.inp* with a 10% root cutout and the built-in twist distribution of the baseline DTU 10-MW RWT rotor. An excerpt from the *XTurb* input file is shown next:

---

**XTurb Example 8.16** Low Induction Rotor – OPTIM = 3 (LIR-10MW-OPTIM-Ch8-3-2.inp)

```
&OPERATION
    PREDICTION = 1,

       BRADIUS   = 101.293,

       RHOAIR    = 1.225,

       MUAIR     = 1.8E-05,

       NPRE      = 1,

       VWIND     = 11,

       RPMPRE    = 8.836,

       PITCHPRE  = 0.000,
&END
&OPTI
    OPTIM     = 3,

       CT0 = 0.5900,

  LAGMULT = 0.0,
&END
```

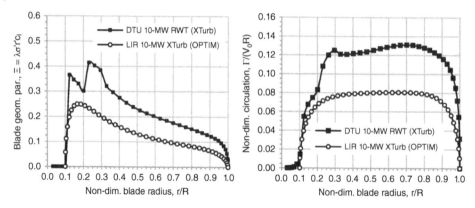

**Figure 8.21** Blade geometry parameter, $\Xi$, and radial circulation distribution, $\Gamma/(V_0R)$ (LIR 10-MW OPTIM, $V_0 = 11$ m s$^{-1}$).

Note the increased blade radius [m] and a wind speed [m s$^{-1}$] chosen close to rated power where both peak rotor thrust and root-flap bending moment are expected. The same rotor speed as for the baseline DTU 10-MW RWT is chosen, thus operating the LIR at a correspondingly higher tip speed and tip speed ratio. The thrust-coefficient constraint is set to the estimated value, and optimization is performed keeping the built-in twist of the DTU 10-MW RWT, though allowing a free (collective) tip pitch angle such that the thrust constraint can be satisfied (OPTIM = 3). Figure 8.21 shows *XTurb* OPTIM results and those for the baseline DTU 10-MW RWT for both the blade geometry parameter, $\Xi$, and the associated radial circulation distribution, $\Gamma/(V_0R)$. Here both optimal $\Xi$ and $\Gamma/(V_0R)$ are lower for the LIR (*XTurb* OPTIM) design compared to the baseline DTU 10-MW RWT, which is expected due to a lower design thrust coefficient.

Integrated performance coefficients as a result of the inverse design are shown next. Indeed, in comparison to Region II data of the baseline DTU 10-MW RWT rotor in Section 8.3.2.3, the ratio of root-flap bending moment coefficients is $C_{Be}/C_{Be,0} = 0.1289/0.1895 \cong 0.68 \leq 1/1.451$, thus satisfying that the LIR design does not exceed the baseline peak root-flap bending moment. This, however, comes at the price of a reduced power coefficient with $C_P/C_{P,0} = 0.4182/0.4989 \cong 0.84$, resulting in a Region II power ratio of $P/P_0 \sim 0.84 \cdot 1.13^2 \cong 1.07$, that is, an approximate Region II power increase of 7%.

---

**XTurb Example 8.17** Low Induction Rotor – OPTIM = 3 (LIR-10MW-OPTIM-Ch8-3-2.inp)

```
8.3.2 OPTIM - LIR 10MW            ***** XTurb V1.9     -   OUTPUT *****
    Blade Number        BN =  3


                                  + Prescribed Wake Method [Chattot, Schmitz] +
PREDICTION
    Blade Radius        BRADIUS  =  101.293 [m]
    Air Density         RHOAIR   =  1.225 [kg/m**3]
    Air Dyn. Visc.      MUAIR    =  0.000018 [kg/(m*s)]
    Number of Cases     NPRE     =  1
```

```
Thrust             T  = 0.5*RHOAIR*VWIND**2.*(pi*BRADIUS**2.)*CT
Power              P  = 0.5*RHOAIR*VWIND**3.*(pi*BRADIUS**2.)*CP
Torque             TO = P / RPM
Bending Moment     BE = 0.5*RHOAIR*VWIND**2.*(pi*BRADIUS**2.)*BRADIUS*CB

Number  VWIND[m/s] RPM[1/min] TSR     PITCH[deg]    CT       CP        CB      ...
     1  11.0000     8.8360    8.5206  0.0000      0.5900   -0.4182   -0.1289   ...
```

The observed reduction in power coefficient is primarily a consequence of the reduced design thrust coefficient. Seen in conjunction with a reduced optimum blade geometry parameter, Ξ, in Figure 8.21, this has some important implications for the LIR blade planform. Keeping the same radial $c_l$ distribution for the LIR design (compared to the baseline DTU 10-MW RWT) would actually result in a reduced blade chord, which is not desirable as the structure is expected to carry the same root-flap bending moment. Here, keeping the built-in twist distribution of the baseline DTU 10-MW RWT turbine (i.e. OPTIM = 3 mode in *XTurb*) is a useful inverse-design strategy as it actually results in a blade planform with a similar chord distribution in the root region compared to the baseline turbine. Figure 8.22 shows *XTurb* OPTIM computed blade chord and pitch distributions, $c/R$ and $\beta$, in comparison to the baseline DTU 10-MW RWT. Note that the built-in twist distribution is the same, while the blade tip pitch is higher for the LIR design as expected for a reduced design thrust coefficient. The interested reader can verify in the corresponding *XTurb* output files that the radial $c_l$ distribution for the LIR design has lower values than those seen on the baseline DTU 10-MW RWT. In the present context, this means in particular that as far as the FFA-W3-### airfoil family is concerned, the LIR design operates at less-than-optimal airfoil $c_l/c_d$, see also discussion in Chaviaropoulos et al. (2013). Consequently, some gain in the LIR power coefficient could be achieved for a different airfoil family with overall lower $c_d$ and higher $c_l/c_d$ at lower angles of attack.

The *XTurb* OPTIM computed blade chord and pitch distributions, $c/R$ and $\beta$, can be recast into a new *XTurb* input file (*LIR-10MW-HVM-Ch8-3-2.inp*) so that power as well as root-flap bending moment can be computed and plotted versus wind speed.

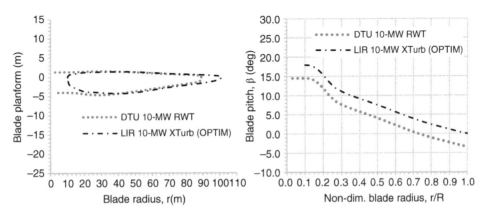

**Figure 8.22** Radial distribution of blade chord, *c*, and blade pitch, *β* (LIR 10-MW OPTIM).

An excerpt from the corresponding *XTurb_Output.dat* is shown next for the design condition:

---

**XTurb Example 8.18  Low Induction Rotor – XTurb_Output.dat**
**(LIR-10MW-HVM-Ch8-3-2.inp)**

```
8.3.3 HVM - LIR 10MW                  ***** XTurb V1.9    -  OUTPUT *****
    Blade Number       BN =  3

                            + Prescribed Wake Method [Chattot, Schmitz] +
PREDICTION
    Blade Radius      BRADIUS   =   101.273 [m]
    Air Density       RHOAIR    =   1.225 [kg/m**3]
    Air Dyn. Visc.    MUAIR     =   0.000018 [kg/(m*s)]
    Number of Cases   NPRE      =   22

    Thrust            T = 0.5*RHOAIR*VWIND**2.*(pi*BRADIUS**2.)*CT
    Power             P = 0.5*RHOAIR*VWIND**3.*(pi*BRADIUS**2.)*CP
    Torque            TO = P / RPM
    Bending Moment    BE = 0.5*RHOAIR*VWIND**2.*(pi*BRADIUS**2.)*BRADIUS*CB

    Number  VWIND[m/s] RPM[1/min] TSR     PITCH[deg]   ...  P[W]            ...  BE[Nm]
    ...
      8  11.0000       8.8360     8.5189  0.0000       ...  -10890519.1513  -31259717.6585
    ...
```

---

Figure 8.23 shows computed rotor power, $P$, and root-flap bending moment, $Be$, versus wind speed, $V_0$, for the LIR design and the baseline DTU 10-MW RWT. Here, the LIR design is analyzed in *XTurb* for both BEMT (METHOD = 1) and HVM (METHOD = 2) modes that agree very well as practically no part of the LIR blade operates in the turbulent wake state where model differences reveal themselves (see e.g. *XTurb* examples on Krogstad turbine). It can be seen in Figure 8.23 that Region II power indeed increases by about 7%, while the peak root-flap bending moment is not exceeded.

While the present example of a LIR design is valid, it is not the absolute optimum. Here the interested reader is once more encouraged to perform individual studies. As far as an estimated increase in AEP is concerned, it is proportional to the wind-speed probability

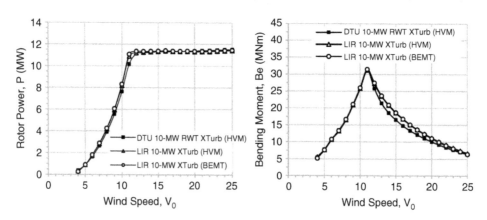

**Figure 8.23** Rotor power, *P*, and root-flap bending moment, *Be*, versus wind speed, $V_0$ (LIR 10-MW OPTIM).

in Region II multiplied by the respective rotor power, $P$. As a (rough) rule-of-thumb, it can be expected that the approximate 7% increase in Region II rotor power will result in an approximately 3% increase in AEP for typical wind speed distributions (see Chapter 1). This is significant, assuming that details of the cost model and overall LCOE support the LIR concept. Another aspect that warrants consideration is how a "larger" LIR design would perform in a wind farm setting, see some first investigations in Quinn et al. (2016).

The LIR concept is one among others considered for future offshore deployment of wind energy. One additional example is that of the Segmented Ultralight Morphing Rotor (SUMR) concept in an effort to reduce rotor mass and hence LCOE (Ananda et al. 2018), with an ultimate goal set by the Advanced Research Projects Agency for Energy (ARPA-E, US Department of Energy) of designing a low-cost 50 MW offshore wind turbine rotor.

### 8.3.2.5 Limitations of VWM Aerodynamic Optimization

As is the case for every physical model, the user has to be cognizant of the assumptions underlying a given theory. For a VWM, in particular, this means the assumption of high aspect ratio blades and the lifting-line model that inherently excludes radial flow. In addition, airfoil section characteristics are determined from external look-up tables that may or may not include rotational augmentation and/or other three-dimensional effects. Further note that, for example, the HVM (METHOD = 2) model in *XTurb* is based on a prescribed wake method (see Section 6.2) that does not take into account vortex rollup effects at blade root and tip. Hence the user has to be cautious when designing (optimal) blade planforms, and it is wise to test/check blade designs using for example, free-wake methods (see Section 6.3), blade-resolved CFD methods (see Chapter 7), and/or actual MS experiments in order to analyze/test whether or not the design performs as desired.

In this context, another interesting aspect is that of the assumed "average" axial induction factor, $a$, which is typically obtained from momentum theory. For the LIR design example (Section 8.3.2.4), $a = 0.18$ was obtained from momentum theory for a design thrust coefficient of $C_T = 0.59$. In reality, however, the local axial induction along the lifting line is higher than this average value due to the proximity of the trailing root-/tip vortices and vortex sheet. While this is not seen in BEM-type rotor designs, computing an actual vortex structure using a VWM reveals larger-than-average flow induction along the lifting line (i.e. blade). This is shown next by looking at the *XTurb_Outout_Method.dat* file associated with the LIR optimization from Section 8.3.2.4 (*LIR-10MW-OPTIM-Ch8-3-2.inp*).

---

**XTurb Example 8.19  Low Induction Rotor – XTurb_Output_Method.dat (LIR-10MW-OPTIM-Ch8-3-2.inp)**

```
8.3.3 OPTIM - LIR 10MW              ***** XTurb V1.9    - OUTPUT   -   METHOD *****
    Blade Number       BN =  3

    Solidity
    0.0340
                                    + Prescribed Wake Method [Chattot, Schmitz] +

   PREDICTION
        VWIND [m/s] RPM [1/min]  PITCH [deg]
```

---

*(Continued)*

**XTurb Example 8.19  (Continued)**

```
    11.0000       8.8360        0.0000

 Number   TSR    PITCH [deg]     CT       CP       CPV      CB       CBV
    1    8.5206  0.0000        0.5900  -0.4182  0.0399  -0.1289  -0.0007

 "Thrust"    >  0    =>     Downwind Direction
 "Torque"    <  0    =>     Energy Extraction (Wind Turbine)   =>   Power = Torque * TSR
 "Bending"   <  0    =>     Flap Bending towards Downwind

 "Axial    Induction Factor"    a      = -ui
 "Angular Induction Factor"   a_prime = +wi/(r/R*TSR)

 r/R    Chord/R Twist[deg] AOA[deg]  PHI[deg] CIRC      ...       a      a_prime

0.1000  0.0000  17.9280  22.7172  40.6452  -0.0000   ...    0.1548   0.1555
0.1014  0.0106  17.9240  22.3709  40.2949   0.0105   ...    0.1558   0.1524
0.1055  0.0201  17.9119  21.3587  39.2706   0.0207   ...    0.1590   0.1438
 ...                       ...                 ...
0.9510  0.0145  0.5006   4.9938   5.4944   0.0560    ...    0.2186   0.0026
0.9657  0.0123  0.3346   5.0775   5.4121   0.0486    ...    0.2184   0.0025
0.9780  0.0099  0.2090   5.1381   5.3471   0.0402    ...    0.2181   0.0025
0.9876  0.0075  0.1140   5.1850   5.2990   0.0310    ...    0.2177   0.0024
0.9945  0.0051  0.0486   5.2197   5.2683   0.0211    ...    0.2168   0.0024
0.9986  0.0026  0.0122   5.2451   5.2573   0.0107    ...    0.2152   0.0023
1.0000  0.0000  0.0000   5.2536   5.2536  -0.0000    ...    0.2147   0.0023
```

It can be seen that the "optimal" distribution of axial-/angular induction factors, $a$ and $a'$, follows closer to a momentum-theory (i.e. no root-/tip losses and profile drag) "average" induction factor for a thrust coefficient of $C_T \cong 0.64$, albeit including wake rotation effects (i.e. reduced local $a$) at the blade root. Here the interested reader is encouraged to perform individual studies and use the HVM "optimal" $a$ and $a'$ distributions in a BEM-type optimization (see Section 8.3.1).

The bottom line is that close-to-optimal designs can be realized relatively fast by reduced-order methods such as BEM and VWM; however, a supposedly small number of final designs should always be verified by both a higher-fidelity computational method and/or a scaled experiment, both of which can be expensive and require orders of magnitude more time and hence cost.

## 8.4  Summary – Scaled Design and Optimization

This chapter was devoted to principles of turbine blade design, scaled design for wind-tunnel testing, and aerodynamic optimization strategies based on both BEM and vortex theories. In the end, the primary constraints for any blade design are postulated by design standards that affect blade (root) thickness for structural integrity as well as noise and transportation constraints, all of which affect the levelized cost-of-energy of a wind project. As such the material presented in this chapter provides the reader with a basic set of skills to conduct optimal blade designs for particular constrained conditions with respect to blade number, tip speed ratio, rotor diameter, and chosen airfoils. In particular, scaling rotor aerodynamics from utility-scale applications to small-scale testing is not as straightforward as it may seem at first, and is primarily a consequence of scaled

Reynolds number and associated reduction in airfoil performance. Hence, different airfoils and blade planform designs may be necessary to resemble the utility-scale (or FS) aerodynamics in a scaled testing environment. Here the reader was provided with a set of strategies and examples, considering scaled blade and wake aerodynamics by means of the radial circulation distribution. Computational examples using the *XTurb* code were added to give the reader the opportunity for further individual studies; these included basic rotor performance comparisons for a selection of MS turbines (NREL Phase VI, MEXICO, Krogstad) considered for computational model comparisons in the wind energy science community. As for aerodynamic optimization, strategies and methodologies for both BEM and VWMs were derived, with appropriate references to material presented in earlier chapters. Here the objective was to provide the reader with a basic understanding of the separate effects of rotation, airfoil drag, and tip effects on optimized blade planforms and for two different physical models. The added examples of the DTU 10-MW RWT and a LIR provide the reader with a set of skills to design the next generation of large wind turbine blades. While an emphasis was given to aerodynamics, choosing appropriate constraints as, for example, in the case of the maximum root-flap bending moment for the LIR, enables a fairly quick design of a baseline rotor blade that can be taken to the next level using advanced system-oriented optimization strategies such as WISDEM (Dykes et al. 2015).

At this stage, the book comes to an end and the author hopes that the reader has developed some of the same passion the author feels for aerodynamics. It had been impossible to capture every aspect of aerodynamics within the limitations of the present book; however, the basic principles were taught and now lie in the hands of the next generation to guarantee the continued evolution of wind energy to continue to grow globally as a viable, reliable, cost-effective, and most of all zero-carbon source of energy.

# References

Adaramola, M.S. and Krogstad, P.-Å. (2010). Wind tunnel simulation of wake effects on wind turbine performance. In: *Proceedings of European Wind Energy Conference, Warsaw, Poland*, 64–67.

Ananda, G. K., Bansal, S., and Selig, M. S. (2018) Aerodynamic Design of the 13.2 MW SUMR-13i Wind Turbine Rotor. AIAA-2018-0994.

Bak, C., Zahle, F., Bitsche, R., Kim, T., Yde, A., Henriksen, L. C., Natarajan, A., and Hansen, M. H. (2013) Description of the DTU 10 MW Reference Wind Turbine. Report-I-0092, Technical University of Denmark (DTU).

Betz, A. (1926). *Windenergie und ihre Ausnützung durch Windmühlen*. Göttingen: Vandenhoeck und Ruprecht.

Björck, A., Ronsten, G., and Montgomerie, B. (1995) Aerodynamic section characteristics of a rotating and non-rotating 2.648m wind turbine blade. *Technical Report* FFA TN 1995-03, Bromma, Sweden.

Boorsma, K. and Schepers, J. G. (2014) New MEXICO Experiment, Preliminary overview with initial validation. Technical Report ECN-E–14-048, Energy Research Center of the Netherlands.

Bottasso, C.L., Campagnolo, F., and Petrovic, V. (2014). Wind tunnel testing of scaled wind turbine models: beyond aerodynamics. *Journal of Wind Engineering and Industrial Aerodynamics* 127: 11–28.

Buckingham, E. (1914). On physically similar systems; illustrations for use of dimensional equations. *Physical Review* 4: 345–376.

Burton, T., Jenkins, N., Sharpe, D., and Bossanyi, E. (2001). *Wind Energy Handbook*. Chichester: Wiley.

Canet, H., Bortolotti, P., and Bottasso, C. L. (2018) Gravo-elastic scaling of very large wind turbines to wind tunnel size. The Science of Making Torque from Wind Conference (Torque 2018), Milan, Italy. doi :10.1088/1742-6596/1037/4/042006.

Chamorro, L.P., Arndt, R.E.A., and Sotiropoulos, F. (2012). Reynolds number dependence of turbulence statistics in the wake of wind turbines. *Wind Energy* 15: 733–742.

Chattot, J.J. (2003). Optimization of wind turbines using helicoidal vortex model. *ASME Journal of Solar Energy Engineering* 125 (4): 418–424.

Chattot, J.J. (2006). Helicoidal vortex model for steady and unsteady flows. *Computers & Fluids* 54: 733–741.

Chaviaropoulos, P. and Sieros, G. (2014) Design of low induction rotors for use in large offshore wind farms. *Proceedings of the Scientific Track*, European Wind Energy Association (EWEA), 2014, Barcelona, Spain.

Chaviaropoulos, P., Beurskens. H. J. M., and Voutsinas, S. (2013) Moving towards large(r) high speed rotors – is that a good idea? *Proceedings of the Scientific Track*, European Wind Energy Association (EWEA), 2013, Vienna, Austria.

Cory, K. and Schwabe, P. (2009) *Wind Levelized Cost of Energy: A Comparison of Technical and Financing Input Variables*, NREL/TP-6A2–46671.

Dexin, H. and Thor, S. E. (1993) The execution of wind energy projects 1986–1992. *Technical Report* FFA TN 1993–19, Bromma, Sweden.

Drela, M. (1989). X-foil: an analysis and design system for low Reynolds number airfoils. In: *Low Reynolds Number Aerodynamics*, vol. 54, Springer-Verlag Lecture Notes in Engineering (ed. T.J. Mueller). New York: Springer.

Dykes, K., Graf, P., Scott, G., Ning, A., King, R., Guo, Y., Parsons, T., Damiani, R. Felker, F., and Veers, P. (2015) Introducing WISDEM TM: An Integrated System Model of Wind Turbines and Plants. National Renewable Energy Laboratory, PR-5000-63564.

Griffith, D. T. and Ashwill, T. D. (2011) The Sandia 100-meter All-Glass Baseline Wind Turbine Blade: SNL100–00. *Tech. Rep. SAND 2011–3779*, Sandia National Laboratories.

Griffith, D. T. and Richards, P. W. (2014) The SNL 100–03 Blade: Design Studies with Flatback Airfoils for the Sandia 100-m Blade. *Tech. Rep. SAND 2014–18129*, Sandia National Laboratories.

Hand, M. M., Simms, D. A., Fingersh, L. J., Jager, D. W., Cotrell, J. R., Schreck, S., and Larwood, S. M. 2001 Unsteady aerodynamics experiment phase VI: wind tunnel test configurations and available data campaigns. *Tech. Report* NREL/TP-500-29955. National Renewable Energy Laboratory, Golden, CO, USA.

Hassanzadeh, A., Naughton, J.W., Kelley, C.L., and Maniaci, D.C. (2016). Wind turbine blade design for subscale testing. The science of making torque from wind (TORQUE 2016). *Journal of Physics: Conference Series* 753: 022048.

Hunt, G. K. (1973) Similarity Requirements for Aeroelastic Models of Helicopter Rotors. RAE Technical Report 72005 – ARC 33730.

IEC (2005). *IEC 1400-1 Wind Turbine Generator Systems*, 3e. IEC, 2005–2008.

Jonkman, J., Butterfield, S., Musial, W., and Scott, G. 2009 Definition of a 5-MW reference wind turbine for offshore system development. *Technical Report* NREL/TP-500-38060. National Renewable Energy Laboratory, Golden, CO, USA.

Kelley, C. L., Maniaci, D. C., and Resor, B. R. (2016) Scaled Aerodynamic Wind Turbine Design for Wake Similarity. AIAA-2016-1521, AIAA SciTech Forum, San Diego CA, USA.

Krogstad, P.A. and Eriksen, P.E. (2013). "Blind test" calculations of the performance and wake development for a model turbine. *Renewable Energy* 50: 325–333.

Krogstad, P.A. and Lund, J.A. (2012). An experimental and numerical study of the performance of a model turbine. *Wind Energy* 15: 443–457.

Lee, S. (2015). Inverse design of horizontal axis wind turbine blades using a vortex line method. *Wind Energy* 18 (2): 253–266.

Manwell, J.F., McGowan, J.G., and Rogers, A.L. (2009). *Wind Energy Explained*, 2e. Chichester: Wiley.

Miller, M., Kiefer, J., Nealon, T., Westergaard, C., and Hultmark, M. (2017) Horizontal Axis Wind Turbine Experiments at Full-Scale Reynolds Numbers. APS Meeting Abstracts.

Munk, M. M. (1921) The Minimum Induced Drag of Aerofoils. NACA Report 121.

Nanos, E. M., Kheirallah, N., Campagnolo, F., and Bottasso, C. L. (2016) Design of a multipurpose scaled wind turbine model. The Science of Making Torque from Wind Conference (Torque 2018), Milan, Italy. doi: 10.1088/1742-6596/1037/5/052016.

Quinn, R., Schepers, G., and Bulder, B. (2016). A parametric investigation into the effect of low induction rotor (LIR) wind turbines on the levelised cost of electricity for a 1 GW offshore wind farm in a North Sea wind climate. *Energy Procedia* 94: 164–172.

Rabl, A. (1985). *Active Solar Collectors and their Applications*. Oxford: Oxford University Press.

Ramsay, R. R., Hoffmann, M. J., and Gregorek, G. M. (1995) Effects of grit roughness and pitch oscillations on the S809 airfoil. *NREL Report* No. NREL/TP-442-7817, National Renewable Energy Laboratory.

Reich, D., Elbing, B., Berezin, C., and Schmitz, S. (2014). Water tunnel flow diagnostics of wake structures downstream of a model helicopter rotor hub. *Journal of the American Helicopter Society* 59 (3): 33–41.

Ronsten, G. (1992). Static pressure measurements on a rotating and a non-rotating 2.375 m wind turbine blade. Comparison with 2D computations. *Journal of Wind Engineering and Industrial Aerodynamics* 39: 105–118.

Ronsten, G. (1994) Geometry and Installation in Wind Tunnels of a STORK 5.0 WPX Wind Turbine Blade Equipped with Pressure Taps. *Technical Report* FFAP-A 1006.

Rosenberg, A. and Sharma, A. (2016). A prescribed-wake vortex lattice method for preliminary design of co-axial, dual-rotor wind turbines. *ASME Journal of Solar Energy Engineering* 138 (6): https://doi.org/10.1115/1.403435.

Schepers, J. G. and Snel, H. (2007) MEXICO, Model experiments in controlled conditions. ECN-E-07-042, Energy Research Center of the Netherlands.

Schepers, J. G., Boorsma, K., Cho, T. et al. (2012) Final report of IEA Task 29, Mexnext (Phase 1): Analysis of MEXICO wind tunnel measurements. ECN-E-12-004, Energy Research Center of the Netherlands.

Schepers, J. G., Boorsma, K., Cho, T. et al. (2014) Final report of IEA Task 29, Mexnext (Phase 2). ECN-E-14060, Energy Research Center of the Netherlands.

Schmitz, S. (2012). *XTurb-PSU: A Wind Turbine Design and Analysis Tool*. The Pennsylvania State University.

Schmitz, S., Bhagwat, M., Moulton, M.A. et al. (2009). The prediction and validation of hover performance and detailed blade loads. *Journal of the American Helicopter Society* 54: 32004.

Selig, M.S. and Tangler, J.L. (1995). Development and application of a multipoint inverse design method for horizontal axis wind turbines. *Wind Engineering* 19 (2): 91–105.

Sheng W., Galbraith, R. A., Coton, F. N., and Gilmour, R. (2006a) The collected data for tests on an S809 airfoil, volume I: pressure data from static, ramp and triangular wave tests. *G.U. Aero Report* 0606, University of Glasgow.

Sheng W., Galbraith, R. A., Coton, F. N., and Gilmour, R. (2006b) The collected data for tests on an S809 airfoil, volume II: pressure data from static and oscillatory tests. *G.U. Aero Report* 0607, University of Glasgow.

Sheng W., Galbraith, R. A., Coton, F. N., and Gilmour, R. (2006c) The collected data for tests on the sand stripped S809 airfoil, volume III: pressure data from static, ramp and oscillatory tests. *G.U. Aero Report* 0608, University of Glasgow.

Simms, D., Schreck, S., Hand, M., and Fingersh, L. J. (2001) NREL Unsteady Aerodynamics Experiment in the NASA-Ames Wind Tunnel: A Comparison of Predictions to Measurements. *Tech Rep.* NREL/TP-500-29494. National Renewable Energy Laboratory, Golden, CO, USA.

Singleton, J. D. and Yeager, W. T. (1998) Important Scaling Parameters for Testing Model-Scale Helicopter Rotors. AIAA-1998–2881.

Somers, D. M. (1997) *Design and Experimental Results for the S809 Airfoil*. Airfoils Inc., State College PA. Published as: National Renewable Energy Laboratory, NREL/SR-440-6918.

Somers, D. M. (2005) The S825 and S826 Airfoils. Airfoils Inc., State College PA. Published as: National Renewable Energy Laboratory, NREL/SR-500-36344.

Sørensen, N.N., Zahle, F., Boorsma, K., and Schepers, G. (2016). CFD computations of the second round of MEXICO measurements. The science of making torque from wind (TORQUE 2016). *Journal of Physics: Conference Series* 753: 022054.

Yeager, W. T. and Kvaternik, R. G. (2001) A Historical Overview of Aeroelasticity Branch and Transonic Dynamics Tunnel Contributions to Rotorcraft Technology and Development. NASA TM-2001–211054.

## Further Reading

Brønsted, P. and Nijssen, R.P.L. (eds.) (2013). *Advances in Wind Turbine Blade Design and Materials*. Woodhead Publishing Ltd.

Burton, T., Jenkins, N., Sharpe, D., and Bossanyi, E. (2011). *Wind Energy Handbook*, 2e. Chichester: Wiley.

Chattot, J.J. and Hafez, M.M. (2015). *Theoretical and Applied Aerodynamics – and Related Numerical Methods*. Dordrecht: Springer.

Eggleston, D.M. and Stoddard, F.S. (1987). *Wind Turbine Engineering Design*. New York: Van Nostrand Reinhold Co.

Hansen, M.O.L. (2008). *Aerodynamics of Wind Turbines*. London: Earthscan.

Jamieson, P. (2011). *Innovation in Wind Turbine Design*. Chichester: Wiley.

Sørensen, J.N. (2016). *General Momentum Theory for Horizontal Axis Wind Turbines*. London: Springer.

# Index

*Aerodynamics of Wind Turbines: A Physical Basis for Analysis and Design,* First Edition. Sven Schmitz.
© 2020 John Wiley & Sons Ltd. Published 2020 by John Wiley & Sons Ltd.
Companion website: www.wiley.com/go/schmitz/wind-turbines